AF805731

Series in BioEngineering

The Series in Bioengineering serves as an information source for a professional audience in science and technology as well as for advanced students. It covers all applications of the physical sciences and technology to medicine and the life sciences. Its scope ranges from bioengineering, biomedical and clinical engineering to biophysics, biomechanics, biomaterials, and bioinformatics.

Indexed by WTI Frankfurt eG, zbMATH.

Elena García-Gareta • José Manuel García-Aznar

Contemporary Tissue Engineering and Regenerative Medicine

From Organ Regeneration to Bioengineered Tissue Models

Springer

Elena García-Gareta
Department of Science and Technology of Materials and Fluids
School of Engineering and Architecture
University of Zaragoza
Zaragoza, Spain

José Manuel García-Aznar
Department of Mechanical Engineering
School of Engineering and Architecture
University of Zaragoza
Zaragoza, Spain

ISSN 2196-8861 ISSN 2196-887X (electronic)
Series in BioEngineering
ISBN 978-3-032-03440-3 ISBN 978-3-032-03441-0 (eBook)
https://doi.org/10.1007/978-3-032-03441-0

© The Editor(s) (if applicable) and The Author(s), under exclusive license to Springer Nature Switzerland AG 2026

This work is subject to copyright. All rights are solely and exclusively licensed by the Publisher, whether the whole or part of the material is concerned, specifically the rights of translation, reprinting, reuse of illustrations, recitation, broadcasting, reproduction on microfilms or in any other physical way, and transmission or information storage and retrieval, electronic adaptation, computer software, or by similar or dissimilar methodology now known or hereafter developed.
The use of general descriptive names, registered names, trademarks, service marks, etc. in this publication does not imply, even in the absence of a specific statement, that such names are exempt from the relevant protective laws and regulations and therefore free for general use.
The publisher, the authors and the editors are safe to assume that the advice and information in this book are believed to be true and accurate at the date of publication. Neither the publisher nor the authors or the editors give a warranty, expressed or implied, with respect to the material contained herein or for any errors or omissions that may have been made. The publisher remains neutral with regard to jurisdictional claims in published maps and institutional affiliations.

This Springer imprint is published by the registered company Springer Nature Switzerland AG
The registered company address is: Gewerbestrasse 11, 6330 Cham, Switzerland

If disposing of this product, please recycle the paper.

Foreword

A new textbook on tissue engineering and regenerative medicine is both timely and essential, offering a comprehensive and balanced introduction to a rapidly evolving field that reaches across biology and engineering to address clinical challenges and advance therapeutic solutions.

Tissue engineering and regenerative medicine are dynamic disciplines at the intersection of life sciences and engineering, focused on restoring functions to damaged tissues and organs. While previous texts exist, recent breakthroughs in areas such as biofabrication, 3D bioprinting, and regenerative cellular therapies ensure these topics remain in flux, and students, practitioners, and researchers require up-to-date, accessible information. This textbook's structure reflects these advances and provides a logical, satisfying organizational scheme, supplemented with ample references for those seeking greater depth on any particular topic.

The unique co-authorship by both a biologist and an engineer offers readers perspectives that are both broad and deep, bridging traditional divides within regenerative medicine. This disciplinary synergy creates an environment where readers can appreciate cellular mechanisms and biological principles, as well as the engineering design, biomaterials, and fabrication techniques necessary for successful tissue reconstruction. The book's approach ensures accessibility for newcomers while providing enough technical rigor for advanced learners, making it suitable for undergraduates, graduate students, clinical researchers, and practicing professionals.

This text is intentionally crafted to attract and enrich readers from varied scientific and engineering backgrounds. Whether trained in biology, chemistry, materials science, mechanical engineering, medicine, or another field, users will find treatments of the major elements: cell sources, scaffold fabrication, biomaterial development, and the ever-expanding catalog of chemical and physical stimuli that guide tissue growth. Examples throughout the book demonstrate rising clinical applications in skin, cartilage, blood vessels, bladders, and challenging organs like heart, liver, and kidney, making the work directly relevant to clinicians, scientists, and engineers alike.

Readers are provided with questions at the end of each chapter, which serve as both summaries and informal exams to confirm understanding and readiness to advance. This active-learning feature supports reflection and gradual mastery, enabling the book to function as both a core text and a reference for self-driven exploration. With chapters devoted to the triad of tissue engineering—cells, scaffolds, and stimuli—as well as modern manufacturing methods such as 3D printing and additive manufacturing, users are encouraged to appreciate both foundational concepts and emerging technologies.

The field's rapid pace is acknowledged—the book incorporates established knowledge and recent advances yet recognizes developing areas that may not be fully mature for introductory discussion. Innovations such as artificial intelligence are expected to influence tissue engineering yet remain largely unexplored, leaving room for future editions to explore their impact as it becomes clearer.

Whether a student just beginning this journey, a postdoctoral researcher seeking current methods, a clinician looking for translational advances, or an engineer imagining the next wave of biofabrication, this book welcomes all. By combining practical techniques, theoretical background, case studies, and emerging research, the authors empower readers to contribute to the multidisciplinary community that propels tissue engineering forward.

In summary, this book stands as an essential, balanced, and forward-looking guide—both a foundation and a springboard for anyone seeking to enter or advance in the vital field of tissue engineering and regenerative medicine.

Department of Biological Engineering and Mechanical Engineering
Massachusetts Institute of Technology,
Cambridge, MA, USA

Roger D. Kamm

Competing Interests The authors have no competing interests to declare that are relevant to the content of this manuscript.

Contents

Chapter 1
Introduction to Tissue Engineering and Regenerative Medicine

Abstract Tissue Engineering is a regenerative medicine approach, based on engineering and biology principles, that aims to restore the functionality of damaged or injured tissues. This chapter begins with an introduction to regenerative medicine, followed by an overview of the principles of tissue engineering and a description of various successful applications. This chapter aims to describe the potential of tissue engineering and its impact in regenerative medicine, serving as an introduction to this book.

1.1 Introduction to Regenerative Medicine

Our body has the capacity to regenerate and self-heal when it is damaged or injured. For instance, when we have a small cut in our skin, we know that our body is able to heal it. However, when a larger wound is produced, we need to close it using stitches. So, although our tissues can regenerate spontaneously, they do it within a limited size. In addition, the capacity of regeneration of our tissues is not the same (Table 1.1). On the one hand, there are tissues with a high regenerative capacity, like bone or the epidermis of skin. But, on the other hand, there are other tissues with a low regenerative capacity such as nerves, ligaments and others. In all cases, the size of the injury is a crucial aspect.

Medicine traditionally uses different strategies to help our body to restore its functionality mainly by means of drugs or therapies, whose aim is to improve the capacity of cells to heal this injury. Nowadays, this kind of medicine is called regenerative medicine and with the combination of other strategies has opened new approaches to improve our body regeneration capacity.

Tissue repair and regeneration capacities are conserved by all species, allowing the renewal and restoration of damaged cells, tissues, organs and even entire body parts [1]. It is mediated by the migration of adult stem cells and their differentiation and specification capacity, but also by the contribution of cell proliferation of both

© The Author(s), under exclusive license to Springer Nature Switzerland AG 2026

E. García-Gareta, J. M. García-Aznar, *Contemporary Tissue Engineering and Regenerative Medicine*, Series in BioEngineering,
https://doi.org/10.1007/978-3-032-03441-0_1

Table 1.1 Different levels of tissue regenerative capacity

Regenerative capacity	Good	Poor
Connective tissues		
Bone	X	
Articular cartilage, ligament, tendon, others		X
Muscles		
Cardiac, skeletal		X
Smooth	X	
Epithelia (e.g. epidermis)	X	
Nerves		X

progenitor and fully differentiated cells. One example of this phenomenon is observed in amphibian limb regeneration after amputation [2, 3].

In Fig. 1.1 we can visualize a scheme of how spontaneous regeneration normally occurs in mammals. When a wound is produced the blood vessels in direct contact with the injury are also damaged and the injured area is infiltrated by platelets to stop the bleeding resulting from the injury. At the same time, the injury is also infiltrated by immune cells that remove the debris, activate regenerative inflammation and create a new extracellular matrix (ECM) normally known as granulation tissue: the immune cells start to clean the dead tissue and other debris resulting from the injury. Fibroblasts[1] proliferate and produce ECM fibers that fill the cleared injury area to prevent it from collapsing. The formation of new blood vessels takes place. The initial ECM starts to be remodelled as the injury is repopulated by new cells through proliferation. Final regeneration is achieved when the injured area has been completely repopulated and the tissue reconstituted to its homeostatic condition.

From an engineering point of view, we could schematize this process of tissue regeneration in Fig. 1.2 as a complex mechanism where the cell is the main actor of the regenerative process. In fact, cells can sense different kinds of microenvironmental stimuli that activate different events to restore tissue integrity, such as, proliferation, migration, proliferation and matrix production, among others.

Traditionally in medicine different kinds of chemical substances have been introduced in our organism in order to increase the response of our cells. For example, the use of growth factors has been quite common to treat cartilage damage [5], bone fractures [6, 7] or wound healing [8].

In orthopedics, however, it is quite common the application of different kinds of external forces to increase bone regeneration capacities. For example, bone distraction is a technique in which bones can increase their size by the application of tension forces in order to repair critical-sized bone defects [9], which are defined as those that will not heal spontaneously within a patient's lifetime [10]. Another effective example has been presented by Gómez-Benito et al. [11], where it has been demonstrated that the application of external mechanical loads can clearly

[1] Other cells may also be present and participate in the process of matrix production, although here we generalize this role on the fibroblasts.

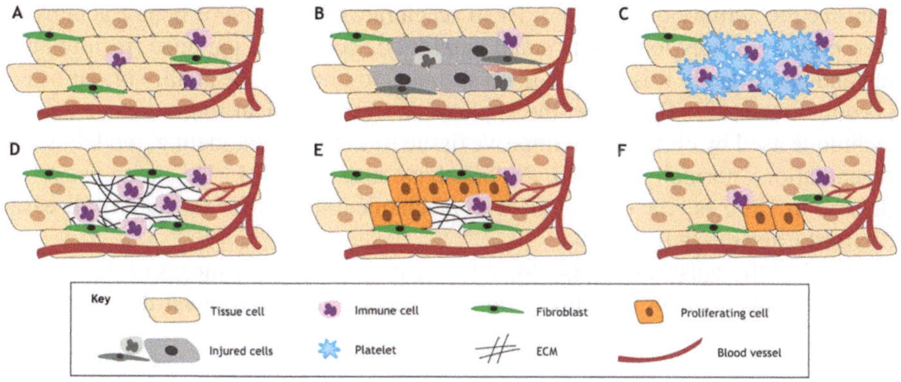

Fig. 1.1 Phases of tissue regeneration. (**a**) Healthy tissue. (**b**) Injury: injured area is shaded in gray, representing damaged and dead cells. The blood vessel in direct contact with the injury is also damaged and is marked in a lighter red colour than the healthy vessels. (**c**) Wound closure and immune response: the injured area is infiltrated by platelets to stop the bleeding resulting from injury to the blood vessels in the damaged area. At the same time, the lesion is also infiltrated by immune cells. (**d**) Clearing of debris and deposition of extracellular matrix (ECM): the immune cells start to clean the dead tissue and other debris resulting from the injury. Fibroblasts proliferate and produce ECM fibres that fill the cleared injury area to prevent the injury from collapsing. The formation of new blood vessels takes place (thinner red lines). (**e**) Resorption of ECM and cell proliferation: the ECM starts to be eliminated as the injury is repopulated by new cells through proliferation. (**f**) Regeneration: the injured area has been completely repopulated and the tissue reconstituted to its homeostatic condition (obtained from [4]). Open access article distributed under the terms and conditions of the Creative Commons Attribution (CC BY) license

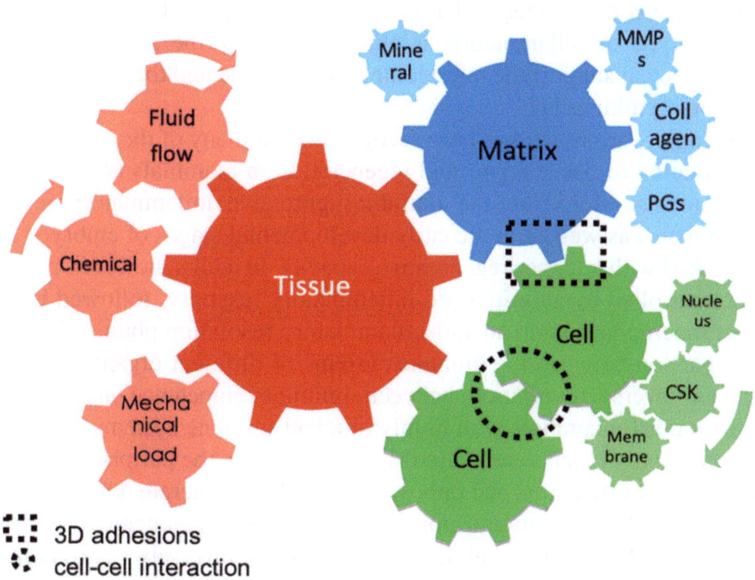

Fig. 1.2 Tissue regeneration can be schematized as a complex mechanism where cells are the main responsible of the regenerative process, according to the multiple and varied surrounding microenvironment

improve the activity of cells in order to accelerate the healing of bone fractures. In fact, high frequency loads with small amplitude accelerated the bone fracture healing and additionally it is really a good stimulus to improve the quality of our osteoporotic bones [12]. Therefore, the application of mechanical forces is an extended technique used by clinicians to improve tissue formation, remodeling, modeling and regeneration, normally termed as "mechanotherapy" [13]. This term was first defined in 1890 as "the employment of mechanical means for the cure of disease" (Oxford English Dictionary).

Physical therapists also apply local forces in our tissues by means of their hands, punches, nails or other systems to improve regeneration and repair during healing [14, 15].

1.2 The Crosstalk Between Inflammation and Regeneration: Regenerative Inflammation

As we have just seen in the previous section, when a tissue is damaged or injured, prior to renewal, immune cells carry out functions essential for proper tissue repair, such as clearance of the injured area and a subsequent inflammation, termed, regenerative inflammation [1]. However, little is known about this regenerative inflammation, because inflammation is commonly known as the process of self-protection to external agents like bacteria or viruses. However, regenerative inflammation is associated with an immunosuppressive and pro-regenerative response generated by monocyte-derived macrophages that promote tissue repair after trauma and injury. This unique type of inflammation is characterized by the secretion of different growth factors that support the remodelling of the tissue and the production of anti-inflammatory cytokines [1].

Interestingly, various studies have recognised that many of the mechanisms linking inflammation to tissue repair and regeneration in mammals are also present in lower organisms. Moreover, the immunosuppressive inflammatory response in lower organisms, as well as in the early developmental stages of embryos and neonates, is linked with a higher regenerative capacity. In adults, however, regeneration is tightly controlled by an initial pro-inflammatory response, followed by the conversion of macrophages into an anti-inflammatory resolution phase [1].

A comparison between the immune systems of different organisms and organs suggests an inverse correlation between immunocompetence and tissue repair (Fig. 1.3). Indeed, mammals with highly efficient immune systems for combatting pathogens can only regenerate injured skeletal muscles, the peripheral nervous system, and the liver efficiently, and only very limitedly regenerate other organs. These findings suggest that the immunosuppressive response of regenerative inflammation is linked to regenerative ability, while also inversely correlating with immunocompetence.

Notably, as in evolution, lower immunocompetence in embryos and neonates correlates with a higher regenerative capacity, as evidenced by scar-free healing in the skin or heart during the early stages of life (Fig. 1.3). Scarring is a consequence of the fibrotic process, which is regulated by regenerative inflammation. In the final phase of regenerative inflammation, macrophages and other cell types, such as fibroblasts and endothelial cells, secrete TGF-β, which promotes the controlled production of the extracellular matrix necessary for rebuilding normal tissue structure. Surprisingly, scarring is not observed in mammals during the embryonic and neonatal stages.

The presence of nerves has also been examined in detail because they play an essential role during limb regeneration [16]. Without nerves, regeneration does not occur (Fig. 1.4).

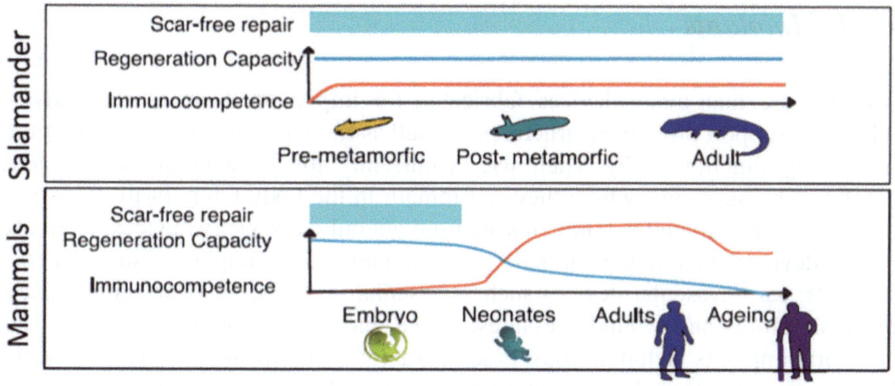

Fig. 1.3 Regenerative inflammation across species (salamander and mammals) at different developmental stages, comparing the scar-free repair, the regeneration capacity and the immunocompetence (obtained from [1]). Open access article distributed under the terms and conditions of the Creative Commons Attribution (CC BY) license

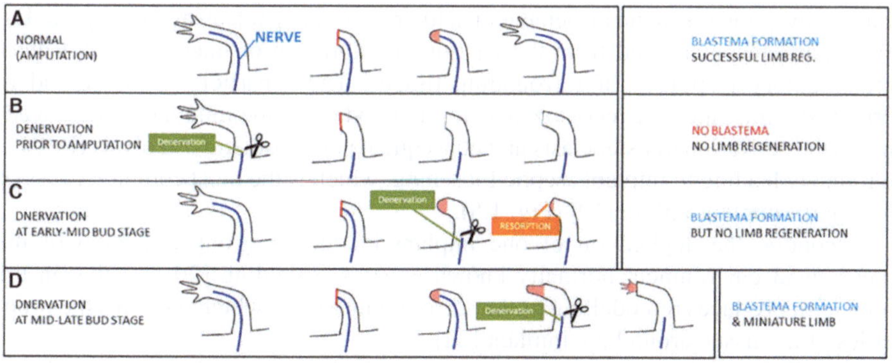

Fig. 1.4 Scheme of denervation effects in limb regeneration. (**a**) The nerve plays an essential role in successful limb regeneration. (**b**) Denervation prior to limb amputation results in no blastema formation. (**c**) Denervation during the early-mid bud stage results in blastema resorption. (**d**) Denervation during the mid-late bud stage results in blastema resorption or the formation of a miniature limb [16]. Open access article distributed under the terms and conditions of the Creative Commons Attribution (CC BY) license

1.3 Different Strategies in Regenerative Medicine: Tissue Engineering

The main aim of Regenerative Medicine consists on the restoring of damaged or injured tissues and organs, but different strategies can be used to achieve this global goal. These strategies involve different technologies from implants, grafts, or cell therapy to tissue engineering. Next, a brief summary is presented of all these technologies.

1.3.1 Implants

Implants are man-made devices fabricated for implantation inside the body to replace or support biological structures, as well as for delivering drugs and monitoring bodily functions [17]. Their use is currently the most common technique employed by surgeons, where they can remain in the body temporarily or permanently. To date, biomedical implants are used not only as sensory devices; brain and neural devices including neuronal, cochlear and retinal implants; subcutaneous implants; cardiovascular devices such as vascular grafts, stent, heart valves, pacemakers; sutures and wound dressings; spinal and dental implants; cosmetic and structural implants including rods, braces, craniofacial, hip and knee replacements; but also as ophthalmic devices including glasses and contact lenses as well as insulin delivery devices [17].

The implantation of a prosthesis or device requires the total replacement of an injured organ; therefore, the functionality and biocompatibility of the implant must be guaranteed. While these implants normally behave well in the short term, they can cause problems in the long term, mainly due to two phenomena. Firstly, wear is caused by poor mechanical behaviour and friction [18] releasing wear particles. Wear particles, generated from the friction and abrasion of materials, can trigger inflammatory responses in surrounding tissues. These particles, often found in orthopedic implants, are recognized as foreign bodies by the immune system, leading to the activation of osteoclasts and consequent periprosthetic osteolysis (PPOL), ultimately leading to implant aseptic loosening, which is the most common cause of long-term implant failure [19] (Fig. 1.5).

Secondly, the application of one implant induces a strong alteration of the mechanical environment normally known as stress-shielding [20, 21] that in the long term produces remodelling of the surrounding tissue, which modifies the properties of the tissue around the implant [22].

Not all the implants remain permanently in the body, recently, an artificial titanium heart has been successfully implanted in a patient during three months, being used as a stopgap while he is waiting for a donor heart [23]. The patient lived with the device for more than 3 months until he underwent surgery to receive a donated human heart. In this case, a device is implanted until a transplant is produced.

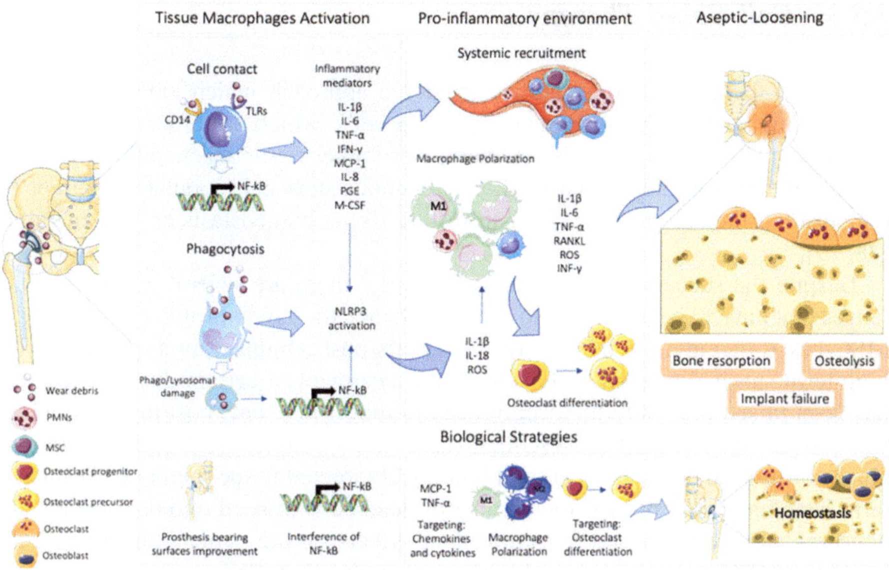

Fig. 1.5 The response of macrophages to implant wear particles and biological strategies to prevent aseptic loosening [19]. Open access article distributed under the terms and conditions of the Creative Commons Attribution (CC BY) license

Therefore, despite the intensive use of implants, they are suboptimal structures that are not adaptive and in the long-term lead to different types of problems as it has been described.

1.3.2 Grafting

Solid organ transplantation has been available to patients for the past half-century; however, it is only within the past 30 years that the development of immunosuppression has allowed the long-term graft survival with minimal risk of rejection [24].

Moreover, in the last years, tissue engineering has provided new technologies like decellularization, 3D bioprinting, cell and organ engineering, and blastocyst complementation, which may offer more possibilities for transplantable organs. Nevertheless, the complexity of solid organs poses a significant challenge, and further research and substantial investments are needed, as well as a synergic collaboration among all stakeholders, namely scientists, academia, industry, funding agencies, and governmental institutions [25].

Currently, the high demand of solid organs from patients, have forced the use of animal organs for humans, normally known as xenotransplantation [26]. In February 2025, the US Food and Drug Administration (FDA) approved the first trial of pig organ transplants as a potential solution to the global shortage of donor organs [27].

1.3.3 Cell-Based Therapy

Cell-based therapy is a regenerative medicine approach which aims to repair or replace damaged tissues and organs by using cells, which are often mesenchymal stem cells (MSCs). So, MSC-based therapy promotes tissue regeneration by stimulating, modulating and regulating the endogenous stem cell population, and/or by replenishing the cell pool, in order to restore tissue homeostasis and regeneration [28, 29].

Despite the success of MSC-based therapies, there are several important challenges that have to be addressed to further enhance the safety profile and efficacy of MSC-based therapy: (1) the fate of these cells post administration, especially the long-term survival of allogeneic cells in the treatment of certain diseases (caution has been raised related to the occurrence of embolism events post infusion, which was proven to be related to MSC-induced innate immune attack), (2) the homing ability of infused cells, as successful homing at targeted tissue might result in long-term benefits to patients, (3) the number of dead cells infused into the patients and concerns are always raised related to their effects on the patient's health, (4) the impact of MSC sources on their downstream application [28].

1.3.4 Tissue Engineering

Finally, tissue engineering (TE) provides a different strategy, which consists on the combination of cell-based therapy with materials or scaffolds. Instead of applying cells in the damaged or injured area, TE proposes using these cells embedded in a material or scaffold that protects them and provides them an architecture to support them. Therefore, scaffolds are three-dimensional (3D) structures that mimic the native ECM of tissues and provide a substrate for cell adhesion and proliferation [17]. These scaffolds can be made of a variety of materials, e.g. synthetic or natural polymers, which can be used to regenerate damaged or deteriorated tissues, or to deliver drugs to specific locations. In the next sections, we explain in more detail the basic concepts and applications in TE.

1.4 Basics Concepts of Tissue Engineering

The term known as Tissue Engineering was officially defined in the 1988 in one National Science Foundation Workshop as "...the application of principles and methods of engineering and life sciences toward fundamental understanding of structure-function relationship in normal and pathological mammalian tissues and the development of biological substitutes to restore, remain or improve tissue function" [30]. A few years later, the famous article by Langer and Vacanti published in

Science in 1993 provided the most utilised definition of tissue engineering: "an interdisciplinary field that applies the principles of engineering and life sciences toward the development of biological substitutes that restore, maintain, or improve tissue function or a whole organ" [31]. Indeed, tissue engineering combines methods and techniques from both life sciences and engineering to build tissues and organs in the laboratory that can then be used for implantation or study of tissue development and disease phenomena. Diverse disciplines such as biology, chemistry, materials science and engineering, mechanical engineering, physics, or medicine come into play to develop tissue-engineered constructs. In order to produce these constructs only three elements are needed, namely cells, scaffolds and stimuli. As already seen in this chapter, cells are considered the main actor in tissue regeneration and therefore, they need to be incorporated into constructs so they can produce the new tissue. Scaffolds are biomaterials that provide a 3D template for cells to carry out their functions. Finally, stimuli, both chemical and physical, provide cells with the necessary cues to direct their functions towards producing the tissue of interest. Thus, these three elements constitute the famous triad of tissue engineering to produce tissue substitutes (Fig. 1.6).

The classic tissue engineering approach (Fig. 1.7) involves isolating primary cells from the patient and seeding them onto a 3D porous scaffold designed to induce cell proliferation and differentiation. A bioartificial system is developed in a specific environment where metabolic, mechanical and electrical stimuli are provided by a bioreactor. In this tissue-specific system, the cells begin to produce an ECM, resulting in the maturation of the tissue in vitro. The tissue-engineered construct is then implanted into the patient and undergoes remodelling in vivo, allowing tissue restoration [32]. This classic tissue engineering approach has evolved and

Fig. 1.6 The three elements of the tissue engineering paradigm

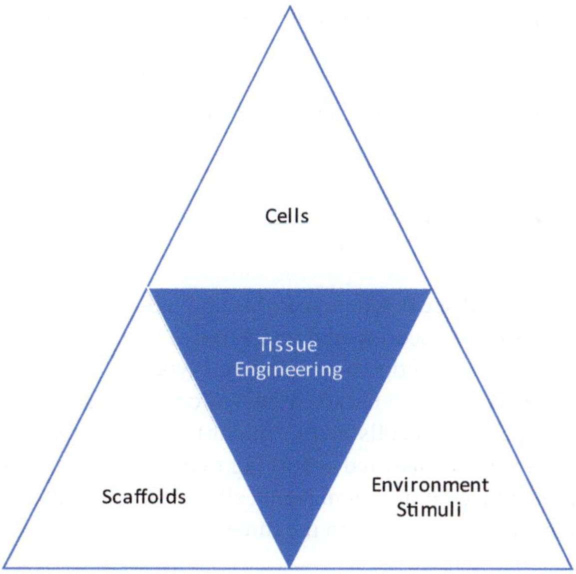

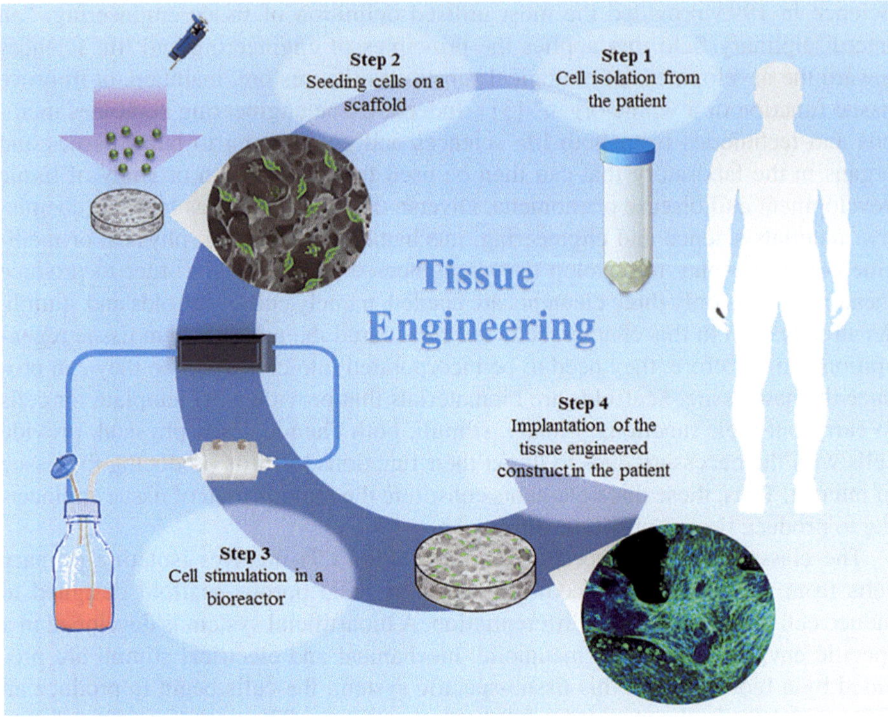

Fig. 1.7 Scheme of the classic tissue engineering approach [32]. Open access article distributed under the terms and conditions of the Creative Commons Attribution (CC BY) license

now the field does not only produce constructs for implantation, but also develops tissue and organ substitutes as models for the study of tissue development events and disease.

Next, each component that defines TE is described briefly, before being covered in detail in different chapters of this book.

1.4.1 Cells

As we will see in Chap. 4 of this book, there are various cell sources that can be used in tissue engineering and regenerative medicine (TERM), with stem cells providing a regenerative edge over other cell types.

Stem cells are undifferentiated cells that possess the ability to differentiate into the specialized cells that form our tissues or self-renew through division [33]. Stem cells can be classified according to their potency, with embryonic stem cells (ESCs) and induced pluripotent stem cells (iPSCs) showing pluripotency, meaning that they can generate cells from the three germ layers, namely the endoderm, mesoderm and ectoderm. Adult stem cells (ASCs) are multipotent and play a crucial role in tissue homeostasis [34, 35]. ASCs have the capacity to differentiate into multiple cell types, generally from a specific group. Ethical considerations are key in the choice of one type of stem cell over other for TERM strategies and research. Chap. 4 will also introduce and discuss the stem cell niche, which is the dynamic in vivo micro-environment where stem cells are found [36]. Therefore, understanding the complex stem cell niche is essential to be able to manipulate and control stem cell fate and behaviour for TERM strategies.

Apart from stem cells, other cells used in TERM include differentiated cells like fibroblasts, chondrocytes, endothelial cells or vascular smooth muscle cells for different tissue regeneration strategies, or cancer cells, either patient-derived or cell lines, for in vitro cancer models [37–39].

1.4.2 Scaffolds

A biomaterial that acts as a scaffold to allow cell attachment, infiltration, growth, function, and ultimately remodelling of its original composition and structure is crucial to produce functional and relevant tissue analogues. As we will see in Chap. 5, TERM scientists possess a biomaterial toolbox for designing and fabricating scaffolds. This toolbox includes polymers, both natural and synthetic, ceramics, metals and carbon-based materials [40–42]. Since all of these materials possess advantages and disadvantages, composites, i.e. combinations of two or more materials, are popular in TERM, as they can be tailored for a specific need [43].

Chapter 5 will discuss the requirements and characteristics necessary in scaffolds for TERM strategies. Some of these characteristics include biocompatibility, bioactivity, biodegradability, suitable mechanical properties, optimal surface properties, or optimal architecture. Some scaffold properties are vital for scaffold-cell interaction, as will be discussed in Chap. 7 [44, 45]. Many of the scaffold properties are dictated by the fabrication method employed to produce the scaffold. Diverse fabrication technologies are available, such as electrospinning, that renders scaffolds composed of ultra-fine fibres, or additive manufacturing techniques, popularly known as 3D printing, capable of producing custom-made scaffolds with controlled architectures. Fabricated scaffolds can have different forms and shapes, as well as formulations. For instance, hydrogels, a particular class of scaffolds, hold much current interest in TERM because they display a degree of flexibility comparable to that of natural tissue due to their significant water content [46]. Due to this intrinsic hydrated nature, hydrogels can be designed to mimic important aspects of the natural ECM of tissues, such as viscoelastic and swelling response.

1.4.3 Stimuli

Stimuli are the third element of the tissue engineering triad and can be biochemical or biophysical. In our bodies, cells inhabit a unique microenvironment where they continuously receive stimuli in the form of growth factors, ions, cytokines, matrikines, exosomes, pH signals, oxygen tension, shear stress, interstitial fluid flow, compression forces, or electrical signals, to name a few [47].

Important biochemical factors in TERM are growth factors, which in vivo are secreted and released into the tissue environment by cells and they act as signalling molecules that bind to specific receptors on the cell surface, triggering a cascade of intracellular events that ultimately influence cell behaviour [48]. Growth factors are incorporated into the scaffold element and particular attention is paid to the sustained release of these macromolecules to elicit an efficient response. Exosomes are another type of biochemical factor that is being investigated for TERM strategies, as discussed in Chap. 4 of this book. These extracellular vesicles are part of the secretome of cells and participate in cell-cell communication. The particular focus on exosomes is on their therapeutic effect, which could be leveraged for TERM applications [49].

Biophysical stimuli can be delivered to tissue-engineered constructs by bioreactors, which are devices that create a controlled and dynamic environment for cell and tissue culture, thereby replicating the physiological conditions of specific tissues or organs in terms of biomechanical forces or electrical stimulation [50]. Bioreactors will be discussed in Chap. 9 of this book.

1.5 Some Preliminary Success Cases in TERM

Tissue engineering, defined as a methodology for the repair and restoration of human tissue, has produced many advances in regenerative medicine in recent years. Indeed, tissue engineering has made it possible to treat a variety of clinical challenges by applying the latest technology to harness the body's own healing and regenerative abilities [51]. Here, some success cases for different tissues and organs are presented here according to the four levels of fabrication complexity described by Shafiee and Atala [52], ranging from the least to the highest complexity: flat tissues and organs such as skin and cartilage; tubular organ structures such as blood vessels and tracheas; hollow nontubular organ structures such as the bladder; and solid organs, including the heart, kidney, bone and liver.

1.5.1 *Flat Tissues and Organs*

Although with strong limitations as presented below, tissue engineering has demonstrated its clinical feasibility in flat organs, such as cartilage and skin.

1.5.1.1 Cartilage

Articular cartilage is one of the tissues that suffers the most damage and deterioration, mainly due to sports injuries, accidental trauma, and ageing. However, as shown in Table 1.1, its regeneration capacity is very low, normally resulting in osteoarthritis. This makes it amenable to repair through tissue engineering. Current surgical interventions offer limited capacity for cartilage tissue regeneration and only temporarily alleviate symptoms [53]. Nevertheless, cartilage tissue engineering strategies are increasingly recognised as promising approaches for cartilage restoration, albeit with considerable limitations [54].

At present, the most commonly used scaffolds for cartilage repair are polyester, polypeptides, polysaccharides, hydrogels and ECM materials (Fig. 1.8), which are mainly used to improve the mechanical properties and biocompatibility of scaffolds [53].

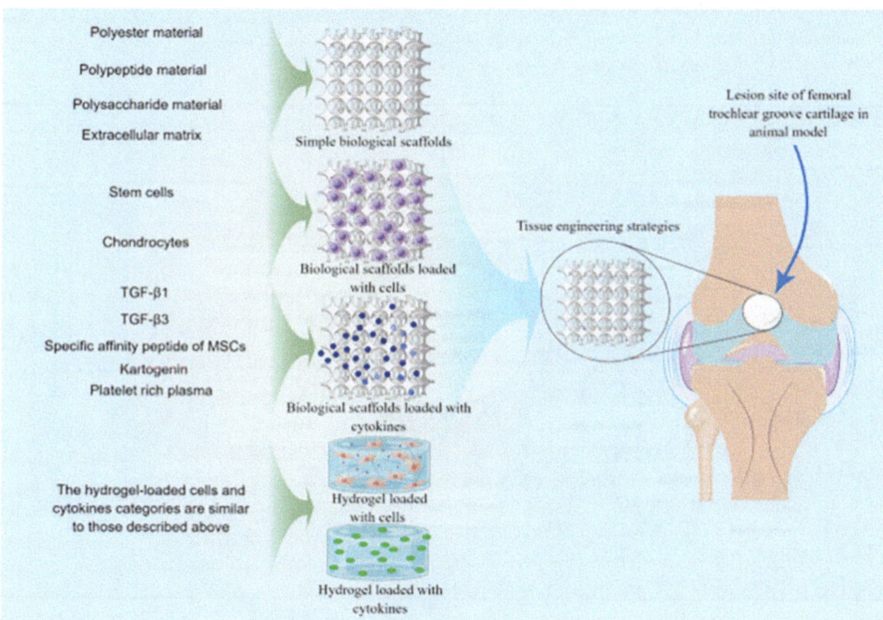

Fig. 1.8 Summary of tissue engineering strategies for cartilage repair [53]. Open access article distributed under the terms and conditions of the Creative Commons Attribution (CC BY) license

Although different strategies and treatments have been developed for the restoring of articular cartilage, all of them often fail to restore its mechanical function [55]. In fact, a key challenge is replicating the exceptional mechanical properties of articular cartilage, which are based on the intricate collagen–proteoglycan network [55]. Therefore, the most recent works focus on understanding how articular cartilage is developed, which arises between late embryonic and early juvenile development [55].

1.5.1.2 Skin

Skin tissue engineering can be used to heal extensive deep tissue injuries, such as large burns, and other conditions involving skin loss. This technique through bioengineered skin substitutes can restore both dermal and epidermal tissue. These substitutes can reduce the time required for stable wound closure with minimal scarring in patients who have insufficient donor sites for autologous partial-thickness skin grafts. However, no commercially available product has yet fully met this goal, particularly in terms on minimal scarring [56, 57]. Tissue-engineered skin substitutes may include cells, biopolymer scaffolds, and drugs and require regulatory approval to demonstrate their safety and efficacy. The clinical translation of these substitutes is not straightforward, nor it is easy to commercialise. Figure 1.9 outlines the main challenges and reasons for the scarcity of commercially available skin substitutes, as detailed in this review work [57].

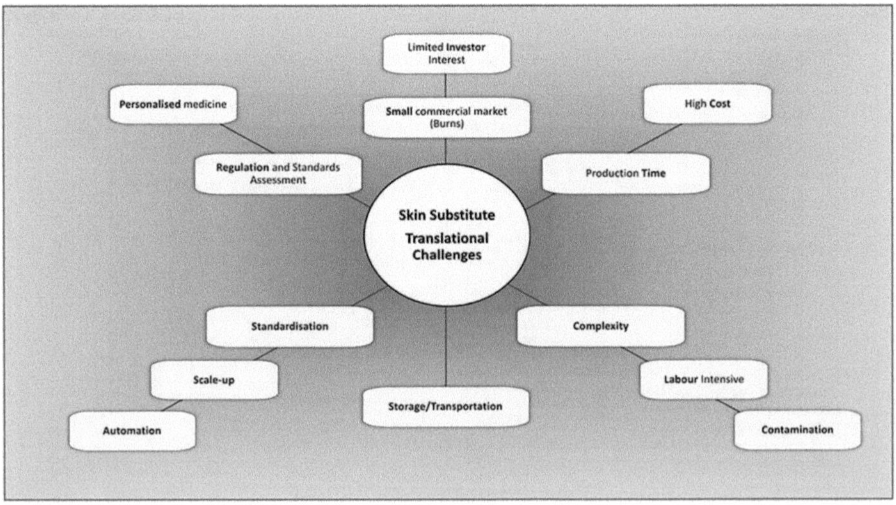

Fig. 1.9 Challenges and considerations in bioengineering of skin substitutes [57]. Open access article distributed under the terms and conditions of the Creative Commons Attribution (CC BY) license

One of the most promising tissue engineering strategies for developing skin grafts to treat large skin injuries is 3D bioprinting [56]. Nevertheless, this technique is still under development, despite preliminary success in fabricating artificial skin constructs using 3D bioprinting [58, 59]. There are still challenges to improve, such as the high cost, limited effusiveness and the inability to reconstitute skin appendages.

1.5.2 Tubular Organs

Tubular organ structures such as blood vessels and tracheas have been fabricated and used in human patients [52].

1.5.2.1 Blood Vessels

Vascular tissue engineering has enormous applications in several vascular diseases, such as coronary artery bypass grafting (CABG) [60], haemodialysis [61], peripheral arterial disease [62] and congenital heart surgery [63].

These tissue-engineered vascular grafts have been tested in large animal preclinical models (Fig. 1.10) showing that after scaffold degradation, the resulting neovessel undergoes growth and remodeling mimicking the native vessel behavior [63].

Despite the wide application of vascular tissue engineering, there are some challenges that need to be addressed [64]. One primary challenge is the high risk of restenosis in small-diameter artificial human blood vessels, which strictly limits their clinical application. Restenosis mainly occurs due to the poor blood compatibility of the materials, which can cause coagulation and thrombosis, as well as the difficulty of achieving endothelialisation of the implanted grafts. Secondly, the immediate availability of engineered blood vessels is also an obstacle. The production of artificial blood vessels with cells takes several months. Although developing smooth muscle or endothelial cells from induced pluripotent stem cells can reduce the time required, the process still takes several weeks to fabricate the graft. Therefore, there is a great need for technology to be developed that can reduce the time required and increase the immediate availability. Thirdly, obtaining blood vessel cells or stem cells from patients is limited. Fourthly, the sterilisation, packaging, storage and transport of grafts greatly hinders the application of artificial blood vessels. To overcome these challenges, the use of novel materials and advanced manufacturing techniques must be investigated [64].

1.5.2.2 Trachea

Patients with long-segment tracheal defects or stenoses cannot undergo conventional tracheal resection and anastomosis surgery; tracheal tissue engineering is therefore an alternative [65]. TE involves choosing the scaffold, selecting the type

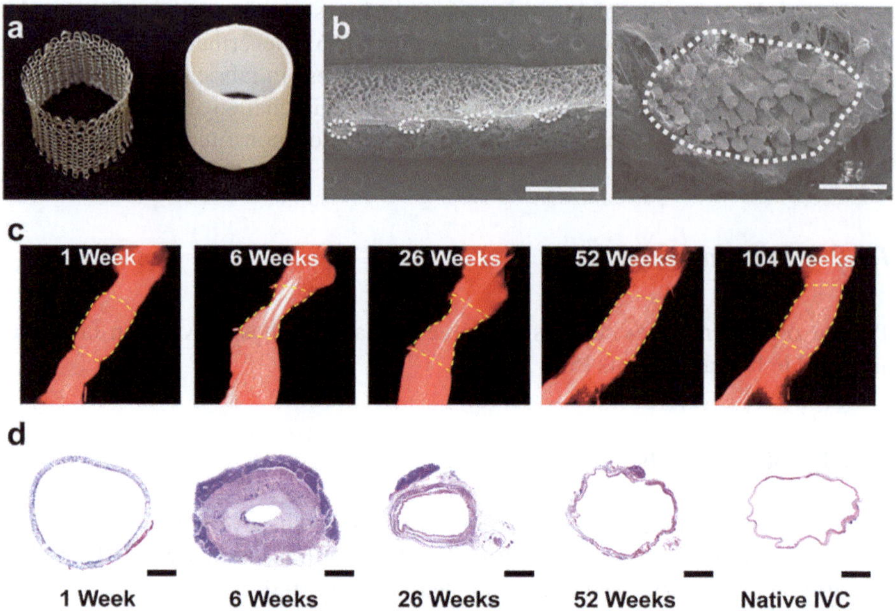

Fig. 1.10 (**a**) Vascular scaffold fabricated from PGA fibers that were knitted into a tube (left) and coated on the inner and outer surface with a PCLA (right). (**b**) Magnified SEM images of the scaffold. Scale bars 1 mm left, 100 µm right. (**c**) Representative 3D reconstructions of a TEVG (outlined in yellow) over its 2-year implantation as an IVC interposition graft in a sheep model. (**d**) Representative histologic H&E images demonstrated characteristic changes in TEVG over the 1-year time course. Scale bar 4 mm. *PGA* polyglycolic acid, *PCLA* polycaprolactone-lactide, *SEM* scanning electron microscope, *IVC* inferior vena cava [63]. Open access article distributed under the terms and conditions of the Creative Commons Attribution (CC BY) license

of stem cells to seed the scaffold with, culturing and expanding the cells, ensuring graft revascularisation at the time of in vivo implantation and using experimental models for tracheal research [65]. Despite these challenges, there have been several successful cases reported in the literature. For instance, the fabrication and transplantation of a trachea into a paediatric patient has been documented, with a follow-up period of 4 years [66]. In this case, a trachea from a deceased donor was decellularised and seeded with the recipient's respiratory epithelium and mesenchymal stromal cells. Computed tomography images of the neck and chest were acquired at various time points post-transplantation. A differentiated respiratory layer was observed and abnormal immune activity was not detected by histocytology. This study marks the first time that computational fluid dynamics has been used to measure tracheal airflow. This method could provide detailed information on how geometrical features correlate with airflow distribution. In a separate study, Zopf et al. [67] used 3D bioprinters to fabricate a tracheal splint that was implanted into a newborn.

1.5.3 Hollow Nontubular Organs

The bladder, an example of a non-tubular hollow organ structure, has been successfully engineered [52, 68]. Autologous biodegradable bladder-shaped scaffolds fabricated from collagen or a collagen-polyglycolic acid composite have been shown to perform better in the long term [69]. Autologous bladder cells were cultured and seeded into the scaffolds. A few weeks later, the bladders were implanted into patients. After a mean of 46 months, the organs were functioning normally [68]. Four girls aged 13–18 with vaginal aplasia successfully received tissue-engineered autologous vaginas [70]. As of the report date, the engineered vaginal organs had been functioning normally for up to 8 years.

Despite the success development of bladder scaffolds, tissue engineering is nowadays marginally influencing urological management [71]. Tissue engineering therapies are mainly limited by their inability to effectively recapitulate the complex cellular, structural, and mechanical environment of native tissues when transitioning from in vitro to in vivo applications [72]. Therefore, new advances and developments in tissue engineering, focusing on manipulation with cells and new biomaterials, will help into the development of reconstructive urology.

1.5.4 Solid Organs

Solid organs are the most complex to fabricate [52]. Although tissue-engineered flat, tubular and non-tubular hollow organs are being used clinically, solid organs have not yet been developed using current technology. Nevertheless, advancements in tissue engineering encompass a variety of approaches, including 3D-printed tissues, cell-based self-assembly and incorporation of cells into synthetic or xenogeneic ECM scaffolds [73].

Tissue engineering has provided some good results with limitations in bones and muscles, however, more difficulties are found in internal organs, such as, livers or kidneys. The next subsections provide a brief overview of these applications.

1.5.4.1 Bone

Bone tissue engineering (BTE) is a clear example of an application where tissue engineering is achieving promising results in addressing large bone defects caused by trauma, infection, congenital malformations and tumours [74, 75]. However, BTE presents some significant clinical challenges, as demonstrated in the review by Hoveidaei et al. [74]. The review focuses on scaffold design, cell sources, growth factors and vascularisation strategies, and highlights their roles in developing effective treatments. While there has been some partial success in the vascularisation of bone defects using the human body as a bioreactor, so Warnke et al. [76]

successfully repaired an extensive mandibular bone defect by growing a custom bone transplant within the latissimus dorsi muscle of an adult male patient. However, vascularisation remains a significant hurdle, with novel strategies such as angiogenic factors, co-culture systems, and bioprinting currently being investigated [74].

Another relevant challenge in tissue engineering in general, and in BTE in particular, as it will be discussed in Chap. 8, is analysing the nervous system's role in bone regeneration. The simultaneous regeneration of bone and nerves, either through the use of materials or by enhancing endogenous neurogenic repair signals, has been proven to promote functional bone regeneration [77].

Despite all the recent advances in BTE, there are still other relevant clinical challenges, such as standardization of protocols, large-scale biomaterial production, personalized medicine approaches, cost-effectiveness, and regulatory issues [75].

1.5.4.2 Heart

Researchers are attempting to engineer cells and tissues in order to treat certain heart disorders, such as myocardial infarction. Various tissue engineering techniques have been employed to combine different cell types with biodegradable and non-degradable biomaterials in an effort to repair diseased or dysfunctional heart muscles [78–80]. Scaffold-free cell sheets, cardiac patches, and injectable hydrogels have been used to partially repair damaged heart tissue (Fig. 1.11).

Scaffold-free cell sheets consist of thin layers of cells, usually cardiomyocytes or stem cells, which are cultivated in vitro and then detached as a cohesive sheet, independent of traditional scaffold materials [81]. They are gaining traction in cardiac tissue engineering as a promising approach for treating heart failure and myocardial infarction [81–83].

Cardiac patches are scaffolds engineered to adhere to the heart's surface after myocardial infarction damage [84] They mimic the native ECM and present mechanical and electrical properties to support the heart's systolic pressure and prevent arrhythmias [78].

Injectable hydrogels are widely used in cardiac tissue engineering due to their strong permeability, biocompatibility, and biodegradability, although they present low mechanical properties and other disadvantages [78]. Nevertheless, despite these limitations, hydrogels are considered an ideal biomaterial for cardiac tissue regeneration in a clinical setting in the future.

Decellularized ECMs are also used in cardiovascular tissue engineering [85]. Heart tissue constructs containing cells have been produced using 3D printers and decellularized ECM [66, 86].

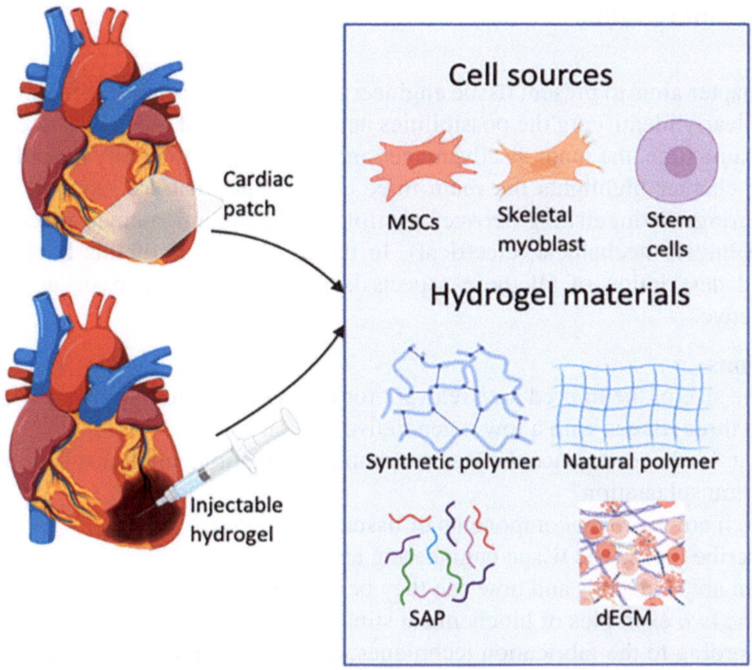

Fig. 1.11 Application of cardiac patches and injectable hydrogels for repairing damaged heart tissue, showing some examples of types of cells and hydrogel materials normally used [78]. *SAP* self-assembling peptide hydrogel. Open access article distributed under the terms and conditions of the Creative Commons Attribution (CC BY) license

1.5.4.3 Other Internal Organs

It is really challenging to recapitulate the native structure, composition and function of these other internal organs using synthetic, printing or self-assembly-based tissue engineering approaches. Therefore, allogeneic or xenogeneic ECM scaffolds are increasingly being used for these organs because they provide the native tissue ECM's structural, compositional and functional properties. However, the degree to which native tissue complexity is preserved varies depending on the antigen elimination/decellularisation/scaffold generation approach used [73]. When native ECM properties are retained, both allogeneic and xenogeneic ECM scaffolds have been shown to influence cell behaviour by directing the differentiation, proliferation, migration, phenotype and function of repopulating cells [87–89].

The liver and the kidney are two of these internal organs that can be regenerated using a whole organ tissue engineering approach [87, 90, 91].

1.6 Conclusions

This chapter aims to present tissue engineering as an approach of regenerative medicine, clearly identifying the possibilities and advantages of using this strategy. But at the same time, the main disadvantages and limitations are briefly described.

The chapter highlights the main three components that are required in tissue engineering, distinguishing between scaffolds, cells and microenvironmental stimuli (chemical, mechanical, electrical). In the next chapters of this book, a more detailed description of all these aspects is presented from a basic and concept perspective.

Questions

- Is the size of the injured area relevant for its regeneration? Why?
- Cite three tissues with a low regenerative capacity.
- What is the difference between xenotransplantation, allotransplantation and autotransplantation?
- Which are the main components of tissue engineering?
- Describe the classic tissue engineering approach.
- What are stem cells and how can they be classified?
- Name two examples of biochemical stimuli used in tissue engineering.
- According to the fabrication techniques, how would you classify human organs or tissues?

References

1. Caballero-Sánchez, N., Alonso-Alonso, S., & Nagy, L. (2024). Regenerative inflammation: When immune cells help to re-build tissues. *The FEBS Journal, 291*(8), 1597–1614.
2. Egger, B., Gschwentner, R., Hess, M. W., Nimeth, K. T., Adamski, Z., Willems, M., et al. (2009). The caudal regeneration blastema is an accumulation of rapidly proliferating stem cells in the flatworm Macrostomum lignano. *BMC Developmental Biology, 9*, 41.
3. Zielins, E. R., Ransom, R. C., Leavitt, T. E., Longaker, M. T., & Wan, D. C. (2016). The role of stem cells in limb regeneration. *Organogenesis, 12*, 16–27.
4. Marques, I. J., Lupi, E., & Mercader, N. (2019). Model systems for regeneration: Zebrafish. *Development, 146*(18), dev167692.
5. Bernard, N. J. (2018). Growth factors respond to cartilage damage. *Nature Reviews Rheumatology, 14*(8), 444–444.
6. Hettiaratchi, M. H., Chou, C., Servies, N., Smeekens, J. M., Cheng, A., Esancy, C., & Krishnan, L. (2017). Competitive protein binding influences heparin-based modulation of spatial growth factor delivery for bone regeneration. *Tissue Engineering Part A, 23*(13-14), 683–695.
7. Ribeiro, F. O., Gómez-Benito, M. J., Folgado, J., Fernandes, P. R., & García-Aznar, J. M. (2015). In silico mechano-chemical model of bone healing for the regeneration of critical defects: The effect of BMP-2. *PLoS One, 10*(6), e0127722.
8. Barrientos, S., Stojadinovic, O., Golinko, M. S., Brem, H., & Tomic-Canic, M. (2008). Growth factors and cytokines in wound healing. *Wound Repair and Regeneration, 16*(5), 585–601.

9. Reina-Romo, E., Gómez-Benito, M. J., Sampietro-Fuentes, A., Domínguez, J., & García-Aznar, J. M. (2011). Three-dimensional simulation of mandibular distraction osteogenesis: Mechanobiological analysis. *Annals of Biomedical Engineering, 39*, 35–43.
10. Roddy, E., DeBaun, M. R., Daoud-Gray, A., Yang, Y. P., & Gardner, M. J. (2018). Treatment of critical-sized bone defects: Clinical and tissue engineering perspectives. *European Journal of Orthopaedic Surgery & Traumatology, 28*, 351–362.
11. Gómez-Benito, M. J., González-Torres, L. A., Reina-Romo, E., Grasa, J., Seral, B., & García-Aznar, J. M. (2011). Influence of high-frequency cyclical stimulation on the bone fracture-healing process: Mathematical and experimental models. *Philosophical Transactions of the Royal Society A: Mathematical, Physical and Engineering Sciences, 369*(1954), 4278–4294.
12. Kiel, D. P., Hannan, M. T., Barton, B. A., Bouxsein, M. L., Sisson, E., Lang, T., & Rubin, C. T. (2015). Low-magnitude mechanical stimulation to improve bone density in persons of advanced age: A randomized, placebo-controlled trial. *Journal of Bone and Mineral Research, 30*(7), 1319–1328.
13. Thompson, W. R., Scott, A., Loghmani, M. T., Ward, S. R., & Warden, S. J. (2016). Understanding mechanobiology: Physical therapists as a force in mechanotherapy and musculoskeletal regenerative rehabilitation. *Physical Therapy, 96*(4), 560–569.
14. Huang, C., Holfeld, J., Schaden, W., Orgill, D., & Ogawa, R. (2013). Mechanotherapy: Revisiting physical therapy and recruiting mechanobiology for a new era in medicine. *Trends in Molecular Medicine, 19*(9), 555–564.
15. McKay, J., Nasb, M., & Hafsi, K. (2022). Mechanobiology-based physical therapy and rehabilitation after orthobiologic interventions: A narrative review. *International Orthopaedics, 46*(2), 179–188.
16. Satoh, A., Mitogawa, K., & Makanae, A. (2018). Nerve roles in blastema induction and pattern formation in limb regeneration. *The International Journal of Developmental Biology, 62*(9–10), 605–612.
17. Goncalves, A. D., Balestri, W., & Reinwald, Y. (2020). Biomedical implants for regenerative therapies. *Biomaterials, 9*, 1–36.
18. Mangan, F., Spece, H., Weiss, A. P. C., Ladd, A. L., Stockmans, F., & Kurtz, S. M. (2024). A review of wear debris in thumb base joint implants. *European Journal of Orthopaedic Surgery & Traumatology, 34*(1), 251–269.
19. Couto, M., Vasconcelos, D. P., Sousa, D. M., Sousa, B., Conceição, F., Neto, E., & Alves, C. J. (2020). The mechanisms underlying the biological response to wear debris in periprosthetic inflammation. *Frontiers in Materials, 7*, 274.
20. Baba, K., Mori, Y., Tanaka, H., Kanabuchi, R., Kuriyama, Y., Kurishima, H., & Aizawa, T. (2025). Effectiveness of stress shielding prevention using a low young's modulus Ti-33.6 Nb-4Sn stem: A 7-year follow-up study. *Medical Science, 13*(2), 51.
21. Kuiper, J. H., & Huiskes, R. (1997). The predictive value of stress shielding for quantification of adaptive bone resorption around hip replacements. *Journal of Biomechanical Engineering, 119*, 166–174.
22. Doblaré, M., & Garcıa, J. M. (2001). Application of an anisotropic bone-remodelling model based on a damage-repair theory to the analysis of the proximal femur before and after total hip replacement. *Journal of Biomechanics, 34*(9), 1157–1170.
23. Mallapaty, S. (2025). Man survives with titanium heart for 100 days—a world first. *Nature, 639*(8056), 848–849.
24. Vacanti, J. P., Otte, J. B., & Wertheim, J. A. (2014). Introduction: Regenerative medicine and solid organ transplantation from a historical perspective. In *Regenerative medicine applications in organ transplantation* (pp. 1–15). Academic.
25. Petrosyan, A., Montali, F., Peloso, A., Citro, A., Byers, L. N., La Pointe, C., & Orlando, G. (2022). Regenerative medicine technologies applied to transplant medicine. *An update. Frontiers in Bioengineering and Biotechnology, 10*, 1015628.
26. Cohen, J. (2025). The organ farm. *Science, 388*(6750), 906–912.

27. Mallapaty, S., & Kozlov, M. (2025). The science behind the first pig-organ transplant trial in humans. *Nature, 638*(8050), 303–304.
28. Hoang, D. M., Pham, P. T., Bach, T. Q., Ngo, A. T., Nguyen, Q. T., Phan, T. T., & Nguyen, L. T. (2022). Stem cell-based therapy for human diseases. *Signal Transduction and Targeted Therapy, 7*(1), 272.
29. Wang, J., Deng, G., Wang, S., Li, S., Song, P., Lin, K., & He, Z. (2024). Enhancing regenerative medicine: The crucial role of stem cell therapy. *Frontiers in Neuroscience, 18*, 1269577.
30. Vacanti, C. A. (2007). The history of tissue engineering. *Journal of Cellular and Molecular Medicine, 10*(3), 569.
31. Langer, R., & Vacanti, J. P. (1993). Tissue engineering. *Science, 260*(5110), 920–926. https://doi.org/10.1126/science.8493529
32. Caddeo, S., Boffito, M., & Sartori, S. (2017). Tissue engineering approaches in the design of healthy and pathological in vitro tissue models. *Frontiers in Bioengineering and Biotechnology, 5*, 40.
33. Tian, Z., Yu, T., Liu, J., Wang, T., & Higuchi, A. (2023). Introduction to stem cells. *Progress in Molecular Biology and Translational Science, 199*, 3–32.
34. Cerneckis, J., Cai, H., & Shi, Y. (2024). Induced pluripotent stem cells (iPSCs): Molecular mechanisms of induction and applications. *Signal Transduction and Targeted Therapy, 9*(1), 112.
35. Kwon, S. G., Kwon, Y. W., Lee, T. W., Park, G. T., & Kim, J. H. (2018). Recent advances in stem cell therapeutics and tissue engineering strategies. *Biomaterials Research, 22*, 1–8.
36. Scodellaro, C., Pina, R. R., Ferreira, F. C., Sanjuan-Alberte, P., & Fernandes, T. G. (2024). Unlocking the potential of stem cell microenvironments in vitro. *Bioengineering, 11*(3), 289.
37. Parkinson, G. T., Salerno, S., Ranji, P., Håkansson, J., Bogestål, Y., Wettergren, Y., & Landberg, G. (2021). Patient-derived scaffolds as a model of colorectal cancer. *Cancer Medicine, 10*(3), 867–882.
38. Pele, K. G., Calderón-Villalba, A., Amaveda, H., Mora, M., Zhang-Zhou, J., Pérez, M. Á., & García-Gareta, E. (2025). Novel hydrogel-based cancer-on-a-chip models for growth of 3D multi-cellular structures and investigation of early angiogenesis in pancreatic ductal adenocarcinoma. *Colloids and Surfaces B: Biointerfaces, 253*, 114736.
39. Rouwkema, J., Rivron, N. C., & van Blitterswijk, C. A. (2008). Vascularization in tissue engineering. *Trends in Biotechnology, 26*(8), 434–441.
40. García-Gareta, E. (2023). Protein-based materials for wound healing. In *Handbook of the extracellular matrix: Biologically-derived materials* (pp. 1–28). Springer International Publishing.
41. Jin, M., Shi, J., Zhu, W., Yao, H., & Wang, D. A. (2021). Polysaccharide-based biomaterials in tissue engineering: A review. *Tissue Engineering Part B: Reviews, 27*(6), 604–626.
42. Kohli, N., Sharma, V., Brown, S. J., & García-Gareta, E. (2019). Synthetic polymers for skin biomaterials. In *Biomaterials for skin repair and regeneration* (pp. 125–149). Woodhead Publishing.
43. Poh, P. S., Woodruff, M. A., & García-Gareta, E. (2020). Polymer-based composites for musculoskeletal regenerative medicine. In *Biomaterials for organ and tissue regeneration* (pp. 33–82). Woodhead Publishing.
44. Frost, O. G., Owji, N., Thorogate, R., Kyriakidis, C., Sawadkar, P., Mordan, N., & Garcia-Gareta, E. (2021). Cell morphology as a design parameter in the bioengineering of cell–biomaterial surface interactions. *Biomaterials Science, 9*(23), 8032–8050.
45. Majhy, B., Priyadarshini, P., & Sen, A. K. (2021). Effect of surface energy and roughness on cell adhesion and growth–facile surface modification for enhanced cell culture. *RSC Advances, 11*(25), 15467–15476.
46. He, Y., DeSart, K., Kubilis, P. S., Irwin, A., Tran-Son-Tay, R., Nelson, P. R., & Berceli, S. A. (2020). Heterogeneous and dynamic lumen remodeling of the entire infrainguinal vein bypass grafts in patients. *Journal of Vascular Surgery, 71*, 1620.
47. Redondo, P. A., Pavlou, M., Loizidou, M., & Cheema, U. (2017). Elements of the niche for adult stem cell expansion. *Journal of Tissue Engineering, 8*, 2041731417725464.

48. Ren, X., Zhao, M., Lash, B., Martino, M. M., & Julier, Z. (2020). Growth factor engineering strategies for regenerative medicine applications. *Frontiers in Bioengineering and Biotechnology, 7*, 469.
49. Kalluri, R., & LeBleu, V. S. (2020). The biology, function, and biomedical applications of exosomes. *Science, 367*(6478), eaau6977.
50. Castro, N., Ribeiro, S., Fernandes, M. M., Ribeiro, C., Cardoso, V., Correia, V., & Lanceros-Mendez, S. (2020). Physically active bioreactors for tissue engineering applications. *Advanced Biosystems, 4*(10), 2000125.
51. Esdaille, C. J., Washington, K. S., & Laurencin, C. T. (2021). Regenerative engineering: A review of recent advances and future directions. *Regenerative Medicine, 16*(5), 495–512.
52. Shafiee, A., & Atala, A. (2017). Tissue engineering: Toward a new era of medicine. *Annual Review of Medicine, 68*(1), 29–40.
53. Yang, C., Chen, R., Chen, C., Yang, F., Xiao, H., Geng, B., & Xia, Y. (2024). Tissue engineering strategies hold promise for the repair of articular cartilage injury. *Biomedical Engineering Online, 23*(1), 92.
54. Nordberg, R. C., Bielajew, B. J., Takahashi, T., Dai, S., Hu, J. C., & Athanasiou, K. A. (2024). Recent advancements in cartilage tissue engineering innovation and translation. *Nature Reviews Rheumatology, 20*(6), 323–346.
55. Pueyo Moliner, A., Ito, K., Zaucke, F., Kelly, D. J., de Ruijter, M., & Malda, J. (2025). Restoring articular cartilage: Insights from structure, composition and development. *Nature Reviews Rheumatology*, 1–18.
56. Balavigneswaran, C. K., Selvaraj, S., Vasudha, T. K., Iniyan, S., & Muthuvijayan, V. (2023). Tissue engineered skin substitutes: A comprehensive review of basic design, fabrication using 3D printing, recent advances and challenges. *Biomaterials Advances, 153*, 213570.
57. Dearman, B. L., Boyce, S. T., & Greenwood, J. E. (2021). Advances in skin tissue bioengineering and the challenges of clinical translation. *Frontiers in Surgery, 8*, 640879.
58. Augustine, R. (2018). Skin bioprinting: A novel approach for creating artificial skin from synthetic and natural building blocks. *Progress in Biomaterials, 7*(2), 77–92.
59. Daikuara, L. Y., Chen, X., Yue, Z., Skropeta, D., Wood, F. M., Fear, M. W., & Wallace, G. G. (2022). 3D bioprinting constructs to facilitate skin regeneration. *Advanced Functional Materials, 32*(3), 2105080.
60. Bao, L., Tang, J., Hong, F. F., Lu, X., & Chen, L. (2020). Physicochemical properties and in vitro biocompatibility of three bacterial Nanocellulose conduits for blood vessel applications. *Carbohydrate Polymers, 239*, 116246.
61. Baba, T., Ohki, T., Maeda, K., Ito, E., Momose, M., Chono, Y., Omori, M., & Shukuzawa, K. (2022). Evaluation of heparin-bonded ePTFE grafts for forearm loop vascular access: Comparison between Gore® PROPATEN vascular graft and ACUSEAL vascular graft. *The Journal of Vascular Access, 23*, 430–435.
62. He, W., Reaume, M., Hennenfent, M., Lee, B. P., & Rajachar, R. (2020). Biomimetic hydrogels with spatial-and temporal-controlled chemical cues for tissue engineering. *Biomaterials Science, 8*(12), 3248–3269.
63. Blum, K. M., Zbinden, J. C., Ramachandra, A. B., Lindsey, S. E., Szafron, J. M., Reinhardt, J. W., & Breuer, C. K. (2022). Tissue engineered vascular grafts transform into autologous neovessels capable of native function and growth. *Communications Medicine, 2*(1), 3.
64. Zhang, Y., Lin, L., Niu, M., Bian, F., Wang, W., & Zu, Y. (2025). Artificial human blood vessels for tissue engineering. *ACS Materials Letters, 7*(4), 1626–1645.
65. Mammana, M., Bonis, A., Verzeletti, V., Dell'Amore, A., & Rea, F. (2024). Tracheal tissue engineering: Principles and state of the art. *Bioengineering, 11*(2), 198.
66. Hamilton, N. J., Kanani, M., Roebuck, D. J., Hewitt, R. J., Cetto, R., Culme-Seymour, E. J., & Birchall, M. A. (2015). Tissue-engineered tracheal replacement in a child: A 4-year follow-up study. *American Journal of Transplantation, 15*(10), 2750–2757.

67. Zopf, D. A., Hollister, S. J., Nelson, M. E., Ohye, R. G., & Green, G. E. (2013). Bioresorbable airway splint created with a three-dimensional printer. *New England Journal of Medicine, 368*(21), 2043–2045.
68. Atala, A., Bauer, S. B., Soker, S., Yoo, J. J., & Retik, A. B. (2006). Tissue-engineered autologous bladders for patients needing cystoplasty. *The Lancet, 367*(9518), 1241–1246.
69. Oberpenning, F., Meng, J., Yoo, J. J., & Atala, A. (1999). De novo reconstitution of a functional mammalian urinary bladder by tissue engineering. *Nature Biotechnology, 17*(2), 149–155.
70. Raya-Rivera, A. M., Esquiliano, D., Fierro-Pastrana, R., López-Bayghen, E., Valencia, P., Ordorica-Flores, R., & Atala, A. (2014). Tissue-engineered autologous vaginal organs in patients: A pilot cohort study. *The Lancet, 384*(9940), 329–336.
71. Ławkowska, K., Rosenbaum, C., Petrasz, P., Kluth, L., Koper, K., Drewa, T., & Trauma and Reconstructive Urology Working Party of the European Association of Urology Young Academic Urologists. (2023). Tissue engineering in reconstructive urology—The current status and critical insights to set future directions-critical review. *Frontiers in Bioengineering and Biotechnology, 10*, 1040987.
72. O'Donnell, B. T., Ives, C. J., Mohiuddin, O. A., & Bunnell, B. A. (2019). Beyond the present constraints that prevent a wide spread of tissue engineering and regenerative medicine approaches. *Frontiers in Bioengineering and Biotechnology, 7*, 95.
73. Somasekhar, L., & Griffiths, L. G. (2023). Current challenges and future promise for use of extracellular matrix scaffold to achieve the whole organ tissue engineering Moonshot. *Stem Cells Translational Medicine, 12*(9), 588–602.
74. Hoveidaei, A. H., Sadat-Shojai, M., Nabavizadeh, S. S., Niakan, R., Shirinezhad, A., MosalamiAghili, S., & Tabaie, S. (2024). Clinical challenges in bone tissue engineering-a narrative review. *Bone, 117363*.
75. Jagadale, S., Damle, M., & Joshi, M. G. (2025). Bone tissue engineering: From biomaterials to clinical trials. *Advances in Experimental Medicine and Biology, 2025*(1479), 73–115. https://doi.org/10.1007/5584_2024_841
76. Warnke, P. H., Springer, I. N. G., Wiltfang, J., Acil, Y., Eufinger, H., Wehmöller, M., & Terheyden, H. (2004). Growth and transplantation of a custom vascularised bone graft in a man. *The Lancet, 364*(9436), 766–770.
77. Sun, W., Ye, B., Chen, S., Zeng, L., Lu, H., Wan, Y., & Guo, X. (2023). Neuro–bone tissue engineering: Emerging mechanisms, potential strategies, and current challenges. *Bone Research, 11*(1), 65.
78. Holme, S., Richardson, S. M., Bella, J., & Pinali, C. (2025). Hydrogels for cardiac tissue regeneration: Current and future developments. *International Journal of Molecular Sciences, 26*(5), 2309.
79. Tenreiro, M. F., Louro, A. F., Alves, P. M., & Serra, M. (2021). Next generation of heart regenerative therapies: Progress and promise of cardiac tissue engineering. *npj Regenerative Medicine, 6*(1), 30.
80. Zhang, D., He, R., Qu, Y., He, C., & Chu, B. (2024). Application of biomaterials in cardiac tissue engineering: Current status and prospects. *MedComm - Biomaterials and Applications, 3*(4), e103.
81. De Pieri, A., Rochev, Y., & Zeugolis, D. I. (2021). Scaffold-free cell-based tissue engineering therapies: Advances, shortfalls and forecast. *npj Regenerative Medicine, 6*(1), 18.
82. Kamata, S., Miyagawa, S., Fukushima, S., Imanishi, Y., Saito, A., Maeda, N., & Sawa, Y. (2014). Targeted delivery of adipocytokines into the heart by induced adipocyte cell-sheet transplantation yields immune tolerance and functional recovery in autoimmune-associated myocarditis in rats. *Circulation Journal, 79*(1), 169–179.
83. Narita, T., Shintani, Y., Ikebe, C., Kaneko, M., Campbell, N. G., Coppen, S. R., & Suzuki, K. (2013). The use of scaffold-free cell sheet technique to refine mesenchymal stromal cell-based therapy for heart failure. *Molecular Therapy, 21*(4), 860–867.
84. Li, M., Wu, H., Yuan, Y., Hu, B., & Gu, N. (2022). Recent fabrications and applications of cardiac patch in myocardial infarction treatment. *View, 3*(2), 20200153.

85. Kafili, G., Kabir, H., Jalali Kandeloos, A., Golafshan, E., Ghasemi, S., Mashayekhan, S., & Taebnia, N. (2023). Recent advances in soluble decellularized extracellular matrix for heart tissue engineering and organ modeling. *Journal of Biomaterials Applications, 38*(5), 577–604.
86. Pati, F., Jang, J., Ha, D. H., Won Kim, S., Rhie, J. W., Shim, J. H., & Cho, D. W. (2014). Printing three-dimensional tissue analogues with decellularized extracellular matrix bioink. *Nature Communications, 5*(1), 3935.
87. Baptista, P. M., Siddiqui, M. M., Lozier, G., Rodriguez, S. R., Atala, A., & Soker, S. (2011). The use of whole organ decellularization for the generation of a vascularized liver organoid. *Hepatology, 53*(2), 604–617.
88. Bowers, S. L., Banerjee, I., & Baudino, T. A. (2010). The extracellular matrix: At the center of it all. *Journal of Molecular and Cellular Cardiology, 48*(3), 474–482.
89. Mazza, G., Al-Akkad, W., Telese, A., Longato, L., Urbani, L., Robinson, B., & Pinzani, M. (2017). Rapid production of human liver scaffolds for functional tissue engineering by high shear stress oscillation-decellularization. *Scientific Reports, 7*(1), 5534.
90. Rayat Pisheh, H., Haghdel, M., Jahangir, M., Hoseinian, M. S., Rostami Yasuj, S., & Sarhadi Roodbari, A. (2024). Effective and new technologies in kidney tissue engineering. *Frontiers in Bioengineering and Biotechnology, 12*, 1476510.
91. Torabinavid, P., Khosropanah, M. H., Azimzadeh, A., & Kajbafzadeh, A. M. (2025). Current strategies on kidney regeneration using tissue engineering approaches: A systematic review. *BMC Nephrology, 26*(1), 66.

Chapter 2
Tissue Mechanobiology

Abstract Tissue engineering aims to restore normal physiological function of damaged or diseased tissues through a specifically tailored combination of cells, extracellular matrices (ECMs), and mechano-chemical stimuli. Therefore, it is crucial to have a good knowledge of our tissues, which are mainly composed of cells, soluble factors, water and the ECM. This matrix not only provides structural support for our cells in a given tissue, but also provides a rich source of signals for the cells. The first part of this chapter will therefore focus on the main mechanical properties, functions and composition of the ECM in different tissues. The properties of these native ECMs, from composition to architecture, are so relevant that it is common to use these decellularised matrices as scaffolds for various tissue engineering applications. The chapter concludes by highlighting the relevant role of cell-matrix adhesions. Finally, the chapter ends with conclusions and questions to reinforce the different concepts and aspects covered in this chapter.

2.1 Introduction

Living tissues are very different from the inert or engineered materials that are normally used to fabricate and develop tissue scaffolds. We need to consider that living tissues are functionally adaptive, updating their internal properties and external geometry according to requirements. Furthermore, living tissues have a hierarchical structure at all size levels [1], which is very difficult to reproduce with engineered materials. In addition to adaptation, growth and remodelling, biological tissues can heal and thus repair themselves.

In general terms, a living tissue consists of a group of cells and their extracellular matrix (ECM) that work together to perform a specific function. This shared functionality, which clearly defines the structural integrity of tissues is based on cell-cell adhesion, cell-extracellular matrix adhesion, and linkages between molecules of the ECM. The ECM also contains water with ions and various soluble factors (growth factors, cytokines, hormones, etc.).

© The Author(s), under exclusive license to Springer Nature Switzerland AG 2026

E. García-Gareta, J. M. García-Aznar, *Contemporary Tissue Engineering and Regenerative Medicine*, Series in BioEngineering,
https://doi.org/10.1007/978-3-032-03441-0_2

Living materials have been studied mainly as passive materials, providing a mechanical function to protect and support cells. But, additionally, living materials have to be able to support the gravitational loads, which are present in every organ of the human body, although other sources of force may occur. Organ or tissue specific loading conditions can be found, for example, in the musculoskeletal system, where muscle loading is responsible for the propulsion of the human body and, together with gravitational forces, joint forces and moments determine locomotion. This leads to mechanical loading and the development of local mechanical stresses and deformations in various tissues such as bone, cartilage, tendons and ligaments. Other examples can be found in the cardiovascular system, where the pumping action of the heart is responsible for the development of blood flow and pressure. Cardiac and vascular tissues are subjected to pulsating hydrostatic pressures and flow-induced shear stresses. Other examples include lung tissue, which is cyclically stretched during respiration, and skin tissue, which is subject to tensile, compressive and shear forces. All living tissues therefore have a clear mechanical function. However, living materials can have many other different functions as active materials, such as ion reservoirs, chemical barriers and catalytic response. All these aspects are going to be studied in this chapter.

2.2 Tissue Mechanics as a Porous Material: Tissues Present a Viscoelastic Behaviour

All our tissues are biphasic porous, i.e. a material that contains pores filled with a fluid with many different mechanical functions: absorption of impact loads, lubrication of joints, its flow ensures the transport (diffusion and convection) of nutrients and waste products and it also includes the transport of ions.

2.2.1 Quantifying the Interstitial Flow in Tissues

The characterization of interstitial flow in tissues requires to measure the interstitial resistance of tissues. For this aim, Darcy's law proposed an equation that describes the flow of a fluid through a porous medium for low Reynolds number (Fig. 2.1) [2]:

$$Q = \frac{KA}{L\eta}\Delta P \tag{2.1}$$

where Q is the volumetric flow rate, A is the cross-sectional area, L is the distance over which the pressure drop is evaluated, h is the dynamic viscosity of the fluid, ΔP is the pressure drop through the porous solid and K is the permeability of this porous solid. Therefore, this permeability is a property of porous materials that quantifies the ability for fluids (gas or liquid) to flow through them. This property is strongly

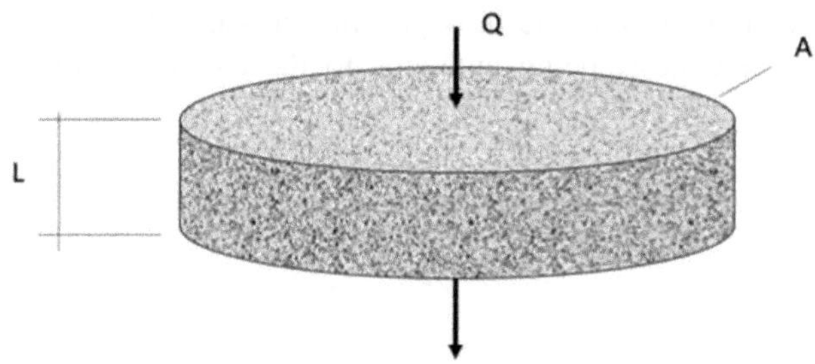

Fig. 2.1 Scheme corresponding to the Darcy's law for describing the flow of a fluid through a porous medium (Levick, 1987)

Table 2.1 Permeability values for different tissues

Tissue	Permeability	Units	References
Cortical bone	1.1×10^{-13}	m^2	Malachanne et al. [3]
Trabecular bone	2.56×10^{-11} to 7.43×10^{-8}	m^2	Rabiatul et al. [4]
Human femoral condyle cartilage	1.18×10^{-15}	$m^4\ N^{-1}\ s^{-1}$	Mow et al. [5]
Cornea	8×10^{-17}	m^2	Stewart et al. [6]
Meniscus fibrocartilage	$1.53–1.87 \times 10^{-15}$	$m^4\ N^{-1}\ s^{-1}$	Kleinhans and Jackson [7]
Brain	$0.5–8 \times 10^{-16}$	m^2	Jamal et al. [8]
Collagen type I hydrogel	$1–6 \times 10^{-12}$	m^2	Moreno-Arotzena et al. [9], Alamán-Díez et al. [10]
Fibrin-based hydrogel	5.73×10^{-13}	m^2	Moreno-Arotzena et al. [9]

dependent on the porosity, pore size and shape, and above all, pore connectivity. The permeabilities of many different tissues are presented in Table 2.1.

In the literature, there are different ways of measuring or quantifying the permeability, which are not identical and, consequently, are expressed in different units (Levick, 1987). The permeability that has been defined in Eq. (2.1) is also known as specific hydraulic conductivity (m^2), which is purely a property of the material, being conductivity to a fluid of unit viscosity:

$$K = \frac{LQ\eta}{A\Delta P}$$

The hydraulic conductance, also called hydraulic resistance or apparent permeability, refers to flow of a specified fluid per unit pressure drop across unit area of material (units, N s m^{-3}):

$$R = \frac{K}{L\eta} = \frac{Q}{A\Delta P}$$

The hydraulic conductivity is the flow per unit pressure drop across unit area of material of unit thickness, being analogue to the electrical conductivity (units, $m^4\,N^{-1}\,s^{-1}$):

$$K' = \frac{K}{\eta} = \frac{LQ}{A\Delta P}$$

2.2.2 Tissue Elasticity

Poroelastic materials exhibit a time-dependent mechanical response, i.e. viscoelastic respose, due to water flow into or out of the ECM porous network when a deformation is induced. However, normally only elasticity, through the elastic modulus or the shear modulus, has been measured in many tissues to compare their different behaviour [11] (Table 2.2).

Mechanical properties of tissues are a very relevant aspect to be considered, because in previous works [12] it was demonstrated that elasticity of substrates controls the differentiation of mesenchymal stem cells. However, traditionally, tissue engineering has focused on the role of chemical regulators (ie. growth factors, bioactive ECM moieties, etc.) within the extracellular environment, and for example, numerous works have confirmed that surface chemistry is a critical parameter contributing to the clinical success of implanted biomaterials and devices [13]. But increasing research works are demonstrating that mechanical properties of the ECM present a relevant role in the regulation of regenerative processes. It is worth to describe some of these works at different scale levels in more detail.

For example, it is very well known by surgeons that bone fracture healing is strongly dependent on the stiffness of the external fixation. Thus, a very low stiffness of the fixator delays fracture healing and causes a larger callus, while a very high stiffness of the fixator induces more bone remodeling than bone regeneration

Table 2.2 Elasticity properties of human tissues (these values have been summarized from many other works in the review work of [11])

Tissue	Modulus	Modulus value
Cortical bone	Elastic	10.4–20.7 GPa
Ligament	Elastic	25–93 MPa
Cartilage	Shear	5.7–6.2 MPa
Nerve	Elastic	5 MPa
Skin	Elastic	60–850 kPa
Bladder	Shear	50 kPa
Kidney	Shear	4–8 kPa
Liver	Elastic	4–6.5 kPa
Pancreas	Elastic	2.9 kPa
Brain	Shear	1–3 kPa

and absence of bone callus. Therefore, a medium stiffness is required to achieve a good regeneration with the corresponding adequate bone callus [14].

At the cellular level, it is really interesting to mention two experimental works. In a first work, the effect of matrix stiffness on the morphogenesis of multicellular epithelial tissues was studied, concluding that under compliant matrices, cells organise themselves with a morphology typical of a mammary duct in a compliant gland (167 Pa), but that increasing stiffness (170–1200 Pa) induces an organisation more similar to carcinoma [15]. In a second study, Engler et al. [12] found that only by tuning the elasticity of the matrix microenvironment could the differentiation of mesenchymal stem cells be regulated. Thus, by changing the stiffness of the substrate, human mesenchymal stem cells could be directed along neuronal, muscle or bone lineages.

Accordingly, the elasticity of most soft tissues has been found to be typically less than 10 kPa, whereas in pathological conditions such as fibrosis, elasticity exceeds 20 kPa [16] (Table 2.3). Excessive ECM production, remodelling and cross-linking contribute to increased matrix stiffness in some healing soft tissue disorders. While the acute wound healing cascade is tightly regulated and coordinated, chronic wounds (as in diabetes or ulcers) develop when the repair process is trapped, most commonly in the inflammatory response phase, unable to trigger the repair programme and progress to the next phase [17]. In the case of tumours, increased matrix stiffness is also produced, but in reality tumours are different from chronic wounds and are more similar to non-healing wounds [17].

However, some recent works highlighted that matrix viscoelasticity could be the main regulator of cell fate: proliferation, migration and differentiation [18–20]. In fact, as it is indicated in this chapter, most of previous works focused on the material elastic (storage) properties, underestimating the viscous (loss) properties, which is now considered to highly influence cell behavior.

2.2.3 *Tissue Viscoelasticity*

A short summary of the main concepts of viscoelasticity is here presented, given that viscoelastic constitutive law is the best approach to understanding the ECM mechanical behaviour. Viscoelastic constitutive laws are characterized by a

Table 2.3 Elasticity properties of human healthy tissues vs diseased tissues [16]

Tissue	Healthy tissue (kPa)	Disease	Diseased tissue (kPa)
Brain	0.5–1	Alzheimer	2–2.3
Lung parenchyma	0.5–9	Idiopatic pulmonary fibrosis	0.4–110
Skin	7–130	Sclerotic lesion	10–325
Breast fat tissue	0.4–25	Breast carcinoma	1.4–75
Liver	4–7	Hepatitis C	13–43
Colon	0.4–7.3	Colorectal cancer	1–68

combination of storage of elastic energy, as a solid, and loss of mechanical energy, as a fluid. This is reflected by hysteresis in the stress–strain relationship during loading and unloading. Viscoelastic materials exhibit stress relaxation in response to a constant deformation (see Fig. 2.2a), and increased strain, or creep, in response to a constant stress (Fig. 2.2b). Hence, relaxation and creep tests are normally used to quantify the response of viscoelastic materials.

Another common methodology to characterize viscoelastic materials is based on the application of an oscillatory load to the material. In this case, for a purely elastic material (stress is proportional to strain) the maximum stress occurs at maximum strain (when deformation is greatest) and both stress and strain are said to be in phase (Fig. 2.3a). For a purely viscous material (stress is proportional to strain rate) the maximum stress occurs when the strain rate is maximum (when flow rate is greatest) and stress and strain are out of phase by 90° or $\pi/2$ radians (quarter of a cycle) (Fig. 2.3c). For a viscoelastic material the phase difference between stress and strain will fall somewhere between the two extremes (Fig. 2.3b). This is illustrated in Fig. 2.3, being δ the phase angle.

Therefore, if we apply a sinusoidal shear deformation:

$$\gamma(\omega,t) = \gamma_0 \sin(\omega t)$$

The shear stress response of a viscoelastic material will be:

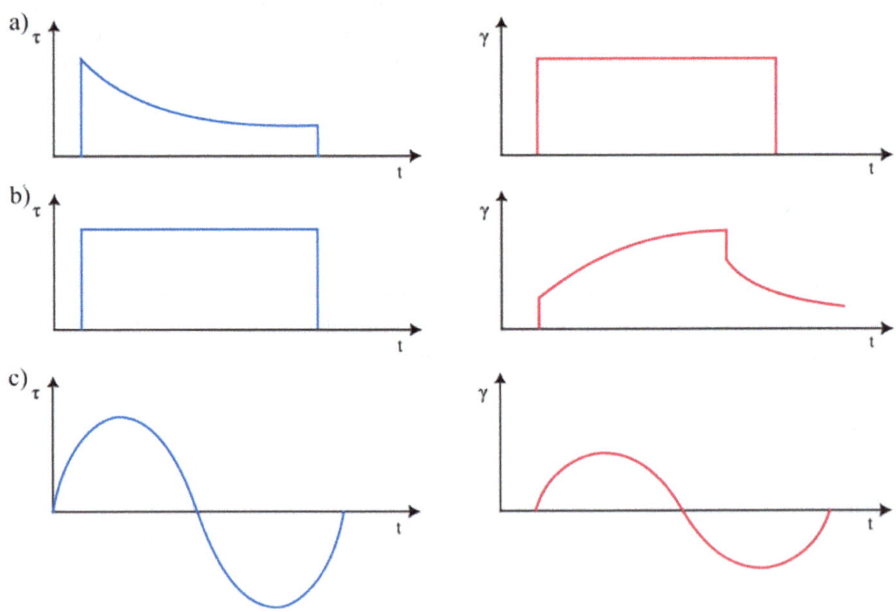

Fig. 2.2 Typical viscoelastic test: (**a**) Stress relaxation test, (**b**) Creep test and (**c**) oscillatory test [21]

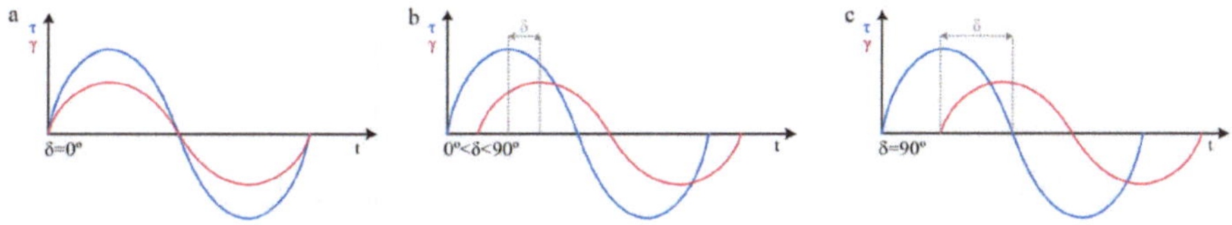

Fig. 2.3 Oscillatory load test normally used in viscoelastic materials, indicating in red the strain and in blue the stress: (**a**) Response of the material when the behaviour of the material is elastic, (**b**) Response of the material when the behaviour of the material is viscoelastic, (**c**) Response of the material when the behaviour is viscous [21]

$$\tau(\omega,t) = G' \cdot \gamma_0 \cdot \sin(\omega t) + G'' \cdot \gamma_0 \cdot \cos(\omega t)$$

where G′ is the storage modulus (it represents the storage of energy) and G″ is the loss modulus, (it represents energy loss).

It is also important to compute the angle phase delta as:

$$tg(\delta) = \frac{G''}{G'}$$

And the dynamic viscosity:

$$\eta = \frac{G''}{\omega}$$

Therefore, with these two parameters G′ and G″ we are able to characterize the viscoelastic properties of matrices.

In conclusion, relaxation, creep and oscillatory tests are the three testing methods normally used to characterize the behaviour of viscoelastic materials. The most appropriate method of measurement may depend on multiple factors that may be determined by external constraints ranging from the available facilities to the sample size.

In order to model this viscoelastic behaviour is common to use mathematical approaches of the constitutive equations based on power-laws. Traditionally, they are represented by springs and dampers in parallel or in series, as seen in the models by Maxwell and Kelvin/Voigt [22] or fitted through the Prony series [23]. A common approach that is clearly improving the viscoelasticity characterization relies on the use of fractional analysis [24], which is normally known as fractional viscoelasticity [22]. In this approach, an additional viscoelastic element is used, the springpot, which is a hybrid approach between a spring and a dashpot (see Fig. 2.4).

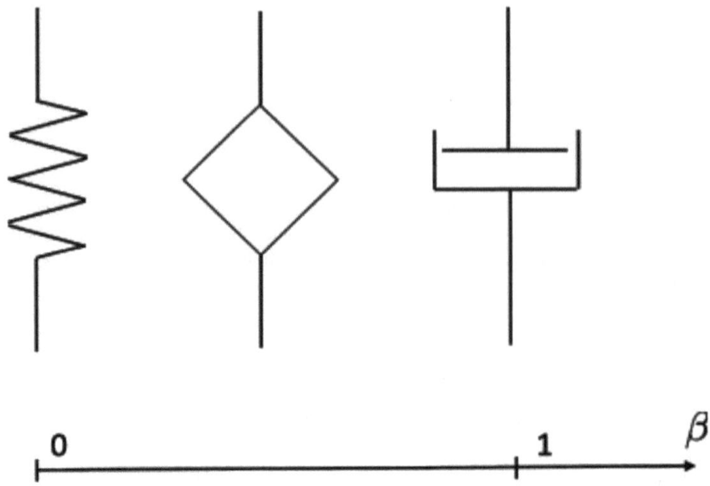

$$\sigma(t) = k\epsilon(t) \qquad\qquad \sigma(t) = \eta\frac{d\epsilon(t)}{dt}$$

$$\sigma(t) = c_\beta\frac{d^\beta\epsilon(t)}{dt^\beta}$$

Fig. 2.4 Scheme of the springpot, which behaves between a spring and a dashpot (based on [22])

2.3 Functions of Extracellular Matrix (ECM)

There are many functions of the ECM, some of which are general, while others depend largely on the tissue in which it is involved. Some of these general functions are described below, with a few particulars noted in each description.

2.3.1 Structural Support to Maintain Tissues and Organs

The most fundamental role of the ECM is to provide structural support and protection for our cells and to ensure the functionality of the organ/tissue involved. Therefore, as a material, the ECM provides sufficient elasticity and strength to guarantee its functionality and to allow the various cell types involved to perform their function [25]. For example, some tissues require high elasticity and extensibility for developing their physiological function, such as arteries, skin, lungs, or the heart which undergo reversible and repetitive deformations. However, other tissues require low deformability and high capacity for support and absorb loads, like our bones.

2.3.2 *Scaffold for Cells-Matrix Interaction: Mechanical Signaling*

Most cells are adherent and, therefore, they require a support to attach to, exert forces and develop their corresponding function. The ECM exactly provides this stable microenvironment that supports the adhesion, migration, proliferation, and differentiation of cells [26]. In fact, many cells die in suspension and they must adhere to a surface. Therefore, cell–ECM interactions play crucial roles in embryonic development, differentiation, tissue remodeling, and diseases including fibrosis and cancer [27].

The coordination between cell contractility, cell-matrix adhesion and matrix stiffness regulates cell mechano-transduction as will be widely explained in the next chapter. Therefore, cells are active elements that not only are able to remodel and alter the ECM, but also cells are able to deform it. Hence, cells receive multiple mechanical stimuli, which could come from different sources: external loads that support the matrix (such as the gravity), loads exerted by surrounding cells and by the cell itself. Then, cell will transduce these mechanical signals in a biological response as it can be cell proliferation, differentiation and matrix synthesis.

Although mechanical properties of the ECM have been the most cited in the research works as regulators of the cell response, there are other biophysical characteristics of the ECM that are really relevant, such as, porosity and pore connectivity (normally measured through the permeability [28]). Pores are essential for cell nutrition, tissue vascularization, proliferation, and migration [29], when this porosity is very low, cells can experience mechanical confinement. Cells within confinement undergo different structural changes, inducing cytoskeletal, chromatin and nuclear membrane proteins reorganization. All these alterations affect to cell morphology, migration, proliferation and differentiation responses [30].

2.3.3 *Exchange and Signalling Medium*

In addition to the regulatory role of the mechanical properties of the ECM, the ECM can provide further signalling information through its specific biochemical composition and the local concentrations of its constituents (e.g. for gene regulation), and can serve as a reservoir and source of signalling molecules, such as cytokines [31] and growth factors [32].

Moreover, the ECM molecules can bind to many soluble factors to regulate their activity, for example, Wang et al. [33] reported that BMP signaling molecule, decapentaplegic (Dpp), was bound to type IV collagen to form a Dpp gradient pattern for the correct development in Drosophila.

Interestingly, degradation by proteolysis releases ECM molecules that can act in a free soluble form. For example, laminins contain repeated epidermal growth factor (EGF)-like domains. When these domains are released proteolytically by matrix metalloproteinase (MMP), they can interact with EGF receptors to activate a downstream signaling pathway [34].

2.4 Composition of the ECM

The ECM is a non-cellular network of several types of macromolecules, in combination with smaller molecules, ions and water, that surrounds cells and is present in almost all biological tissues. As we have previously described, this ECM provides mechanical and biochemical support for cells in tissues, supplying cells with nutrients, local and systemic factors, and removing wastes. All these functions are essential for driving key cellular events, such as, migration, proliferation and differentiation. Certainly, different levels of mechanical and chemical support are provided depending on tissue specificity. In fact, the ECM is produced and assembled by the cells themselves as they proliferate, differentiate and migrate, in response to various regulatory microenvironmental stimuli, making it extraordinarily adaptable and dynamic.

ECM is composed of fibrillar molecules, such as, collagens, elastins, fibronectins and laminins (which give tissue strength and resilience), hydrophilic proteoglycans (PGs) and glycosaminoglycans (GAGs) (which form hydrated gels that act as a cushion), and other proteins/glycoproteins such as matricellular proteins [35, 36] (see Fig. 2.5). Next, a brief description of each component and its functionality is presented with respect to the ECM functions.

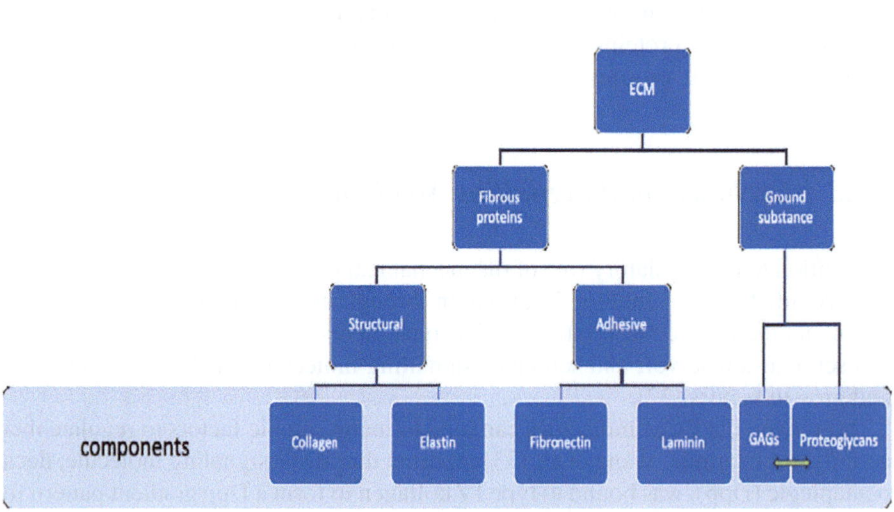

Fig. 2.5 Main components of the ECM and its functional organization

2.4.1 Collagen

Collagen is the most abundant fibrous protein within the interstitial ECM, constituting up to 30% of the total protein mass, being in particular collagen types I, II and III the most common, which account for 80–90% of total body collagen [35]. In fact, the collagen superfamily includes 28 members that contain at least one triple-helical domain and provide 3D structures for mechanical support of tissues and organs [37]. Collagens can be divided into two subgroups: fibril-forming and nonfibril-forming collagens (see Table 2.4).

Collagens, which constitute the main structural element of the ECM, provide tensile strength, regulate cell adhesion, support chemotaxis and migration, and direct tissue development.

2.4.2 Elastin

Elastin is one of the major structural proteins found in the native ECM, in fact, elastin fibres give resiliance. The elastic property is crucial for many tissues that goes from connective tissues to various organs, including vascular walls, lungs, skin, and so on.

The elastin network is produced by specific cells through tropoelastin (TE), which is cross-linked by fibrillins and other microfibril proteins [40]. Fibrillins also participate in the binding of proteins to elastin, but are also involved in cell signalling through interaction with syndecans and integrins, and the storage of the transforming growth factor β (TGFβ) family of GFs in the matrix. Elastin is mainly formed during development and childhood and gradually degrades during adulthood and ageing [35].

Lysyl oxidase (LOX) and LOX-like (LOXL) proteins initiate the covalent cross-linking of TE and collagen fibrils, which stabilizes the corresponding networks [41].

2.4.3 Glycosaminoglycans (GAGs)

The 'ground substance' of the extracellular matrix is made up mainly of glycosoaminoglycans (GAGs), which link together to form even larger molecules called proteoglycans. This ground substance is an amorphous, gelatinous, transparent, colourless material that fills the spaces between fibres and cells.

GAGs are long, unbranched polysaccharide chains composed of repeating sugar units [42]. They are highly negatively charged due to sulphate and carboxyl groups, so they attract a lot of cations (e.g. sodium ions), which in turn causes a lot of water to be drawn into the matrix. They are highly hydrophilic and inflexible because the long sugar chains cannot fold like proteins, so they remain extended.

Table 2.4 Collagen types showing their main location and organization [38, 39] (*FACITs* fibril-associated collagens with interrupted triple helix)

Collagen type	Forming	Location
I	Fibrillar	bone, skin, tendons, cornea, blood vessel walls and other connective tissues
II	Fibrillar	Cartilage, vitreous body, nucleus pulposus
III	Fibrillar	Blood vessels, intestinal organs, cartilage, skin
IV	Network-forming	Basement membranes
V	Fibrillar	Bone, cornea, lung, fetal membranes
VI	Network-forming	Ubiquitous tissue distribution, all connective tissues except bone
VII	Anchoring fibrils	Skin, dermal-epidermal junctions, oral mucosa, cervix
VIII	Network-forming	Cornea, endothelium
IX	FACITs	Cartilage, vitreous body, cornea
X	Network-forming	Hypertrophic cartilage
XI	Fibrillar	Articular cartilage, vitreous body
XII	FACITs	Skin, tendon, perichondrium, periosteum, lungs, liver, placenta, vessel walls
XIII	Transmembrane	Epidermis, hair follicle, endomysium, intestine, chondrocytes, lungs, liver
XIV	FACITs	Skin, tendon, perichondrium, periosteum, lungs, liver, placenta, vessel walls, bone, cornea
XV	Multiplexins	Heart and skeletal muscle, kidney, pancreas
XVI	FACITs	Hyaline cartilage, skin, basement membranes, tissue junctions
XVII	Transmembrane	Dermal-epidermal junctions
XVIII	Multiplexins	Lung, kidney, brain, smooth muscle, eye, liver
XIX	FACITs	Basement membranes, tissue junctions
XX	FACITs	Corneal epithelium, embryonic skin, sternal cartilage, tendon
XXI	FACITs	Basement membranes, tissue junctions, blood vessels
XXII	FACITs	Basement membranes, tissue junctions
XXIII	Transmembrane	Metastatic tumor cells
XXIV	Fibrillar	Bone and cornea
XXV	Transmembrane	Neurons
XXVI	Other collagens	Testis and ovaries
XXVII	Fibrillar	Cartilage
XXVIII	Other collagens	Cartilage, eye, ear, lung

These properties imply that GAGs form a sponge-like matrix that absorbs a lot of water, like a porous hydrated gel, and are good at resisting compressive stress. This is particularly important, for example, in cartilage, a form of connective tissue found in joints that have to withstand large compressive forces when we walk. The

Table 2.5 Tissue distribution and engineering applications for each GAG [43]

Glycosaminoglycan	Tissue distribution	Tissue Engineering application
Hyaluronic acid	Skin, cartilage, vitreous body	Chondrogenesis, osteogenesis, muscle and liver regeneration, wound healing
Chondroitin sulphate	Cartilage, bone, skin, cornea and arteries	Chondrogenesis, osteogenesis, nerve regeneration, wound healing
Dermatan sulphate	Skin, blood vessels and heart	Nerve regeneration and heart implant coatings
Heparan sulphate	Lung, liver, skin and arteries	Chondrogenesis and osteogenesis
Keratan sulphate	Cartilage, cornea, intervertebral disc	Non-existent

extracellular matrix consists of 10% GAGs and most of the rest is water. This also provides nice big spaces for molecules to diffuse and for cells to move around in.

There are four main groups of GAGs, distinguished by the sugar residues, the type of linkage between residues, and the number and location of sulphate groups (see Table 2.5). Hyaluronic acid is the simplest, as it does not contain sulphated sugars and is not covalently linked to other proteins to form proteoglycans. The other three are sulphated, complex, small, secreted by the cell, and always attached to a protein to form a proteoglycan (see next subsection): chondroitin sulphate and dermatan sulphate, heparan sulphate (heparin) and keratan sulphate.

Glycosaminoglycans play a key role driving cell behavior and controlling the microenvironment surrounding cells. Therefore, it is common its use to drive stem cell fate and promote cell proliferation in regenerating tissues [43].

2.4.4 Proteoglycan

Proteoglycans are GAGs covalently attached to a 'core' glycosylated protein [44]. As mentioned above, all GAGs except hyaluronic acid are attached to a protein in this way. These GAG-protein mixtures are very heterogeneous, and this variability means that they can form pores of different sizes to regulate the size of molecules that can move through the matrix. Many forms of proteoglycans are present in virtually all extracellular matrices of connective tissues. The primary biological function of proteoglycans is derived from the physicochemical properties of the glycosaminoglycan component of the molecule, which provides hydration and swelling pressure to the tissue, enabling it to withstand compressive forces.

Proteoglycans are vital components of the ECM in articular cartilage, providing biomechanical properties that are critical for its proper functioning [45]. They are key players in chondral diseases, particularly in the degradation of the extracellular matrix. The evaluation of proteoglycan molecules can serve as biomarkers of joint degradation in osteoarthritis patients, as well as assessing the quality of repaired tissue following different treatment strategies for chondral injuries. Proteoglycans, such as aggrecan, biglycan, decorin, perlecan and versican, are important in maintaining articular cartilage health and the cartilage repair process.

Many different types of proteoglycans are found in many connective tissues, on cell surfaces and in intracellular compartments, being very difficult to characterize and classify [46]. In addition to their purely structural function, some of them have other functions related to the regulation of enzyme and growth factor activity. For example, the perlecan may play an important role as the major anionic site responsible for the charge selectivity in glomerular filtration [44].

2.4.5 Glycoproteins: Laminin and Fibronectin

Collagen and elastin are the main structural components of the ECM, however, glycoproteins are the main actor to make the ECM a cohesive and integrative network of molecules [47]. In fact, they are intermediaries that link structural molecules between each other, and also connect structural molecules and cells. In each glycoprotein molecule, there are several domains binding different molecules that altogether forms cross-linked molecular networks. Although cohesion is a main function, glycoproteins perform many other functions.

Laminin and fibronectin are the main glycoproteins, although there are others like tenascin (it is released into the ECM of tendons, bones and cartilage during the embryonary development, although, it can also be found in other tissues) and osteopontin (mainly found in bones).

Fibronectins present several molecular domains that recognise and bind glycosaminoglycans, proteoglycans and some transmembrane proteins such as integrins [48]. Fibronectins are therefore intermediaries that provide links between different ECM molecules, but also between the ECM and cells. Fibronectins are present in almost all tissues. They are found as insoluble fibres in connective tissue and as soluble molecules in body fluids such as blood. In addition to maintaining the cohesion of the ECM, fibronectins have many other functions. For example, during embryonic development, cells move from one side to the other along adhesion pathways made of fibronectin. They are also important in tissue remodelling.

Laminins are a major component of the basal lamina, the ECM layer that separates epithelia and muscle cells from connective tissues [47]. Laminins are synthesized by epithelial cells, muscle cells, neurons, and bone marrow cells. Besides the structural function in the ECM, laminins also influence cell behavior and differentiation through cell interactions with integrins. That is why mutation of laminins usually leads to pathological processes. During the embryonary development, laminins are the first type of glycoprotein to be released to the extracellular matrix [47].

In addition, there are other glycoproteins that have domains for binding collagen, fibronectin and some cell receptors, but their main function is not to maintain the structural integrity of the ECM, but to modify cell activity and facilitate ECM remodelling. They are released into the ECM at specific times and locations by many cell types. They are therefore not constitutive but transient proteins in the

ECM that occur during embryonic development, normal tissue development or during tissue repair after pathological damage. Some of these proteins are tenascin, tromospondin, osteopontin, fibulin and SPARC protein [49].

2.5 ECM as a Biologic Scaffold for TERM

The ECM from native tissues has been widely used as scaffold for numerous TERM applications [50]. The ECM from which these scaffold materials are derived from a variety of tissues, including heart valves, blood vessels, skin, nerves, skeletal muscle, tendons, ligaments, small intestinal submucosa, urinary bladder and liver [50].

The scaffold material selected for a tissue engineering application must promote in vivo remodelling, i.e. the degradation and replacement of the scaffold by new host tissue, and the organisation and differentiation of the new host tissue in relation to surrounding structures. The remodelling process is strongly influenced by the microenvironment of each tissue, including factors such as blood supply, pH, O_2 and CO_2 concentrations, mechanical stressors and the host-scaffold interface. These microenvironmental factors will ultimately determine the fate of the scaffold material and the type of remodelling that will occur, i.e. constructive remodelling, destructive remodelling, scarring or something in between these extremes.

Although there is a debate about the importance of the composition vs. structure of these materials in scaffold design, the use of native tissue scaffolds (or derived from cultured cells) guarantees at least its biocompatibility, biodegradability and favourable immune respose [51].

Nevertheless, the composite structure of these ECM molecules, as well as their in vivo degradability, has marked effects upon the host response and the remodeling events that determine the eventual clinical outcome.

Therefore, biological scaffolds composed of ECM have a complex composition with a variety of diverse molecules that are perfectly suited to support the cellular processes necessary to regenerate tissues or organs. However, it is very important to note that the ability of an ECM from one tissue to function as a biological scaffold material for the same or a different tissue can be highly variable [50].

Normally the utilization of decellularized organs/tissues and cell-derived ECM presents many possibilities in TERM and a brief description of these applications is presented in these subsections.

2.5.1 ECM Scaffolds for Organ Transplantation

Solid organ transplantation is currently the only curative treatment option available for patients with end-stage organ failure. Although transplantation has improved patients' quality of life and survival, the shortage of organs means that only a small number of patients can benefit [52].

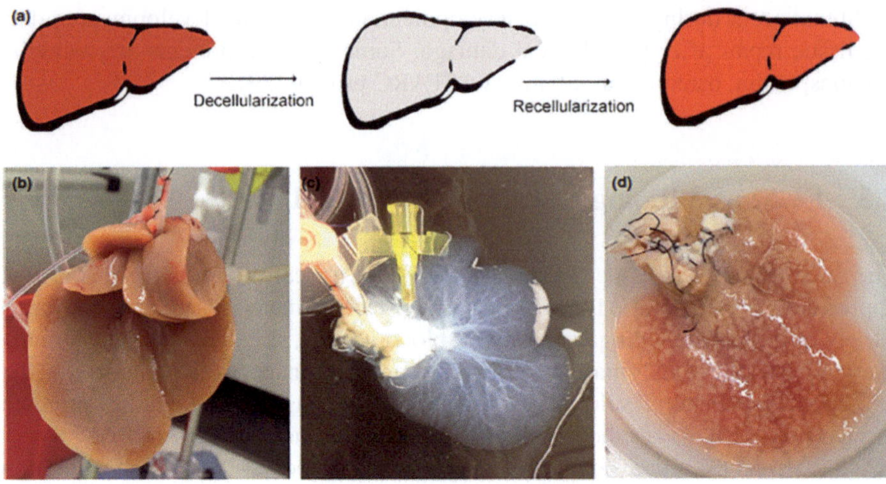

Fig. 2.6 Example of the approach used for organ transplantation using de- and recellularization: (**a**) Scheme of a liver, (**b**) Native rat liver, (**c**) Decellularized rat liver and (**d**) Recellularized rat liver. Image from Hillebrandt et al. [52], open access article distributed under the terms and conditions of the Creative Commons Attribution (CC BY) license

This approach of organ transplantation by means of another organ using ECM scaffolds is based on the technique of decellularisation and recellularization (see Fig. 2.6). In a future clinical setting, a xenogenic human (e.g. porcine) or allogenic organ will be used for decellularisation, followed by recellularisation using cells from the patient suffering from organ failure [e.g. induced pluripotent stem cells (iPSCs)] (isolated, organ-specifically differentiated and expanded). After recellularisation and maturation of this neo-organ, the personalised graft is implanted without the need for long-term immunosuppression [52].

2.5.2 ECM Scaffolds for Tissue Regeneration

Allogenic/xenogeneic sourced tissue-/organ-derived materials have been widely studied for use in tissue repair [53].

The native ECM offers many opportunities for tissue repair as it contains a composition and architecture that incorporates the complex microenvironment of proteins and matrix components of the target tissue/organ, such as collagen, proteoglycans, fibronectin, laminin, elastin and other glycoproteins [54]. These properties of native ECM are often difficult to achieve with conventional fabrication methods and biomaterials, which has motivated its utilization for tissue repair. Nevertheless, they have to be decellularized in order to remove cellular material and antigens, preserving and minimizing the loss of ECM properties. Maintaining the intrinsic biochemical and biophysical cues of the native ECM provides structural and chemical signalling cues to regulate cell behaviour in terms of supporting cell adhesion, migration, proliferation and differentiation, events critical for tissue repair [53].

This decellularised native ECM can therefore be used for direct transplantation or for the manufacture of scaffolds for subsequent clinical use. Its versatility allows the development of different biomaterial forms such as powders, gels, sheets, 3D structures through additive manufacturing [54].

2.5.3 *Scaffolds from Cell-Derived ECM*

Cell-derived matrices (CDMs) are emerging as an alternative strategy to conventional decellularised ECM (dECM), mainly to overcome the potential inflammatory/pathogenic risk associated with dECM tissues/matrices. In this approach, CDMs are produced in vitro by cell-assembled ECM deposition of a fibrillar network composed of a variety of proteins. The three common methods of obtaining CDMs include monolayer cell sheet culture, template culture and pellet culture (Fig. 2.7). These CDMs generally require milder methods for accelerated decellularisation with greater preservation of ECM components and bioactivity with effective removal of cellular debris [54]. Decellularised CDMs can mostly be used as

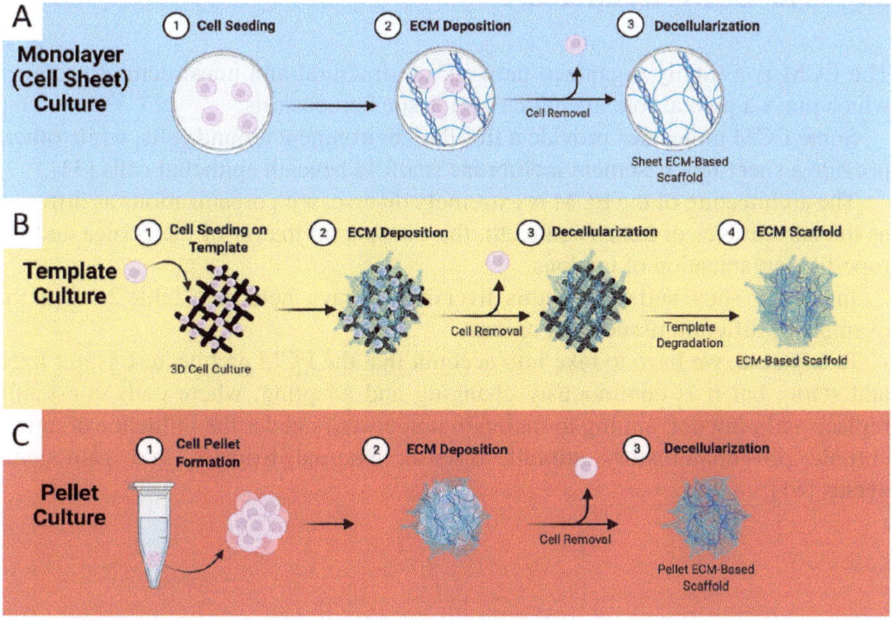

Fig. 2.7 Methodological strategies to fabricate CDMs: (**a**) monolayer cell sheet culture, (**b**) template culture and (**c**) pellet culture [54] (open access article distributed under the terms and conditions of the Creative Commons Attribution (CC BY) license)

in vitro models to study cellular behaviour and allow application-specific tailoring by cell type and culture method (2D/3D, static/dynamic, culture medium, etc.) [55], mainly due to the low complexity of the structural architecture and consequently lower mechanical properties.

2.5.4 *Decellularization*

The goal of the decellularization process is to completely remove cellular material while preserving the architecture and biochemical composition of the ECM. The three classifications of common decellularization methods include: physical, chemical, and enzymatic. Tissue or organ decellularization methods rely on immersion or perfusion of solutions containing chemical or biological agents or physical stresses to disrupt cell membranes to induce cell death and removal. The choice of decellularization agent, treatment time, and protocol complexity depends on the characteristics of the target tissue/organ, such as tissue thickness, density, structure, origin, and lipid content, as well as its intended use. A good review of the different methods can be found in the recent work by Golebiowska et al. [54].

2.6 The ECM Architecture

The ECM is a highly organized network of structural and nonstructural proteins, which plays a pivotal role in cellular and tissue homeostasis.

Some ECM molecules provide a fibrillar environment around cells, while others provide a sheet-like basement membrane scaffold beneath epithelial cells [31].

The architecture of the ECM is extremely diverse, with organizations as different as the hard tissues of bones and teeth, the transparent matrix of the cornea and the rope-like organization of tendons.

In order to show and classify this diversity, we have included a Table 2.6 with the example of different tissues and organs.

In any case, we have to take into account that the ECM architecture is not fixed and static, but it is continuously changing and adapting, where cells constantly replace and remodel, aiming to maintain homeostasis under the influence of aging, chronic pro-inflammatory stimuli, disease, external irritants, and pathogenic agents [61].

Table 2.6 List of tissues/organs with a different architecture and composition

Tissue/ organ	Organization	Composition	References
Cortical bone	concentrically arranged cylindrical structures called osteons (or Haversian systems)	Collagen I and bone mineral	Martin et al. [56]
Cornea	It is avascular and is composed of three layers	Fibril forming collagens (I, III, V), fibril associated collagens with interrupted triple helices (XII and XIV), network forming collagens (IV, VI and VIII) as well as small leucine-rich proteoglycans (SLRP) expressed in the stroma: decorin, biglycan, lumican, keratocan, and fibromodulin	Espana and Birk [57]
Tendon	Bundles of parallel fibrillar ECM	Collagen (65–80% of its dry mass, primarily collagen I) and elastin (1–2% of its dry mass) embedded in a well hydrated, proteoglycan-rich matrix	Mouw et al. [47]
Cartilage	Four main layers: the superficial zone, the middle zone, the deep zone, and the calcified zone	principally composed of water, collagen type II, and proteoglycans, with other noncollagenous proteins and glycoproteins present in lesser amounts	Fox et al. [58]
Liver	The liver is made of two major lobes, a large right and a smaller left lobe. Both are made up of 8 segments that consist of 1000 lobules (small lobes). These lobules are connected to small ducts (tubes) that connect with larger ducts to form the common hepatic duct	Main components: collagens (type I, III, IV, V, VI), glycoproteins, proteoglycans and elastin	Arriazu et al. [39]
Basal lamina	It is a set of cell-adherent, organized layers of ECM molecules that form the supporting structure for epithelial, endothelial, muscle and fat cells, the central nervous system and the Schwann cells of peripheral nerves	Collagen IV and VII, laminins bound to nidogen glycoproteins and heparan sulphate proteoglycans	Mouw et al. [47]

(continued)

Table 2.6 (continued)

Tissue/ organ	Organization	Composition	References
Lung	Lung is organized in parenchymal and non-parenchymal structures. The lung parenchyma resembles a sponge, and occupies 90% of total lung volume. Non-parenchymal structures consist of the bronchial tree, pulmonary vessels, and interlobular septa	Main components: collagens (types I, II, III, V and XI), large elastic fibres, glycoproteins (fibrillin-1, -2 and -3). elastin microfibril interface-located proteins and members of the elastin-crosslinking lysyl oxidase (LOX) family	Burgstaller et al. [59], Itoh et al. [60]
Brain	Neural ECM aggregates form a latticed network that varies in its composition and organization throughout development. These loose networks provide enough space for the developing brain to grow and to extend the neural network; tight networks in later life restrict growth and reinforce existing connections	It predominantly consists of hyaluronic acid, proteoglycans and GAGs, although it contains fibrous components such as collagens, fibronectins and other molecules in a small percentage	Mouw et al. [47]

2.7 Cell-Matrix Interactions

All cells are hosted in the ECM of the tissues they form, and are active inhabitants capable of responding to signals received from the matrix, as well as modifying and remodeling the matrix. A cell is capable of sensing multiple cues from its surrounding microenvironment, including signals bound to the ECM and to neighboring cells, mechanical stimuli, and soluble signals from neighboring and distant cells. The cell then responds to these signals through multiple pathways, each involving different cascades of internal molecular interactions [62]. Thus, the ECM exerts a clear guidance role for our cells through various multiphysical mechanisms in the form of biochemical signals (e.g., composition), biophysical factors (e.g., stiffness, porosity, and permeability), and topographical cues (e.g., surface roughness). Therefore, advancing our understanding of how cells sense and respond to each other to the multitude of extrinsic signals present in their local microenvironment is crucial in tissue regeneration and tissue engineering. There is a homeostatic balance of cell-ECM interactions in native and healthy tissues that can be perturbed by aging, irritants, or disease, resulting in aberrant cell behavior [61]. Tissue regeneration and tissue engineering aim to restore this homeostatic balance, which is not easy to achieve due to the complexity of the healing process.

The mechanical interactions between the cells and the extracellular matrices are bi-directional. On the one hand, cells interact with ECMs mechanically, including by pulling, often through actomyosin-based contractility coupled to the ECM through integrin-based adhesions, and by pushing, often through actin polymerization and microtubules. The mechanical properties of ECMs mediate these

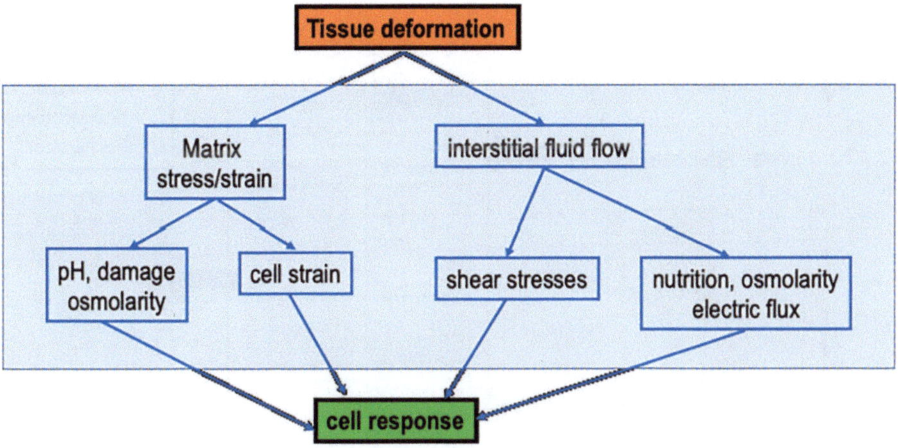

Fig. 2.8 Secondary effects due to tissue strain and how they affect the cell response

interactions, resulting in cell mechanotransduction and affecting cell behaviours. A more detailed explanation of the mechanobiology of how cells interact with the matrix is provided in the next chapter. On the other hand, when tissues are deformed due to external loads they cause changes at cell level, such as, shear strain, osmolarity, pH, etc. (see Fig. 2.8). As example, it is interesting to see the review work in bone tissue by García-Aznar et al. [63] (see Fig. 2.9).

2.8 Conclusions

This chapter presents the most relevant concepts associated with native tissue ECM, focusing on tissue repair strategies through tissue engineering approaches. The relevance of quantitatively characterizing the composition of this ECM, as well as its architectural properties associated with relevant biophysical properties such as stiffness, viscoelasticity and permeability, has been discussed. However, a more extensive analysis could have been performed on both properties of the ECM (the composition and its architecture), being current topics of research, although this is beyond the scope of this chapter.

Nevertheless, it is important to highlight some aspects of current work in these areas. For example, mass spectrometry techniques are currently being used to determine the detailed composition of the ECM from many mouse and human tissues, and the results are indexed in a "matrisome" database (http://matrisomeproject.mit.edu), which has been enormously useful in understanding ECM physiology [64]. In addition, the architecture of the ECM is beginning to be studied, the so-called "matricture", which attempts to quantify the spatiotemporal organization of all ECM proteins [61].

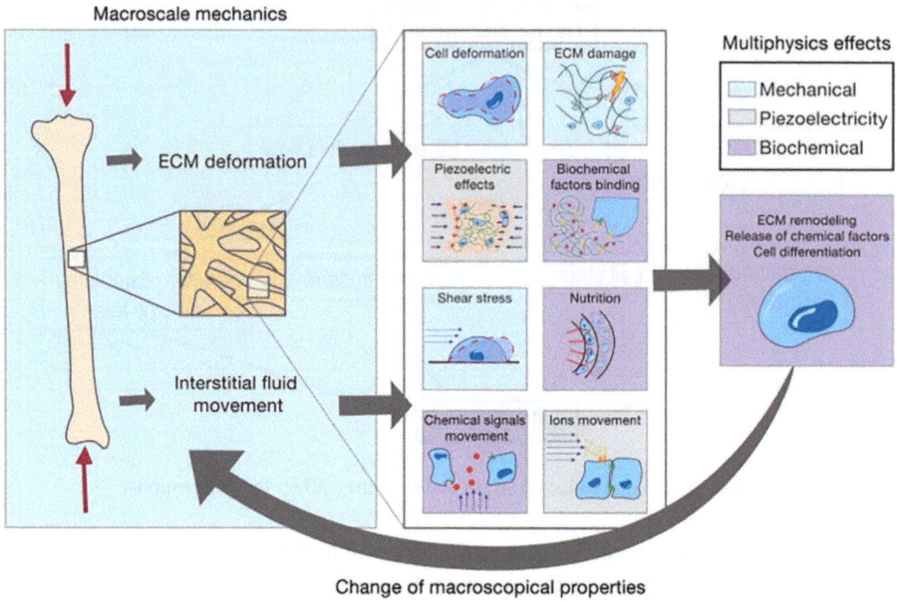

Fig. 2.9 Interaction between load forces at bone scale and multiphysics at tissue and cellular level. Under loading conditions, bone tissue is deformed causing the deformation of the extracellular matrix (ECM) and the flux of interstitial fluid. Matrix deformation can induce different multiphysics effects, such as, cell deformation, damage breaking the connections between cells, electric charges formation within the matrix and alteration of biochemical factors binding to the matrix. The movement of interstitial fluid also induces multiphysics effects like shear stresses on the cell, improving the nutrition, streaming potentials and movement of biochemical factors. All multiphysics stimuli have an impact at cell level with the corresponding biochemical cell response [63], Open access article distributed under the terms and conditions of the Creative Commons Attribution (CC BY) license)

It is also very important to keep in mind that the ECM is more than a passive material, because when it supports and transmits loads, other biophysical effects are activated. In fact, under mechanical stress, collagen (the main component of the ECM) is biochemically active like a giant mechanical "enzyme", producing radicals and subsequently reactive oxygen species (essential biological signaling molecules) [65]. This effect clearly presents implications for aging, inflammation and pain.

This knowledge of the ECM is crucial given the regulatory role of the ECM on cell shape, fate and metabolism. Tissue regeneration and tissue engineering strategies aim to restructure altered ECM consequence of a disease and revert to a normal native ECM.

Questions

- The following figure [37] shows the elastic and viscoelastic properties of the endometrium in different areas, with E′ as the solid line and E″ as the dashed line.

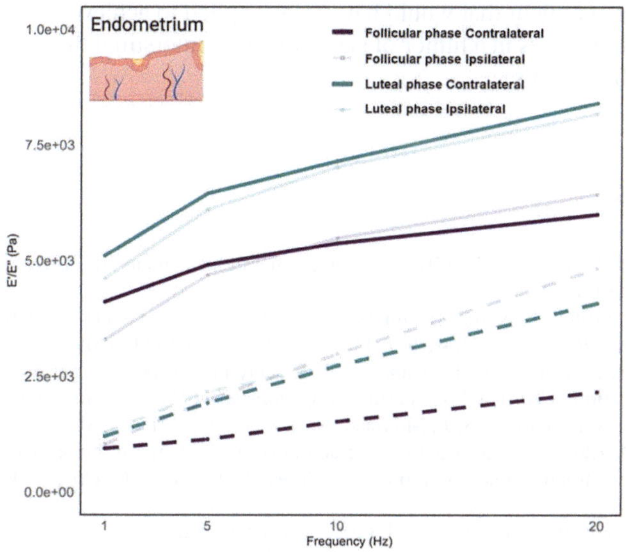

- Is it possible to say that at low frequencies it is an elastic, viscoelastic or viscous material? Why? What happens with increasing frequency?
- Can be used the same decellularization method for native ECMs as for cell-derived ECMs?
- A network of fibrous proteins, adhesion proteins, and proteoglycans make up the extracellular matrix (ECM). Name at least two of each type:
  - Fibrous proteins
  - Adhesion proteins
  - Proteoglycans
- What is a decellularized tissue and why is it attractive for a good tissue engineering scaffold?
- How could you measure the permeability of a hydrogel that is used in this microfluidics-based experiment [9]? Assume that all the data you need is known.

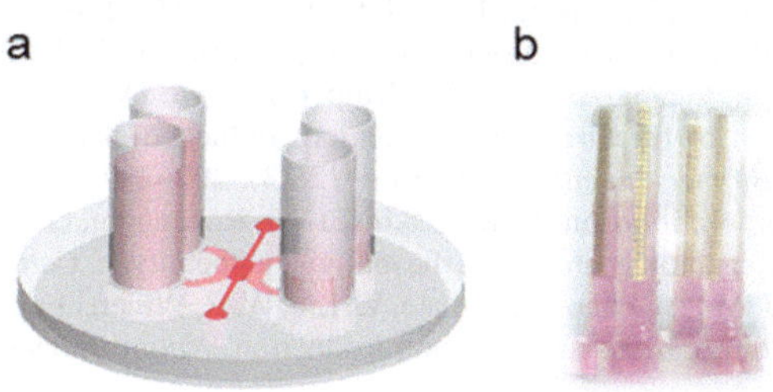

- In the work of Yang et al. [66] the permeability of a brain tumor (glioblastoma) was measured, knowing that its value is 2.81×10^{-15} m^2. We are asked to estimate the pressure gradient that would have to be applied to achieve an interstitial fluid velocity of 10 mm/s in a tumor of size 1 mm (the interstitial fluid with a viscosity of 6.9×10^{-4} kg/m/s).

References

1. Fratzl, P., & Weinkamer, R. (2007). Nature's hierarchical materials. *Progress in Materials Science, 52*(8), 1263–1334.
2. Levick J. R. (1987). Flow through interstitium and other fibrous matrices quarterly. *Journal of Experimental Physiology, 72*(4), 409–437. https://doi.org/10.1113/expphysiol.1987.sp003085
3. Malachanne, E., Dureisseix, D., Cañadas, P., & Jourdan, F. (2008). Experimental and numerical identification of cortical bone permeability. *Journal of Biomechanics, 41*(3), 721–725.
4. Rabiatul, A. A. R., Fatihhi, S. J., Md Saad, A. P., Zakaria, Z., Harun, M. N., Kadir, M. R. A., & Syahrom, A. (2021). Fluid–structure interaction (FSI) modeling of bone marrow through trabecular bone structure under compression. *Biomechanics and Modeling in Mechanobiology, 20*, 957–968.
5. Mow, V. C., Gu, W. Y., & Chen, F. H. (2005). Structure and function of articular cartilage and meniscus. *Basic Orthopaedic Biomechanics and Mechano-biology, 181*, 258.
6. Stewart, J. M., Schultz, D. S., Lee, O. T., & Trinidad, M. L. (2009). Collagen cross-links reduce corneal permeability. *Investigative Ophthalmology & Visual Science, 50*(4), 1606–1612.
7. Kleinhans, K. L., & Jackson, A. R. (2018). Hydraulic permeability of meniscus fibrocartilage measured via direct permeation: Effects of tissue anisotropy, water volume content, and compressive strain. *Journal of Biomechanics, 72*, 215–221.
8. Jamal, A., Mongelli, M. T., Vidotto, M., Madekurozwa, M., Bernardini, A., Overby, D. R., et al. (2020). Infusion mechanisms in brain White matter and their dependence on microstructure: An experimental study of hydraulic permeability. *IEEE Transactions on Biomedical Engineering, 68*(4), 1229–1237. https://doi.org/10.1109/TBME.2020.3024117
9. Moreno-Arotzena, O., Meier, J. G., Del Amo, C., & García-Aznar, J. M. (2015). Characterization of fibrin and collagen gels for engineering wound healing models. *Materials, 8*(4), 1636–1651.
10. Alamán-Díez, P., Borau, C., Guerrero, P. E., Amaveda, H., Mora, M., Fraile, J. M., García-Gareta, E., García-Aznar, JM., & Pérez, M. Á. (2023). Collagen-laponite nanoclay hydrogels for tumor spheroid growth. *Biomacromolecules*, 24(6), 2879–2891.
11. Guimarães, C. F., Gasperini, L., Marques, A. P., & Reis, R. L. (2020). The stiffness of living tissues and its implications for tissue engineering. *Nature Reviews Materials, 5*(5), 351–370.
12. Engler, A. J., Sen, S., Sweeney, H. L., & Discher, D. E. (2006). Matrix elasticity directs stem cell lineage specification. *Cell, 126*(4), 677–689.
13. Handorf, A. M., Zhou, Y., Halanski, M. A., & Li, W. J. (2015). Tissue stiffness dictates development, homeostasis, and disease progression. *Organogenesis, 11*(1), 1–15.
14. Gómez-Benito, M. J., García-Aznar, J. M., Kuiper, J. H., & Doblaré, M. (2006). A 3D computational simulation of fracture callus formation: Influence of the stiffness of the external fixator. *Journal of Biomechanical Engineering, 128*(3), 290–299.
15. Paszek, M. J., Zahir, N., Johnson, K. R., Lakins, J. N., Rozenberg, G. I., Gefen, A., & Weaver, V. M. (2005). Tensional homeostasis and the malignant phenotype. *Cancer Cell, 8*(3), 241–254.
16. Du, H., Bartleson, J. M., Butenko, S., Alonso, V., Liu, W. F., Winer, D. A., & Butte, M. J. (2023). Tuning immunity through tissue mechanotransduction. *Nature Reviews Immunology, 23*(3), 174–188.
17. Hua, Y., & Bergers, G. (2019). Tumors vs. chronic wounds: An immune cell's perspective. *Frontiers in Immunology, 10*, 2178.
18. Carvalho, E. M., & Kumar, S. (2023). Lose the stress: Viscoelastic materials for cell engineering. *Acta Biomaterialia, 163*, 146–157.

19. Chaudhuri, O., Cooper-White, J., Janmey, P. A., Mooney, D. J., & Shenoy, V. B. (2020). Effects of extracellular matrix viscoelasticity on cellular behaviour. *Nature, 584*(7822), 535–546.
20. Elosegui-Artola, A., Gupta, A., Najibi, A. J., Seo, B. R., Garry, R., Tringides, C. M., & Mooney, D. J. (2023). Matrix viscoelasticity controls spatiotemporal tissue organization. *Nature Materials, 22*(1), 117–127.
21. García-García, D. E. (2024). *Computational modeling to evaluate the mechanical properties of hydrogels used as scaffolds in tissue engineering*. Master Thesis Universidad de Zaragoza.
22. Bonfanti, A., Kaplan, J. L., Charras, G., & Kabla, A. (2020). Fractional viscoelastic models for power-law materials. *Soft Matter, 16*(26), 6002–6020.
23. Pacheco, J. E. L., Bavastri, C. A., & Pereira, J. T. (2015). Viscoelastic relaxation modulus characterization using prony series. *Latin American Journal of Solids and Structures, 12*(2), 420–445.
24. Garcia, P. D., Guerrero, C. R., & Garcia, R. (2020). Nanorheology of living cells measured by AFM-based force–distance curves. *Nanoscale, 12*(16), 9133–9143.
25. Hoshiba, T., & Yamaoka, T. (2019). Extracellular matrix scaffolds for tissue engineering and biological research. *Current Stem Cell Research & Therapy, 12*(3), 233–246.
26. Zhang, W., Liu, Y., & Zhang, H. (2021). Extracellular matrix: An important regulator of cell functions and skeletal muscle development. *Cell & Bioscience, 11*, 1–13.
27. Yamada, K. M., Doyle, A. D., & Lu, J. (2022). Cell–3D matrix interactions: Recent advances and opportunities. *Trends in Cell Biology, 32*(10), 883–895.
28. Olivares, V., Cóndor, M., Del Amo, C., Asín, J., Borau, C., & García-Aznar, J. M. (2019). Image-based characterization of 3D collagen networks and the effect of embedded cells. *Microscopy and Microanalysis, 25*(4), 971–981.
29. Saif Ur Rahman, M., Wu, J., Chen, H., Sun, C., Liu, Y., & Xu, S. (2023). Matrix mechanophysical factor: Pore size governs the cell behavior in cancer. *Advances in Physics: X, 8*(1), 2153624.
30. Doolin, M. T., Moriarty, R. A., & Stroka, K. M. (2020). Mechanosensing of mechanical confinement by mesenchymal-like cells. *Frontiers in Physiology, 11*, 365.
31. Walma, D. A. C., & Yamada, K. M. (2020). The extracellular matrix in development. *Development, 147*(10), dev175596.
32. Sérandour, A.-L., Loyer, P., Garnier, D., Courselaud, B., Théret, N., Glaise, D., Guguen-Guillouzo, C., & Corlu, A. (2005). *Hepatology, 41*, 478.
33. Wang, X., Harris, R. E., Bayston, L. J., & Ashe, H. L. (2008). *Nature, 455*, 72.
34. Schenk, S., Hintermann, E., Bilban, M., Koshikawa, N., Hojilla, C., Khokha, R., & Quaranta, V. (2003). *The Journal of Cell Biology, 161*, 197.
35. Karamanos, N. K., Theocharis, A. D., Piperigkou, Z., Manou, D., Passi, A., Skandalis, S. S., & Onisto, M. (2021). A guide to the composition and functions of the extracellular matrix. *The FEBS Journal, 288*(24), 6850–6912.
36. Sutherland, T. E., Dyer, D. P., & Allen, J. E. (2023). The extracellular matrix and the immune system: A mutually dependent relationship. *Science, 379*(6633), eabp8964.
37. Franko, R., Franko, Y., Ribes Martinez, E., Ferronato, G. A., Heinzelmann, I., Grechi, N., & Ferraz, M. A. M. M. (2024). Mechanical properties of native and decellularized reproductive tissues: Insights for tissue engineering strategies. *Scientific Reports, 14*(1), 7347.
38. Karsdal, M. (2023). *Biochemistry of collagens, laminins and elastin: Structure, function and biomarkers*. Elsevier.
39. Arriazu, E., Ruiz de Galarreta, M., Cubero, F. J., Varela-Rey, M., Perez de Obanos, M. P., Leung, T. M., & Nieto, N. (2014). Extracellular matrix and liver disease. *Antioxidants & Redox Signaling, 21*(7), 1078–1097.
40. Vindin, H., Mithieux, S. M., & Weiss, A. S. (2019). Elastin architecture. *Matrix Biology, 84*, 4–16.
41. Vallet, S. D., & Ricard-Blum, S. (2019). Lysyl oxidases: From enzyme activity to extracellular matrix cross-links. *Essays in Biochemistry, 63*, 349–364.

42. Perez, S., Makshakova, O., Angulo, J., Bedini, E., Bisio, A., de Paz, J. L., & Ricard Blum, S. (2023). Glycosaminoglycans: what remains to be deciphered? *JACS Au, 3*(3), 628–656.
43. Sodhi, H., & Panitch, A. (2020). Glycosaminoglycans in tissue engineering: A review. *Biomolecules, 11*(1), 29.
44. Yanagishita, M. (1993). Function of proteoglycans in the extracellular matrix. *Acta Patholigica Japonica, 43*(6), 283–293.
45. Alcaide-Ruggiero, L., Cugat, R., & Domínguez, J. M. (2023). Proteoglycans in articular cartilage and their contribution to chondral injury and repair mechanisms. *International Journal of Molecular Sciences, 24*(13), 10824.
46. Iozzo, R. V., & Schaefer, L. (2015). Proteoglycan form and function: A comprehensive nomenclature of proteoglycans. *Matrix Biology, 42*, 11–55.
47. Mouw, J. K., Ou, G., & Weaver, V. M. (2014). Extracellular matrix assembly: a multiscale deconstruction. *Nature Reviews. Molecular Cell Biology, 15*, 771–785.
48. Pankov, R., & Yamada, K. M. (2002). Fibronectin at a glance. *Journal of Cell Science, 115*, 3861–3863.
49. Murphy-Ullrich, J. E., & Sage, E. H. (2014). Revisiting the matricellular concept. *Matrix Biology, 37*, 1–14.
50. Badylak, S. F., Freytes, D. O., & Gilbert, T. W. (2009). Extracellular matrix as a biological scaffold material: Structure and function. *Acta Biomaterialia, 5*(1), 1–13.
51. Zhu, M., Li, W., Dong, X., Yuan, X., Midgley, A. C., Chang, H., & Kong, D. (2019). In vivo engineered extracellular matrix scaffolds with instructive niches for oriented tissue regeneration. *Nature Communications, 10*(1), 4620.
52. Hillebrandt, K. H., Everwien, H., Haep, N., Keshi, E., Pratschke, J., & Sauer, I. M. (2019). Strategies based on organ decellularization and recellularization. *Transplant International, 32*(6), 571–585.
53. Bonnans, C., Chou, J., & Werb, Z. (2014). Remodelling the extracellular matrix in development and disease. *Nature Reviews. Molecular Cell Biology, 15*(12), 786–801. https://doi.org/10.1038/nrm3904
54. Golebiowska, A. A., Intravaia, J. T., Sathe, V. M., Kumbar, S. G., & Nukavarapu, S. P. (2024). Decellularized extracellular matrix biomaterials for regenerative therapies: Advances, challenges and clinical prospects. *Bioactive Materials, 32*, 98–123.
55. Assunção, M., Dehghan-Baniani, D., Yiu, C. H. K., Später, T., Beyer, S., & Blocki, A. (2020). Cell-derived extracellular matrix for tissue engineering and regenerative medicine. *Frontiers in Bioengineering and Biotechnology, 8*, 602009.
56. Martin, R. B., Burr, D. B., Sharkey, N. A., & Fyhrie, D. P. (1998). *Skeletal tissue mechanics* (Vol. 190). Springer.
57. Espana, E. M., & Birk, D. E. (2020). Composition, structure and function of the corneal stroma. *Experimental Eye Research, 198*, 108137.
58. Sophia Fox, A. J., Bedi, A., & Rodeo, S. A. (2009). The basic science of articular cartilage: Structure, composition, and function. *Sports Health, 1*(6), 461–468.
59. Burgstaller, G., Oehrle, B., Gerckens, M., White, E. S., Schiller, H. B., & Eickelberg, O. (2017). The instructive extracellular matrix of the lung: Basic composition and alterations in chronic lung disease. *European Respiratory Journal, 50*(1), 1601805.
60. Itoh, H., Nishino, M., & Hatabu, H. (2004). Architecture of the lung: Morphology and function. *Journal of Thoracic Imaging, 19*(4), 221–227.
61. Reuten, R., Mayorca-Guiliani, A. E., & Erler, J. T. (2022). Matritecture: Mapping the extracellular matrix architecture during health and disease. *Matrix Biology Plus, 14*, 100102.
62. Johnson, A. W., & Harley, B. (Eds.). (2011). *Mechanobiology of cell-cell and cell-matrix interactions*. Springer Science & Business Media.
63. García-Aznar, J. M., Nasello, G., Hervas-Raluy, S., Pérez, M. Á., & Gómez-Benito, M. J. (2021). Multiscale modeling of bone tissue mechanobiology. *Bone, 151*, 116032.

64. Naba, A., Clauser, K. R., Hoersch, S., Liu, H., Carr, S. A., & Hynes, R. O. (2012). The matrisome: in silico definition and in vivo characterization by proteomics of normal and tumor extracellular matrices. *Molecular & Cellular Proteomics, 11*(4), 014647.
65. Zapp, C., Obarska-Kosinska, A., Rennekamp, B., Kurth, M., Hudson, D. M., Mercadante, D., & Gräter, F. (2020). Mechanoradicals in tensed tendon collagen as a source of oxidative stress. *Nature Communications, 11*(1), 2315.
66. Yang, Y., Yuan, T., Panaitescu, C., Li, R., Wu, K., Zhou, Y., & Zhan, W. (2024). Exploring tissue permeability of brain tumours in different grades: Insights from pore-scale fluid dynamics analysis. *Acta Biomaterialia, 190*, 398–409.

Chapter 3
Introduction to Cell Mechanobiology and Its Impact on Tissue Engineering and Regenerative Medicine

Abstract In tissue engineering the main actors are the cells, as they define how tissue regeneration is going to occur. Therefore, a primary challenge in tissue engineering is controlling the cellular function in order to achieve reliable and repeatable tissue regeneration strategies. If tissue engineering provides commercial products, they should be reliable and safe for patients under all pathologies. For this challenge, it is also crucial to advance in the understanding of how cells sense and respond to their microenvironment, in particular to mechanical conditions in a process normally known as mechano-transduction. Hence, this chapter offers a brief description of how cells behave according to different microenvironmental conditions, specially focused on how cells interact with material and with other cells. In addition, a brief explanation is presented of how cells are able to sense and transduce these signals in a cellular response. Normally, cell nucleus is constantly exposed to biochemical signals and mechanical deformations, which together can modulate many cellular and nuclear functions. Finally, this chapter ends with the most relevant conclusions and, additionally, some questions are suggested to reinforce the different concepts and aspects covered in this chapter.

3.1 Cells as Passive and Active Materials

Although cells have been extensively studied from a biological point of view, they also play an important role as mechanical elements that can behave as both active and passive materials. On the one hand, cells can deform, support and transmit stresses to the environment in which they are anchored. Therefore, as a passive material, cells exhibit complex mechanical behaviour. In fact, cells exhibit viscoelastic behaviour similar to that of a viscous fluid [1]. The viscoelasticity of cells is commonly evaluated by the complex shear modulus $G^* = G' + i\,G''$, which is defined as the complex ratio in the frequency domain between the applied stress and the resulting strain. The real part (G') is known as the storage modulus and represents the elastic contribution, while the imaginary part (G'') is known as the loss

© The Author(s), under exclusive license to Springer Nature Switzerland AG 2026

E. García-Gareta, J. M. García-Aznar, *Contemporary Tissue Engineering and Regenerative Medicine*, Series in BioEngineering,
https://doi.org/10.1007/978-3-032-03441-0_3

modulus and represents the dissipative or viscous contribution. Dynamic measurements of G* have shown that the viscoelastic behaviour of living cells is timescale dependent [2, 3]. Furthermore, cell stiffness shows a high variability depending on conditions and measurement techniques, even varying by several orders of magnitude, in the order of tens to thousands of pascals [4].

On the other hand, cells are sources of force and can exert forces or stresses in the medium to which they are attached (see Fig. 3.1) [5, 6]. Cells therefore behave like an active material, which can even be a strained/stressed structure. Indeed, there are filaments within the cell body that maintain a pre-existing tension even in the absence of external loading. Recent results have confirmed that there is a filamentary network under tension inside the cell: when these fibres are cut with a laser, they snap back [7]. This active behaviour of the cell that generates this internal tension is due to molecular motors that generate forces that are transmitted from the cell to the extracellular matrix (ECM) [8]. This internal bias strongly modifies the stiffness of the cell and its viscoelastic behaviour. Thus, the cell is characterised by a dual and interactive behaviour: as a passive material and as an active contractile system. The question that remains unanswered is: what is the factor that regulates the value of this internal prestress? It seems that the concentration of certain solutes, especially calcium ions, could control the level of pre-stress in smooth muscle cells [9].

But the mechanical characteristics of cells are very dynamical and can adapt their behaviour according to the needs that are required. In fact, cells can act both as solids and fluids: some structural functions require solid cells and whereas many other cell functions require fluidity [10, 11] (Fig. 3.2).

This dynamic and adaptive cell behaviour, which can vary from fluid to solid or vice versa, and with very different values of stiffness and viscosity, is largely due to the cytoskeleton, which will be clearly defined in the next section.

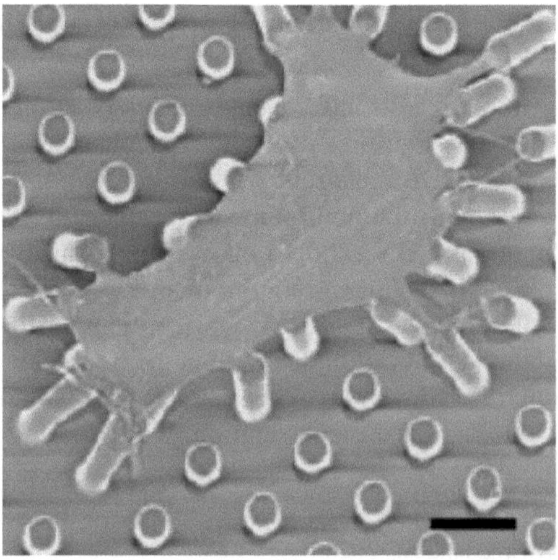

Fig. 3.1 Cell exerting traction forces in a micropillar-based substrate [5]. (Open access article distributed under the terms and conditions of the Creative Commons Attribution (CC BY) license)

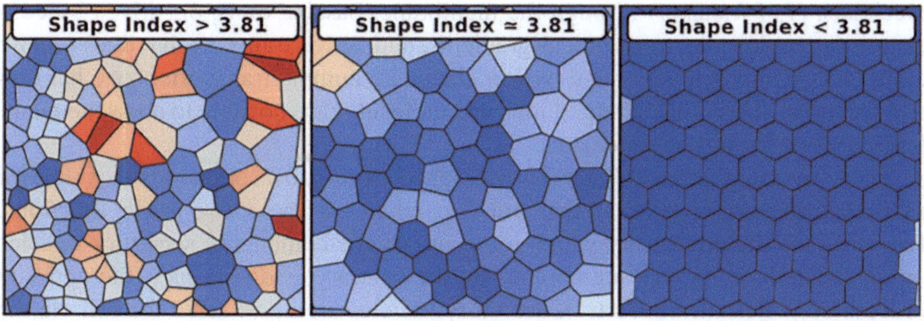

Fig. 3.2 Numerical analysis in function of the shape index (adimensional number that is calculated by the ratio between the cell perimeter and the square root of the cell area) of the cell transition (middle) between a fluid-like state (unjammed, top) and solid-like state (jammed, bottom) (adapted from [12])

3.2 Main Mechanical Components of the Cell: The Cytoskeleton

The mechanical properties of the cell are largely determined by four main components with different contributions: the membrane, the cytosol, the cytoskeleton (CSK) and the nucleus. Although each component will be discussed separately in each subsection, they all work together as a system to give each cell dynamic and adaptive mechanical properties.

3.2.1 Cell Membrane

The cell membrane is a thin, flexible, selectively permeable layer that separates the cell interior from the extracellular environment. It is formed by a double layer of phospolipid molecules in which proteins are embedded. Two different types of membrane proteins can be distinguished (1) integral proteins, which partly or entirely penetrate the membrane, and (2) peripheral proteins, which are laterally attached to one side of the membrane. In fact, cell membranes are typically asymmetric, their interior and exterior faces can carry different proteins and present different properties. Moreover, these membrane proteins can diffuse laterally or sideways throughout the membrane, giving them their characteristic appearance of a fluid rather than a solid. Certainly, these proteins can regulate the mechanical behaviour of the membrane.

The cell membrane develops two main functions. On the one hand, the cell membrane serves as a structural and mechanical element, maintaining and protecting the physical integrity of the cell body, by resisting bending and regulating cell shape. On the other hand, the cell membrane regulates the molecular transport between the cell interior and the extracellular environment. Normally there are two main mechanisms of transport through the membrane: passive transport driven by concentration- and/or electrochemical-gradients, and active transport that does require extra energy and it is regulated by ion channels, pumps, transporters, exchangers and receptors. Therefore, charged ions and molecules are not able to cross the membrane except through specific sites.

There are many other collateral processes like exocytosis or endocytosis (known as bulk transport processes) that are dependent on these two main functions. Exocytosis is the process by which a cell moves the contents of secretory vesicles out of the cell via the cell membrane. On the contrary, endocytosis is the opposite process by which the contents of secretory vesicles are moved into the cell via the cell membrane.

The cell membrane has a very low shear stiffness, both in-plane and out-of-plane. However, it has very high in-plane tensile stiffness and strength [13] (Table 3.1). The low shear stiffness of the membrane and the ability of the membrane to adapt to the presence of shear stresses indicate that the membrane will always seek to minimise (reduce to near zero) the shear stresses in the membrane. This behaviour of high tensile strength and virtually no shear strength has been described and explained thanks to the model known as the Fluidic Mosaic Model. The fluid mosaic model of the cell membrane is so called because the cell membrane is made up of different parts that work together, so it is made up of a bunch of phospholipids, cholesterol, proteins and carbohydrates [14]. The phospholipids are attracted to each other, but they are also constantly moving and bouncing around a bit. The spaces created by the membrane's fluidity are incredibly small, so it is still an effective barrier.

The four main properties of cell membrane are determined by their molecular composition: fluidity, permeability, heterogeneity and asymmetry [15]. The cell membrane acts like a fluid, because molecules can move inside the membrane, where higher fluidity means that movements are more frequent. Membrane permeability is selective, which is a consequence of the internal hydrophobic environment generated by the lipid fatty acid chains, which is difficult to cross by molecules having a net electric charge. Membrane heterogeneity and asymmetry are a consequence of fluidity, because molecules can move inside the membrane.

Membrane fluidity and stiffness are regulated by temperature, cholesterol and lipid composition. Decreasing the temperature induces a lower movement of lipids creating an ordered and regular structure with a solid-like behaviour, on the contrary, increasing the temperature produces a more disorganized structure with a fluid-like response. Cholesterol serves as a buffer, preventing lower temperatures

Table 3.1 Cell membrane stiffness [13]

Stiffness	N/m
Axial	0.45
Shear	6–9, 10–9
Bending	Approx. 10–19

from inhibiting fluidity and preventing higher temperatures from increasing fluidity. Finally, lipid composition is also determinant. If saturated fatty acids are compressed by decreasing temperatures, they press in on each other, making a dense rigid membrane. If unsaturated fatty acids are composed, the kinks push adjacent phospholipid molecules away, which helps maintain fluidity in the membrane.

The cell surface properties are not only determined by the cell membrane, because actin cortex is also contributing. In fact, cell surface is normally considered as an adaptive composite material [16], formed by the plasma membrane, the membrane-cortex-attachment and the adjacent cell cortex. It is very relevant to know that this lipid-cortex interface largely relies on non-covalent interactions. In consequence, the cytoskeleton continuously can de- and reattach from the lipid bilayer. These transient detachments not only open up the possibility to readily adjust interactions with material and cells, but also the ability to transport and fuse vesicles (i.e., large objects) without permanent detachment of the plasma membrane from the underlying cortex.

The cell membrane and cell cortex exert a crucial role in mechanotransduction as it will be detailed in Sect. 3.4 [17]. In fact, mechanical forces can deform the cortex and the membrane as also the mechano-sensitive proteins that are attached in the cell surface, such as, ion channels, piezo channels, TREK channels, G-protein couple receptors.

Moreover, the cell membrane is also crucial in cell-matrix and cell-cell adhesions as it will be explained in the next Sect. 3.3. In particular, cell-matrix adhesions are large protein complexes that consist of integrins and a large array of adaptor proteins that can grow or not in response to mechanical signals. In the case of cell-cell junctions, these are specialized regions of the plasma membrane consisting of protein complexes that couple adjacent cells.

3.2.2 *Cell Cytoplasm: The Cytosol*

The cytoplasm is formed by the cell content enclosed within the cell membrane and outside the nucleus. Apart from the CSK, the cytoplasm is formed by the cytosol, which is an aqueous solution formed by a myriad of proteins and molecules that fill the compartments of the cytoplasm.

Cytoplasmic mechanics plays a crucial role in cell organisation [18]. When cells organise themselves spatially or divide, they translocate many organelles in their cytoplasm. This organelle movement is driven by cytoskeletal forces, under some frictional viscoelastic forces due to the surrounding crowded cytoplasmic medium.

3.2.3 Cell Nucleus

Cell mechanics is mainly determined by both the mechanics of the largest organelle in the cell, the nucleus, and the cytoskeletal architecture of the cell [19]. Hence, cytoskeletal and nuclear mechanics are the major contributors to cell mechanics.

The nucleus is constituted by two concentrated lipid membranes containing the DNA molecules that encode the genetic information. The CSK biopolymers connect the nucleus by linker of nucleoskeleton and cytoskeleton (LINC) complexes that span the nuclear envelop, regulating the force transfer to the nucleus with important consequences in gene expression (see also below in mechanotransduction section). The cell nucleus has been reported to be ten-fold stiffer than the surrounding cytoplasm. Therefore, the nucleus could play an important role in the mechanical stabilization of the cell [20].

From a TERM perspective, the nucleus develops a crucial role in determining cell fate and function in response to various signals and stimuli. In addition to biochemical inputs, the nucleus is constantly exposed to intrinsic and extrinsic mechanical forces that modulates many of the cellular and nuclear functions [21]. These functions have long been considered to be downstream of cytoplasmic signalling pathways and dictated by gene expression, when however, nuclear deformation is determinant. Therefore, it is critical to improve our understanding of how the physiological forces can induce nuclear deformation.

3.2.4 Cytoskeleton

The cytoskeleton (CSK) is a complex, heterogeneous and filamentous structure that extends from the nucleus to the cell membrane providing a continuous and dynamic connection between almost all cellular structures, defining the most significant mechanical characteristics of a cell. In fact, the CSK constitutes the dynamic skeleton of the cell from which the cell is able to change its shape, coordinate its movements, exert mechanical forces and sense the extracellular environment. It consists of a biopolymer network consisting of three major components (Table 3.2): filamentous actin (F-actin), intermediate filaments and microtubules. These cytoskeletal polymers are at length-scales (a few microns at most), all corresponding to semiflexible polymers. The thermal fluctuation of one-dimensional semiflexible polymers or filaments is governed by their bending energy and can be characterized using the concept of persistence length L_p. In the absence of thermal fluctuations at zero temperature filaments are straight because of their bending rigidity (K_b). Sufficiently large and thermally fluctuating filaments lose their straight conformation. Only subsystems with contour length $L < <L_p$ appear rigid and maintain an average straight conformation. Larger filaments $L> > L_p$, on the other hand, appear flexible. In the 'semiflexible' regime for which L is comparable to L_p, statistical mechanics is governed by the competition of the thermal energy T and the bending

Table 3.2 Geometrical and mechanical properties of cytoskeletal components [22]

	Diameter (nm)	Persistence length (μm)	Bending stiffness (Nm^2)	Young's modulus (Pa)
Actin filament	6–8	15	7×10^{-26}	$1.3–2.5 \times 10^9$
Microtubule	25	6000	2.6×10^{-23}	1.9×10^9
Intermediate filament	10	1–3	$4–12 \times 10^{-27}$	$2–5 \times 10^6$

Table 3.3 Functions of the cytoskeleton main components

Category	Main functions
Actin filament	• Resist tension • Source of protrusive forces • Mechanical support to cells. • To allow cell motility and cell division • It is the scaffold on which myosin proteins generate force to support cell contractility
Microtubules	• Maintain cell shape by resisting compression • Motility via flagella or cilia • Move chromosomes during cell division • Move organelles • Pushing forces to deform cell membrane
Intermediate filaments	• Resist tension • Anchor nucleus and some other organelles • Provide mechanical stability to cells • Define some inter-cellular unions
CSK motor proteins	• Cell contractility • Intracellular organelle transport • Membrane association

rigidity. In Table 3.3, we list the persistence lengths and bending rigidities associated to the polymers that constitute the cell CSK [22].

F-actin is the main component that regulates the mechanical behaviour of the cell. In fact, its depolymerization implies a significant decrease in cell stiffness [2, 23, 24]. In vitro experiments of reconstituted F-actin networks showed that tension sustained by the filaments plays a critical role in the network rheology [25]. Actin bundles can bind to myosin, a motor protein able to move the bundles relative to each other by hydrolyzing adenosine triphosphate (ATP), creating what is known as a stress fibre, which is a structure able to support forces in the cell. Therefore, actin filaments in conjunction with myosin are the main force-generating mechanisms of the actin CSK, playing a crucial role in the active behaviour of the cell. Microtubules and intermediate filaments define the main passive behaviour of the cell. Ingber [26] proposed that cells are prestressed tensegrity structures with internal molecular struts and cables, with microtubules being effective at withstanding compression (the struts), and actin filaments being more adequate for working under tension. He hypothesized this theory based on the fact that microtubules often appear to be curved in living cells, whereas intermediate filaments are almost always linear. This is consistent with the engineering rule that tension straightens and compression buckles or bends the bar elements. These elements that constitute the CSK create a crowded network of structural proteins that regulates cell shape and drives cell motions, being able to modify and orient this filamentous structure in function of the mechanical and functional needs of the cell in a process known as CSK remodelling [27].

3.3 Actin Filaments

Actin filaments (F-actin) are linear polymers of globular actin (G-actin) subunits and occur as microfilaments in the cytoskeleton and as thin filaments that are part of the contractile system in muscle and non-muscle cells. Actin filaments are semiflexible and highly dynamic during polymerisation, allowing cells to rapidly change shape and respond to intracellular and extracellular forces. Actin filaments are semiflexible on the scale of cell length (10 microns). Therefore, shorter filaments behave like rigid rods and longer filaments can bend. Bending of actin filaments is accompanied by twisting due to the helical structure of actin filaments. Actin filaments sometimes form bundles that can withstand higher compressive forces. Actin crosslinking proteins play an important role in the formation of actin networks and bundles, and thus in the mechanical properties of cells [17].

Actin filaments are polarized and present a minus end (depolymerization is more frequent) and a plus end, where the addition of new actin proteins, is more frequent than depolymerization.

3.4 Microtubules

Microtubules are rigid, long and straight cytoskeletal filaments, being the stiffest ones. They are hollow tubes made of polymers of tubulin, with an outer diameter of 25 nm and with a 17 nm interior space diameter [28]. Microtubules present dynamic instability and they can switch rapidly between growing and shrinking processes to reorganize quickly and they can be assembled and disassembled rapidly. Therefore, they exhibit cycles of growth, shortening and regrowth [29].

Microtubules follow a polar organization, where they have directionality throughout their length and their two ends are different. Polarity is essential for different microtubule functions, mainly providing tracks for long-distance, motor-driven transport of vesicles and organelles [29].

The role of microtubule dynamics is crucial for generating forces to drive chromosome motions required for cell division [30].

3.5 Cytoskeletal Motor Proteins

These proteins convert chemical energy into motion. Chemical energy is derived from ATP hydrolysis and motion is generated by conformational changes depending on the bound nucleotide. Coordinated with their cycle of nucleotide hydrolysis and conformational change, the proteins cycle between states in which they are tightly bound to their filament tracks and states in which they are unbound. Through a mechanochemical cycle of filament binding, conformational change, filament release, conformational relaxation and filament rebinding, the motor protein and its associated cargo move one step at a time along the filament (typically a distance of a few nanometres). The identity of the track and the direction of movement along it is determined by the motor domain (head), while the identity of the cargo (and therefore the biological function of the individual motor protein) is determined by the tail of the motor protein [31].

Dozens of different motor proteins coexist in every eucaryotic cell. They differ in the type of filament they bind to (either actin or microtubules), the direction in which they move along the filament, and the "cargo" they carry. Many motor proteins carry membrane-enclosed organelles—such as mitochondria, Golgi stacks, or secretory vesicles—to their appropriate locations in the cell. Other motor proteins cause cytoskeletal filaments to slide against each other, generating the force that drives such phenomena as cell contraction, as for example, the myosin II.

Next, some cytoskeletal motors are cited, indicating the filaments where they move and their main function:

- The nonmuscle myosin II slides between actin filaments clearly determining the contractile properties of the actomyosin cytoskeleton [32]
- The myosin I molecular motors are comprised of a motor domain that binds to and interacts with actin in response to ATPase cycling, regulating membrane tension, cell adhesion, and actin architecture [33]
- Myosin V is a highly processive motor that spends most of its time strongly associated with actin, being responsible of intracellular transport of multiple cargoes over long distances within the cell [34].
- Kinesin-1 is the major anterograde motor for transport along microtubules, exerting a crucial role in the intracellular transport and the nervous system development [35].
- Dyneins are motors that move toward the minus end of microtubules, being responsible for transporting intracellular cargos in interphase cells and mediating spindle assembly and chromosome positioning during cell division [36].

3.6 Intermediate Filaments

Intermediate filaments are highly flexible and extensible filaments, being shorter than microtubules and actin fibers.

Intermediate filaments are not polarised because both ends of the filament are equivalent and their dimers lack polarity. This is a major difference from actin filaments and microtubules, which depend on their polarity for many functions, including the unidirectional movement of motor proteins. In addition, these filaments are much more heterogeneous than actin filaments or microtubules because the number of protofilaments along a single filament can vary. In fact, no energy is required for filament assembly. They are therefore the more stable components of the cell.

The main function of IFs is to create cell cohesion and prevent the acute fracture of epithelial cell sheets under tension. This is due to the extensive and complex interactions between the protofilaments that make up an intermediate filament, which increase its resistance to compression, torsion, stretching and bending forces. These features also enable intermediate filaments to help stabilise the extended axons of nerve cells and to line the inside of the nuclear envelope, where they help to store and protect the cell's DNA [37].

Five types of intermediate filaments can be found:

- Type I and II: keratins.
- Type III: desmin and vimentin
- Type IV: neurofilaments
- Type V: lamins

3.7 Cell-Cell and Cell-Matrix Interactions

One of the most relevant cellular processes for TERM applications is how cells interact with each other and with the ECM (see Fig. 3.3). These interactions are essential for many biological processes as cell adhesion, cell migration, tissue organization, and differentiation, and as a result they play central roles in embryonic development, remodelling, regeneration and homeostasis of tissue and organ systems. Most of cells require of these interactions to develop their functions successfully. Certainly, these interactions are also dependent on the cell type.

This interaction between cells or with the matrix is mediated by specialised molecules on the cell surface: cell adhesion molecules (CAMs) [38] or cell adhesion receptors [39]. Cell adhesion is generally defined as the binding of a cell to a substrate, which may be another cell, a surface or an organic matrix. These CAMs are typically transmembrane receptors consisting of an intracellular domain that interacts with cytoplasmic proteins, including the cytoskeleton, and an extracellular domain that specifically binds to other adhesion components. The major classes of CAMs in cells include cadherins, selectins, integrins and Ig-CAMs (cell adhesion molecules of the immunoglobulin superfamily). Cell adhesion receptors therefore

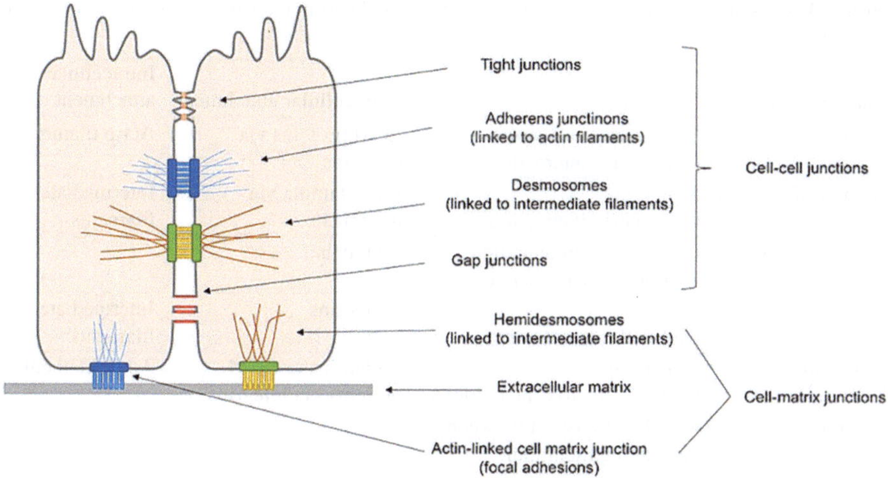

Fig. 3.3 Schematic representation of different cell-matrix and cell-cell junctions of how epithelial cells can interact (source from internet https://commons.wikimedia.org/wiki/File:Cell_junctions_types_shown_on_epithelial_cells_including_cell-cell_and_cell-matrix_junctions.jpeg)

recognise and interact either with other cell adhesion receptors on neighbouring cells or with proteins of the ECM. ECM proteins are typically large glycoproteins, including the collagens, fibronectins, laminins and proteoglycans, which assemble into fibrils or other complex macromolecular arrays [39].

Next, it is shortly presented a summary of the main functions that these adhesions exert on cell processes:

- Structural integrity to the tissues: it means that cells and tissues work together with a structural objective as it is supporting and transmitting loads. In the case of contractile cells, this adhesion provides the means of force transmission from the cells to the surrounding matrix to allow contraction. In non-contractile cells, it still provides the needed mechanical coupling to the surrounding tissue, necessary for a variety of cellular functions.
- Mechano-sensing: cell adhesive and contractile properties work together to sense the mechanical environment that surrounds cells. The adhesions are crucial to connect the cell body to other cells or to the matrix.
- Cell signaling: cellular interactions with the ECM and with neighboring cells widely regulate many different signaling events including those involved in proliferation, survival, and differentiation [40]. For example, recently in neurons it has been studied that CAMs regulate protein synthesis by modulating the activity of transcription factors, gene expression, protein translation, and the structure and distribution of organelles involved in protein synthesis and transport [41].
- Cell migration: certainly, cell adhesion regulates cell migration and requires a strong spatial and temporal coordination of multiple mechano-chemical signals [42].

Table 3.4 Classification of different cell interactions, indicating their functions and how they are interacting

Junction type	Function	Extracellular attachment	Intracellular attachment
Focal adhesions	Structural Cell-matrix interactions	ECM proteins via integrins	Actin filaments
Hemidesmosomes	Structural Cell-matrix interactions	Basal lamina via integrins	Intermediate filaments
Gap junctions	Cell-cell communication Ion/molecule exchange	Channels	
Desmosomes	Structural Cell-cell interactions	Cadherins	Intermediate filaments
Adherent junctions	Structural Cell-cell/-matrix interactions	Cadherins or ECM proteins via integrins	Actin filaments
Tight junctions	Regulate transport between cells along membrane		

These junctions are dependent on the type of filament that is using for attachment that normally can be the actin or the intermediate filaments. We can see in the next Table 3.4 how these junctions can be classified.

3.7.1 Focal Adhesions

Focal adhesions consist of a complex network of trans-membrane integrins and cytoplasmic proteins that form a <200-nm plaque connecting mechanically the ECM to the actin cytoskeleton [43]. These structures are dynamical and their evolution is mechano-sensitive, meaning that they can grow or shrink in response to the mechanical forces that are transmitting [44]. Focal adhesions begin as nascent adhesions with only a few integrins. While some nascent adhesions are short-lived, other nascent adhesions mature into stable focal adhesions, depending on the mechanical loads that they support. Normally, the mechanical dependency of the stability of these FAs follow a catch-bond law [45], and there are many different models simulating these adhesion with the known clutch model [46]. In fact, some authors classify these adhesions: nascent, maturation, fibrillar and rear adhesions [47]. The principal components of the FAs are integrins, a family of transmembrane proteins that connect the cytoskeleton with the extracellular matrix (Fig. 3.4).

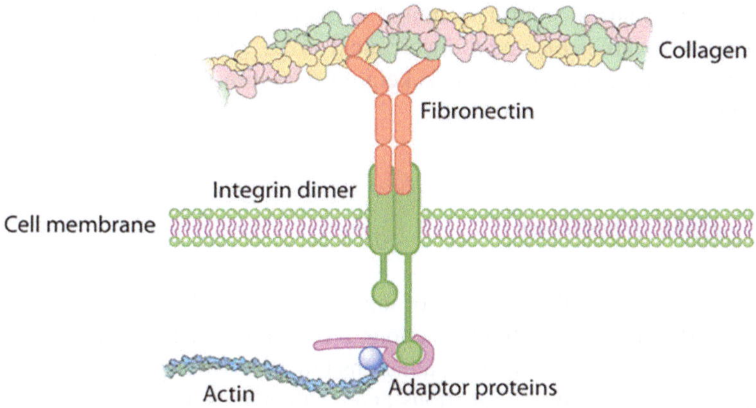

Fig. 3.4 Integrin connects the extracellular matrix with the actin cytoskeleton inside the cell (obtained from Nature Education 2010)

3.7.2 *Hemidesmosomas*

Hemidesmosomes (HDs) are highly specialised integrin-mediated epithelial attachment structures that anchor cells to the ECM by providing a link between the underlying basement membrane (BM) and the internal mechanical stress resistant intermediate filament (IF) network. Modulation of hemidesmosome function is of crucial importance for a variety of biological processes, such as terminal differentiation of basal keratinocytes and keratinocyte migration during wound healing and carcinoma invasion [48].

3.7.3 *Gap Junctions*

Gap junctions are clusters of intercellular channels that allow the direct diffusion of ions and small molecules between neighbouring cells. The ability of cells to exchange ions is fundamental to the function of electrically excitable cells such as neurons, heart and smooth muscle. Nevertheless, all cells in solid tissues are connected by gap junctions [49].

3.7.4 *Adherens Junctions and Desmosomes*

Adherens junctions (AJs) and desmosomes link the actin and intermediate filament networks of neighbouring cells into a mechanical unit. While AJs are involved in mechanosensing and transducing mechanical forces between the plasma membrane and the actomyosin cytoskeleton, desmosomes and intermediate filaments (IFs) provide the mechanical stability required to maintain tissue architecture and integrity when tissues are subjected to mechanical stress. Desmosomes are essential for stable intercellular cohesion, whereas keratins determine cell mechanics but are not involved in tension generation [50].

3.7.5 *Tight Junctions*

The tight junctions regulate passage of ions and small molecules between cells and establish cell polarity. The tight junctions are arranged nearest the apical (lumen exposed) portion of the cell and the adherens junctions immediately underneath the tight junctions. Both the gap junctions and desmosomes are located more basally. Historically, the adherens and tight junctions have been thought of as discrete complexes. However, it is now clear that a high level of interdependency exists between the two junctional complexes [51].

3.8 Mechanotransduction

Most tissues are constantly supporting external forces, such as mechanical loads, stretch, and shear stress, in addition to intrinsic forces generated by cells through contractile forces. Both extrinsic and intrinsic forces contribute to tissue morphogenesis, homeostasis, and regeneration and regulate cell fate: proliferation, matrix secretion and migration. Sensing and transmitting forces depends to a large extent on tight interactions between cell adhesion complexes and the cytoskeleton. Cellular mechanosensing and mechanotransduction can be defined as the cellular processes that convert mechanical cues into intracellular biochemical signals regarding cell activities and metabolism [52].

Mechanical signals that originate at the level of the ECM (e.g. as a consequence of external, physiological loading) can be transduced to the nucleus through a molecular network that connects the ECM to the nucleus [8]. Therefore, for advance in the understanding of how mechanotransduction works, it is crucial to know how these forces are applied and transmitted into and through the cell. As we have seen in the previous section, cells are anchored to the ECM by means of integrins, a specialised class of transmembrane proteins. Their extracellular domain enables them to bind to the ECM (or substrate). Integrin molecules are heterodimeric transmembrane receptors and composed of an alpha and beta subunit, the combination of which results in a binding affinity which is specific to certain ECM proteins [53]. Their cytoplasmic domain binds to actin-associated proteins, such as talin, vinculin, zyxin and paxillin [8]. Certain integrin proteins can also bind to intermediate filaments. Integrins cluster into specialised anchoring complexes, called focal adhesion complexes (FAs) which are essential to transmit mechanical signals across the cell membrane from the ECM to the cytoskeleton. While integrins constitute molecular bridges between the ECM and the cytoskeleton, cadherins, which mediate cell-to-cell adhesion, form junctional complexes as well, which bind actin filaments and intermediate filaments respectively [8].

To better understand how cellular mechanotransduction process works, normally is convenient to divide this process in other subprocesses [54]:

- Mechanoreception or mechanosensing, it is the process corresponding to the detection of the mechanical stimulus and transmission of the signal from outside of the cell to inside. This process mainly occurs in the cell membrane, where there are different cell membrane receptors. As we have explained, integrins are one of the main actors in mechanotransduction, because they allow to connect the ECM with the cell cytoskeleton through the membrane. But there are others, like G proteins, growth factor receptors, primary cilium and stretch-activated ion channels.
- Intracellular signal transduction, it is the transduction of the stimulus to the nucleus, it occurs in the location in the cell where a molecular response can be generated. Some of these intracellular signalling can be the following:
  - Ca^{2+} changes in the intracellular calcium concentration are known to regulate intracellular signaling and CSK remodeling
  - Rho GTPases regulates many aspects of intracellular actin dynamics, Rho proteins have been described as molecular switches and play a role in cell proliferation, apoptosis, gene expression, and multiple other common cellular functions
  - MAPK mitogen-associated protein kinase pathways convey information to effectors, coordinate incoming information from other signaling cascades, amplify signals, and initiate a variety of response patterns
  - NO nitric oxide acts as a second messenger, it is a free radical that can diffuse through the plasma membrane and affect nearby cells
  - PI3K phosphoinositol-3-kinase is an intracellular signaling pathway regulating apoptosis
- Target activation, all these signals converge to activate transcription factors (it is a protein that controls the rate of transcription of genetic information from DNA to messenger RNA, by binding to a specific DNA sequence), which stimulate gene expression and other nuclear events that may cause alterations in cell behavior. For example, inducing an increased mitotic activity and increased collagen synthesis [54]. Only few mechanosensitive transcription factors have been so far identified that are induced to shuttle from the cytoplasm and activate a given genetic program (see Fig. 3.5). For example, YAP is a recognised a mechanosensitive transcriptional activator with a critical role in cancer, regeneration, and organ size control [56]. In this work, authors show that force applied to the nucleus directly drives YAP nuclear translocation by decreasing the mechanical restriction of nuclear pores to molecular transport. Exposure to a stiff environment leads cells to establish a mechanical connection between the nucleus and the cytoskeleton, allowing forces exerted through focal adhesions to reach the

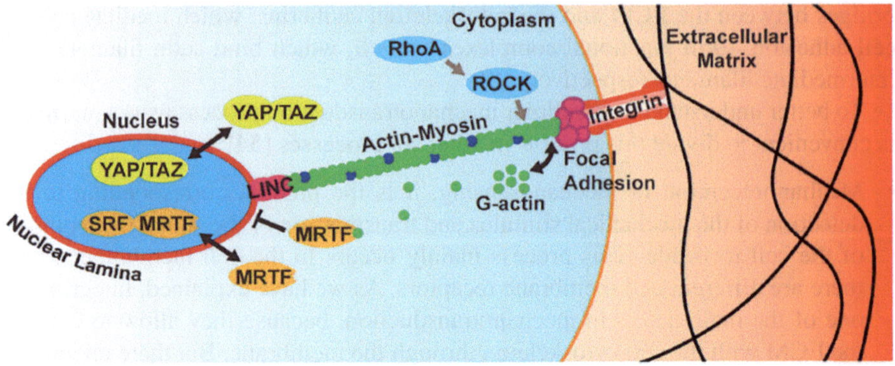

Fig. 3.5 Overview of mechanotransduction pathways. Large protein complexes called focal adhesions form on the intracellular part of integrin clusters and attach to actin-myosin polymers. Actin-myosin chains attach to other focal adhesions or to the LINC complex on the nucleus. RhoA and ROCK play an important, but not exclusive, role in myosin phosphorylation and actin-myosin force generation. The LINC complex passes both nuclear membranes and attaches to the nuclear lamina. YAP and TAZ enter the nucleus upon sufficient cellular tension where they serve as co-transcription factors. MRTFs bind to globular-actin monomers that prevent nuclear entry. When concentrations of G-actin are sufficiently low, MRTF can enter into the nucleus and bind to SRF to initiate target-gene transcription (obtained from [55]) (open access article distributed under the terms and conditions of the Creative Commons Attribution (CC BY) license)

> nucleus. Force transmission then leads to nuclear flattening, which stretches nuclear pores, reduces their mechanical resistance to molecular transport, and increases YAP nuclear import. The restriction to transport is further regulated by the mechanical stability of the transported protein, which determines both active nuclear transport of YAP and passive transport of small proteins. These results clearly show a mechanosensing mechanism mediated directly by nuclear pores, demonstrated for YAP but with potential general applicability in transcriptional regulation

As it is indicated in Fig. 3.5, mechanotransduction requires connection between the nucleus and the cytoskeleton This connection between the inner(INM) and the outer nuclear membrane of the nuclear envelope host the complex responsible for tying together nucleoskeleton, nuclear envelope and cytoskeleton: the Linker of Nucleoskeleton and Cytoskeleton (LINC) [57]. The main components of LINC system so far identified are SUN and nesprin proteins (see Fig. 3.6).

3.9 Conclusions

This chapter presents the most important basic concepts of cell mechanics and its crucial impact on tissue response for regenerative purposes. Control of cell and tissue function is essential for the design and development of various strategies in TERM. Typically, this control has focused on biochemical considerations and mechanics has been underestimated.

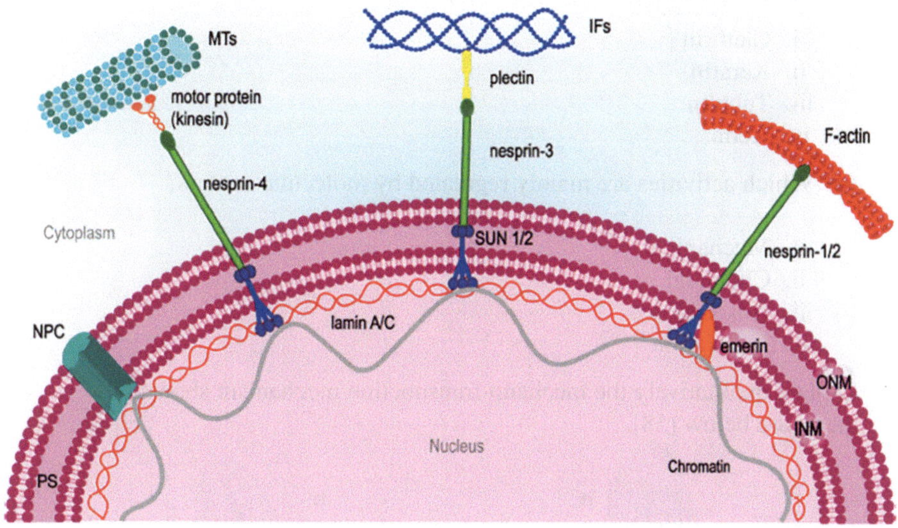

Fig. 3.6 Overview of the main components of the LINC complex. (obtained from [57]) (open access article distributed under the terms and conditions of the Creative Commons Attribution (CC BY) license)

This chapter highlights the importance of how mechanical conditions determine the final cell fate (proliferation, differentiation or matrix synthesis, among others) and thus the control of their behaviour. Indeed, this chapter provides new insights into how different scaffold designs can present different mechanical and architectural properties to control cellular and tissue functions for tissue engineering and regenerative medicine applications, depending on the tissue type and injury.

Questions

1. Which of the following cell structure-function pairs is mismatched? Indicating which would be its main function in the wrong cases.

    (a) Microtubule—compressive strength
    (b) Intermediate filament—tensile strength
    (c) Integrin—organelle transport
    (d) Myosin II—endocytosis
    (e) Ion channel—cilia motility

2. Choose only one correct answer in each case:

    (a) What is the function of the cell membrane?

        i. To control substances entering and leaving the cell
        ii. Cellular respiration
        iii. Contains genetic information
        iv. D Synthesises proteins

(b) Microtubules are made up of the protein:

i. Clathrin
ii. Keratin
iii. Tubulin
iv. Actin

(c) Which activities are mainly regulated by molecular motors?

i. Mechanical
ii. Chemical
iii. Absorption
iv. Regulation

3. Explain qualitatively the mechano-transduction mechanism shown in the figure below [58].

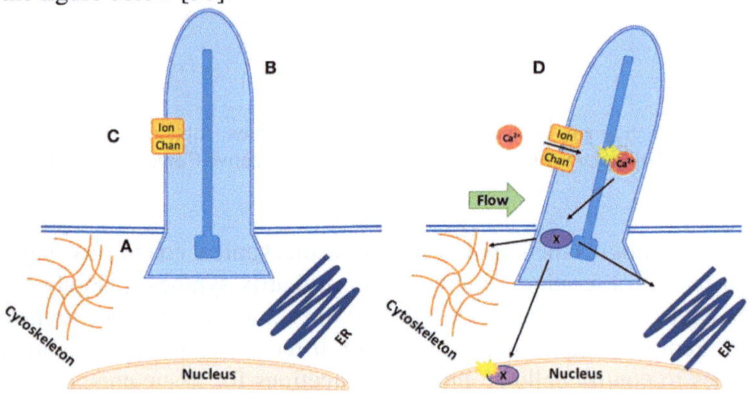

References

1. Desprat, N., Richert, A., Simeon, J., & Asnacios, A. (2005). Creep function of a single living cell. *Biophysical Journal, 88*(3), 2224–2233.
2. Fabry, B., Maksym, G. N., Butler, J. P., Glogauer, M., Navajas, D., & Fredberg, J. J. (2001). Scaling the microrheology of living cells. *Physical Review Letters, 87*(14), 148102.
3. Stamenović, D. (2006). Two regimes, maybe three? *Nature Materials, 5*(8), 597–598.
4. Stamenovic, D., & Wang, N. (2000). Invited review: Engineering approaches to cytoskeletal mechanics. *Journal of Applied Physiology, 89*(5), 2085–2090.
5. Hur, S. S., Jeong, J. H., Ban, M. J., Park, J. H., Yoon, J. K., & Hwang, Y. (2020). Traction force microscopy for understanding cellular mechanotransduction. *BMB Reports, 53*(2), 74.
6. Trichet, L., Le Digabel, J., Hawkins, R. J., Vedula, S. R. K., Gupta, M., Ribrault, C., & Ladoux, B. (2012). Evidence of a large-scale mechanosensing mechanism for cellular adaptation to substrate stiffness. *Proceedings of the National Academy of Sciences, 109*(18), 6933–6938.
7. Kumar S., Maxwell I. Z., Heisterkamp A., Polte T. R., Lele T. P., Salanga M., Mazur E., & Ingber D. E. (2006). Viscoelastic retraction of single living stress fibers and its impact on

cell shape cytoskeletal organization and extracellular matrix mechanics. *Biophysical Journal 90*(10), 3762–3773. https://doi.org/10.1529/biophysj.105.071506.
8. Wang, N., Tytell, J. D., & Ingber, D. E. (2009). Mechanotransduction at a distance: Mechanically coupling the extracellular matrix with the nucleus. *Nature Reviews Molecular Cell Biology, 10*(1), 75–82.
9. Stålhand J., Klarbring A., & Holzapfel G. A. (2008). Smooth muscle contraction: Mechanochemical formulation for homogeneous finite strains. *Progress in Biophysics and Molecular Biology 96*(1–3) 465–481. https://doi.org/10.1016/j.pbiomolbio.2007.07.025.
10. Park, J. A., Fredberg, J. J., & Drazen, J. M. (2015). Putting the squeeze on airway epithelia. *Physiology, 30*(4), 293–303.
11. González-Valverde, I., & García-Aznar, J. M. (2019). An agent-based and FE approach to simulate cell jamming and collective motion in epithelial layers. *Computational Particle Mechanics, 6*, 85–96.
12. Gonzalez-Valverde, I. (2018). Modeling and simulation of multi-cellular systems using hybrid FEM/agent-based approaches(doctoral dissertation, PhD thesis, University of Zaragoza, Zaragoza).
13. Jacobs, C. R., Huang, H., & Kwon, R. Y. (2012). *Introduction to cell mechanics and mechanobiology*. Garland Science.
14. Nicolson, G. L. (2014). The fluid—Mosaic model of membrane structure: Still relevant to understanding the structure, function and dynamics of biological membranes after more than 40 years. *Biochimica et Biophysica Acta (BBA) - Biomembranes, 1838*(6), 1451–1466.
15. Jacobson, K., Liu, P., & Lagerholm, B. C. (2019). The lateral organization and mobility of plasma membrane components. *Cell, 177*(4), 806–819.
16. Lamparter, L., & Galic, M. (2020). Cellular membranes, a versatile adaptive composite material. *Frontiers in Cell and Developmental Biology, 8*, 684.
17. Uray, I. P., & Uray, K. (2021). Mechanotransduction at the plasma membrane-cytoskeleton interface. *International Journal of Molecular Sciences, 22*(21), 11566.
18. Arjona, M. I., Najafi, J., & Minc, N. (2023). Cytoplasm mechanics and cellular organization. *Current Opinion in Cell Biology, 85*, 102278.
19. Fischer, T., Hayn, A., & Mierke, C. T. (2020). Effect of nuclear stiffness on cell mechanics and migration of human breast cancer cells. *Frontiers in Cell and Developmental Biology, 8*, 393.
20. Versaevel, M., Riaz, M., Grevesse, T., & Gabriele, S. (2013). Cell confinement: Putting the squeeze on the nucleus. *Soft Matter, 9*(29), 6665–6676.
21. Kalukula, Y., Stephens, A. D., Lammerding, J., & Gabriele, S. (2022). Mechanics and functional consequences of nuclear deformations. *Nature Reviews Molecular Cell Biology, 23*(9), 583–602.
22. Mofrad M. R. K., Kamm R. D. (2006). *Cytoskeletal mechanics models and measurements in cell mechanics*. Cambridge University Press.
23. Trepat X., Grabulosa M., Buscemi L., Rico F., Farré R., & Navajas D. (2005). Thrombin and histamine induce stiffening of alveolar epithelial cells. *Journal of Applied Physiology, 98*(4), 1567–1574. https://doi.org/10.1152/japplphysiol.00925.2004.
24. Smith B. A., Tolloczko B., Martin J. G., & Grütter P. (2005). Probing the viscoelastic behavior of cultured airway smooth muscle cells with atomic force microscopy: stiffening induced by contractile agonist. *Biophysical Journal, 88*(4), 2994–3007. https://doi.org/10.1529/biophysj.104.046649.
25. Gardel M. L., Nakamura F., Hartwig J., Crocker J. C., Stossel T. P., & Weitz D. A. (2006). Stress-dependent elasticity of composite actin networks as a model for cell behavior. *Physical Review Letters, 96*(8). https://doi.org/10.1103/PhysRevLett.96.088102.
26. Ingber D. E. (2006). Cellular mechanotransduction: putting all the pieces together again. *The FASEB Journal, 20*(7), 811–827. https://doi.org/10.1096/fsb2.v20.7 https://doi.org/10.1096/fj.05-5424rev.

27. Bursac P., Lenormand G., Fabry B., Oliver M., Weitz D. A., Viasnoff V., Butler J. P., & Fredberg J. J. (2005). Cytoskeletal remodelling and slow dynamics in the living cell Nature. *Materials, 4*(7), 557–561. https://doi.org/10.1038/nmat1404.
28. Hawkins, T., Mirigian, M., Yasar, M. S., & Ross, J. L. (2010). Mechanics of microtubules. *Journal of Biomechanics, 43*(1), 23–30.
29. Gudimchuk, N. B., & McIntosh, J. R. (2021). Regulation of microtubule dynamics, mechanics and function through the growing tip. *Nature Reviews Molecular Cell Biology, 22*(12), 777–795.
30. Forth, S., & Kapoor, T. M. (2017). The mechanics of microtubule networks in cell division. *Journal of Cell Biology, 216*(6), 1525–1531.
31. Alberts, B. (2017). *Molecular biology of the cell*. Garland Science.
32. Vicente-Manzanares, M., Ma, X., Adelstein, R. S., & Horwitz, A. R. (2009). Non-muscle myosin II takes Centre stage in cell adhesion and migration. *Nature Reviews Molecular Cell Biology, 10*(11), 778–790.
33. McIntosh, B. B., & Ostap, E. M. (2016). Myosin-I molecular motors at a glance. *Journal of Cell Science, 129*(14), 2689–2695.
34. Sarshad, A. A., & Percipalle, P. (2014). New insight into role of myosin motors for activation of RNA polymerases. *International Review of Cell and Molecular Biology, 311*, 183–230.
35. Gindhart, J. G., & Weber, K. P. (2009). Lysosome and endosome organization and transport in neurons.
36. Canty, J. T., Tan, R., Kusakci, E., Fernandes, J., & Yildiz, A. (2021). Structure and mechanics of dynein motors. *Annual Review of Biophysics, 50*, 549–574.
37. Goldman, R. D., Cleland, M. M., Murthy, S. P., Mahammad, S., & Kuczmarski, E. R. (2012). Inroads into the structure and function of intermediate filament networks. *Journal of Structural Biology, 177*(1), 14–23.
38. Edelman, G. M. (1983). Cell adhesion molecules. *Science, 219*(4584), 450–457.
39. Gumbiner, B. M. (1996). Cell adhesion: The molecular basis of tissue architecture and morphogenesis. *Cell, 84*(3), 345–357.
40. Juliano, R. L. (2002). Signal transduction by cell adhesion receptors and the cytoskeleton: Functions of integrins, cadherins, selectins, and immunoglobulin-superfamily members. *Annual Review of Pharmacology and Toxicology, 42*(1), 283–323.
41. Kozlova, I., Sah, S., Keable, R., Leshchyns, I., Janitz, M., & Sytnyk, V. (2020). Cell adhesion molecules and protein synthesis regulation in neurons. *Frontiers in Molecular Neuroscience, 13*, 592126.
42. Merino-Casallo, F., Gomez-Benito, M. J., Hervas-Raluy, S., & Garcia-Aznar, J. M. (2022). Unravelling cell migration: Defining movement from the cell surface. *Cell Adhesion & Migration, 16*(1), 25–64.
43. Kanchanawong, P., Shtengel, G., Pasapera, A. M., Ramko, E. B., Davidson, M. W., Hess, H. F., & Waterman, C. M. (2010). Nanoscale architecture of integrin-based cell adhesions. *Nature, 468*(7323), 580–584.
44. Cóndor, M., & García-Aznar, J. M. (2017). A phenomenological cohesive model for the macroscopic simulation of cell–matrix adhesions. *Biomechanics and Modeling in Mechanobiology, 16*, 1207–1224.
45. Escribano, J., Sánchez, M. T., & García-Aznar, J. M. (2014). A discrete approach for modeling cell–matrix adhesions. *Computational Particle Mechanics, 1*, 117–130.
46. Chan, C. E., & Odde, D. J. (2008). Traction dynamics of filopodia on compliant substrates. *Science, 322*(5908), 1687–1691.
47. Vicente-Manzanares, M., & Horwitz, A. R. (2011). Cell migration: An overview. In *Cell migration: Developmental methods and protocols* (pp. 1–24). Springer.
48. Walko, G., Castañón, M. J., & Wiche, G. (2015). Molecular architecture and function of the hemidesmosome. *Cell and Tissue Research, 360*, 363–378.
49. Goodenough, D. A., & Paul, D. L. (2009). Gap junctions. *Cold Spring Harbor Perspectives in Biology, 1*(1), a002576.

50. Hatzfeld, M., Keil, R., & Magin, T. M. (2017). Desmosomes and intermediate filaments: Their consequences for tissue mechanics. *Cold Spring Harbor Perspectives in Biology, 9*(6), a029157.
51. Campbell, H. K., Maiers, J. L., & DeMali, K. A. (2017). Interplay between tight junctions & adherens junctions. *Experimental Cell Research, 358*(1), 39–44.
52. Di, X., Gao, X., Peng, L., Ai, J., Jin, X., Qi, S., & Luo, D. (2023). Cellular mechanotransduction in health and diseases: From molecular mechanism to therapeutic targets. *Signal Transduction and Targeted Therapy, 8*(1), 282.
53. Huveneers, S., & Danen, E. H. (2009). Adhesion signaling–crosstalk between integrins, SRC and Rho. *Journal of Cell Science, 122*(8), 1059–1069.
54. Wong, V. W., Longaker, M. T., & Gurtner, G. C. (2012). Soft tissue mechanotransduction in wound healing and fibrosis. *Seminars in Cell & Developmental Biology, 23*(9), 981–986.
55. Zonderland, J., & Moroni, L. (2021). Steering cell behavior through mechanobiology in 3D: A regenerative medicine perspective. *Biomaterials, 268*, 120572.
56. Elosegui-Artola, A., Andreu, I., Beedle, A. E., Lezamiz, A., Uroz, M., Kosmalska, A. J., & Roca-Cusachs, P. (2017). Force triggers YAP nuclear entry by regulating transport across nuclear pores. *Cell, 171*(6), 1397–1410.
57. Martino, F., Perestrelo, A. R., Vinarský, V., Pagliari, S., & Forte, G. (2018). Cellular mechanotransduction: From tension to function. *Frontiers in Physiology, 9*, 824.
58. Pagnozzi, L. A., & Butcher, J. T. (2017). Mechanotransduction mechanisms in mitral valve physiology and disease pathogenesis. *Frontiers in Cardiovascular Medicine, 4*, 83.

Chapter 4
Cells and Growth Factors for Tissue Engineering and Regenerative Medicine

Abstract Cells are one of the cornerstones of the tissue engineering and regenerative medicine (TERM) paradigm and are ultimately responsible for producing the new tissue. Because of their regenerative potential, stem cells are central to TERM, where particular care is taken into recreating their natural niche, which regulates and modulates stem cell fate. Other cells used in TERM include differentiated cells or even cell lines for in vitro models. Growth factors and exosomes, which are both released by cells into the environment, can be incorporated into scaffolds to elicit desired responses in target cells. Cell sheet tissue engineering creates cell patches for clinical repair of various tissues. Chapter 4 offers a comprehensive overview of the cell element of the TERM paradigm.

4.1 Introduction

Cells are one of the three building blocks of the tissue engineering and regenerative medicine (TERM) paradigm, where combined with scaffolds and bio-chemical/physical cues allow the production of tissue analogues or equivalents [1]. These tissue-engineered equivalents can be used clinically for the repair or replacement of lost or damaged tissues and organs, or as models for the study of tissue phenomena including disease. Cells are an important component in TERM since they are ultimately responsible for new tissue formation and remodelling. Therefore, the right choice of cells for a particular TERM application is of vital importance.

Stem cells are undifferentiated cells that possess the ability to differentiate into the specialized cells that form our tissues, or self-renew through division [2]. Stem cells are central to regenerative medicine and are the foundation for growth, development and repair of tissues. With different degrees of potencies, stem cells include embryonic stem cells (ESCs) and induced pluripotent stem cells (iPSCs), which are pluripotent, and adult stem cells (ASCs), which are multipotent and play a crucial role in tissue homeostasis [2–5]. The different stem cell types have their own set of advantages and disadvantages, with ethical considerations being central in the

© The Author(s), under exclusive license to Springer Nature Switzerland AG 2026

E. García-Gareta, J. M. García-Aznar, *Contemporary Tissue Engineering and Regenerative Medicine*, Series in BioEngineering,
https://doi.org/10.1007/978-3-032-03441-0_4

choice of one type over other for TERM strategies and research. The stem cell niche is the dynamic in vivo microenvironment where stem cells are found [6]. A plethora of biochemical and biophysical factors in the niche direct and regulate stem cell fate [6]. Therefore, understanding the complex stem cell niche is essential to be able to manipulate and control stem cell fate and behaviour for TERM strategies. Beside stem cells, there are other cells used in TERM, including differentiated cells like fibroblasts, chondrocytes, endothelial cells or vascular smooth muscle cells for various tissue regeneration strategies, or cancer cells, either patient-derived or cell lines, for in vitro cancer models [7–12].

Growth factors are biochemical cues often incorporated into scaffolds, with special care taken into their sustained release into the target site [13, 14]. Growth factors are signalling molecules released by cells into the environment that eventually trigger a cellular response in the target cell, thereby influencing cell behaviour. Several strategies exist for adding these important cues to scaffolds, including the use of nanocarriers for their delivery [14]. Also released by cells into the environment are the intriguing exosomes, which are nano-sized endosomal derived membrane biovesicles implicated in cell-cell communication [15]. It has been shown that mesenchymal stem cells (MSCs), a type of multipotent ASCs used in multiple TERM strategies, mainly deploy their therapeutic outcomes via paracrine effects derived from their secretome, which includes exosomes [16]. Therefore, many efforts are directed towards including these cell-derived vesicles in TERM approaches.

We talk about scaffold-free tissue engineering when the scaffold component is merely used as a temporary support for cells. This is the so-called cell sheet tissue engineering, which has achieved human clinical use for regeneration and repair of various tissues [17, 18]. In cell sheets cell-to-cell contact is maintained and ECM adhesive components are found on the basal side of the cells. Cell sheets adhere to the application site thereby integrating with the host tissue, which is a clear advantage compared with the injection of cells that have been harvested by enzymatic treatment [17].

This chapter offers a comprehensive overview of the cell element of the TERM paradigm, including stem cells, the stem cell niche, differentiated cells, exosomes, and cell sheet tissue engineering. Additionally, growth factors are discussed. Strategies for engineering stem cell niches, incorporating growth factors and exosomes into scaffolds, and fabricating cell sheets are covered as part of the overview, which ends with questions to underpin the knowledge presented in this chapter.

4.2 Stem Cells

Stem cells are undifferentiated cells, which have the ability to differentiate into specialized cell types and self-renew through division [2]. They serve as the foundation for growth, development, and repair in living organisms and are central to regenerative medicine and research. However, ethical considerations, particularly with ESCs, and technical challenges in their use remain topics of ongoing discussion, research and innovation.

4.2.1 Hierarchy of Stem Cells

Stem cells are typically classified into the stem cell hierarchy based on their differentiation potential or potency. In decreasing order of differentiation potential, stem cells can be totipotent, pluripotent, multipotent, oligopotent, and unipotent (Fig. 4.1).

Totipotent stem cells are the embryonic stem cells present during the first few cell divisions post-fertilization. These totipotent stem cells can give rise to a

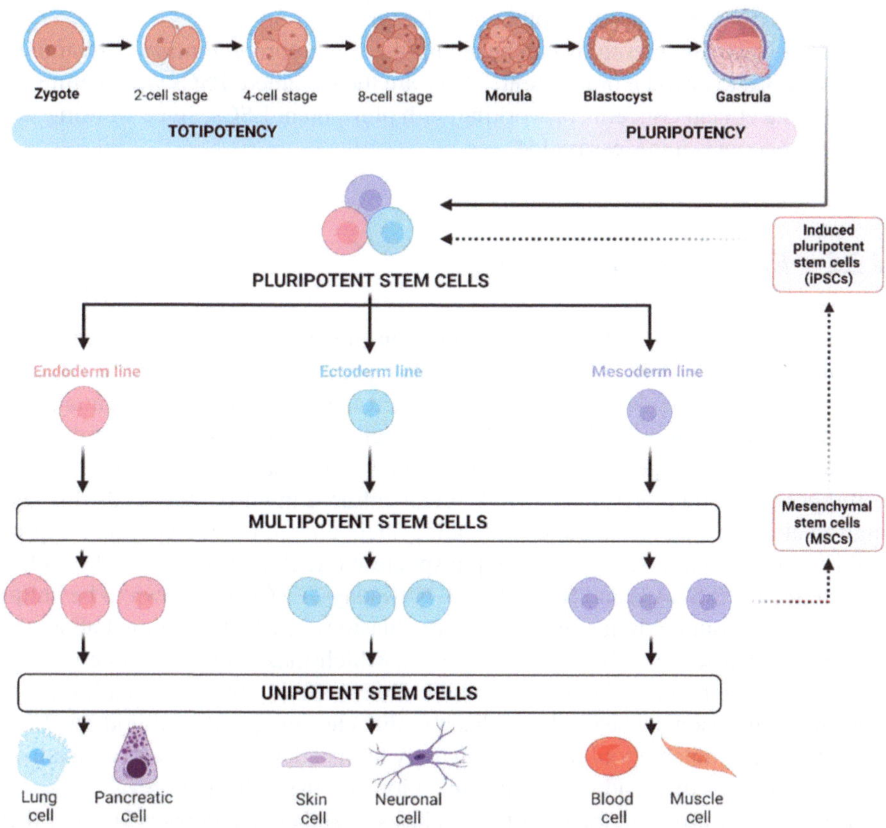

Fig. 4.1 Hierarchy of stem cells. ESCs are embryonic stem cells. ASCs are adult stem cells. iPSCs are induced pluripotent stem cells, which have been produced from differentiated cells and multipotent stem cells. Image adapted from Erceg Ivkošić et al. [19], open access article distributed under the terms and conditions of the Creative Commons Attribution (CC BY) license

complete, viable organism. As cell division continues, a blastocyst is formed where in their inner cell mass ESCs reside. These ESCs are pluripotent stem cells that can generate cells from the three germ layers, namely the endoderm, mesoderm and ectoderm [5].

Next in the stem cell hierarchy we find ASCs, which are multipotent cells with the capacity to differentiate into multiple cell types, generally from a specific group [20]. For instance, hematopoietic stem cells, which are found in the bone marrow, peripheral blood or umbilical cord blood, give rise to the different blood cells, which includes cells from the myeloid and lymphoid lineages [20]. ASCs are part of adult tissues and have a prominent role in tissue homeostasis and healing. Oligopotent cells are a type of stem cell that has a more restricted differentiation potential compared to pluripotent or multipotent stem cells. They are found in adult tissues and have a more restricted differentiation potential than multipotent ASCs [19]. Finally, we find unipotent cells which are capable of differentiating into one particular cell type. At the end of the spectrum, we find differentiated cells, which can renovate themselves but cannot produce other cellular types.

This downstream hierarchy can be manipulated by reprogramming differentiated cells or ASCs to become pluripotent. These are the so-called iPSCs which eliminate some of the ethical concerns of working with pluripotent ESCs, since working with embryos becomes unnecessary [4].

4.2.2 *Embryonic Stem Cells*

ESCs were first isolated from preimplantation mouse embryos in 1981 and from human embryos in 1998 [21, 22]. These cells represent some of the earliest uncommitted cells in tissue development and allowed the study of the differentiation of hundreds of cell types. Early on, ESCs were defined according to their capacity to proliferate for prolonged periods of time while being capable of differentiation into cell types of the three germ layers [2]. When cultured in vitro, these cells display a high nucleus to cytoplasm ratio, indicating their uncommitted state, and express certain surface markers, e.g. alkaline phosphatase (ALP) and stage-specific embryonic antigen-3 (SSEA3) and SSEA4 [22]. Nowadays, ESCs are known to be unique cells derived from the inner cell mass of the blastocyst, which is formed about 4–5 days post-fertilization. ESCs are pluripotent, which makes them a cornerstone of regenerative medicine and developmental biology, offering hope for treating conditions like Parkinson's disease, heart disease, diabetes, or spinal cord injuries. ESCs also serve as a valuable tool for studying early human development and understanding the processes that lead to cell specialization and tissue formation.

Despite their promise, the use of ESCs has sparked ethical debates because their derivation typically involves the destruction of embryos [3]. This controversy has driven researchers to explore alternative sources, such as iPSCs, which are discussed next. Other issue with the clinical use of ESCs is that these cells are allogenic to the patient, and therefore, immune rejection could occur [3]. Also importantly, ESCs present cellular and molecular similarities with tumour cells and cancer cell lines, such as fast proliferation rate, lack of contact inhibition, high telomerase activity, high expression of oncogenes like MYC9 and KLF4, and astonishing similarities in their overall gene expression patterns or microRNA (miRNA) signatures [5]. When human ESCs are injected into immunodeficient mice, they form teratomas, which are tumours characteristic of ESCs. Furthermore, the literature shows that culture-adapted ESCs can form teratocarcinomas, which are the malignant counterparts of teratomas and thus, more aggressive [5]. The field of stem cell research continues to balance the immense therapeutic potential of ESCs with the societal, tumorigenicity, and ethical considerations surrounding their use.

4.2.3 *Induced Pluripotent Stem Cells*

iPSCs are generated by reprogramming adult somatic cells into an embryonic-like state through the introduction of specific transcription factors, typically Oct4, Sox2, Klf4, and c-Myc [23]. These cells display pluripotency and thus, are able to differentiate into any cell type in the body, making them highly valuable for regenerative medicine, disease modelling, and drug development. Unlike ESCs, iPSCs bypass ethical concerns since they are derived from adult tissues rather than embryos. Their discovery by Shinya Yamanaka in 2006 revolutionized stem cell research and earned him a Nobel Prize in 2012, along with John Gurdon. However, challenges such as genetic instability and tumorigenicity remain areas of active research to enhance their therapeutic application [4, 5, 24].

Undoubtedly, iPSCs have opened new frontiers in personalized medicine, as they can be derived directly from a patient's own somatic cells, reducing the risk of immune rejection during transplantation therapies [4]. Additionally, they could be allogenic, thus reaching a large number of patients [4]. Both approaches have advantages and disadvantages, as summarized in Fig. 4.2.

These cells are also invaluable in studying complex diseases by creating patient-specific models, offering insights into disease mechanisms at the cellular level [4]. Furthermore, iPSCs are being explored for their role in high-throughput drug screening, enabling the identification of novel treatments with reduced reliance on animal models. Despite their potential, improving reprogramming efficiency, ensuring genomic integrity, and addressing ethical concerns about genetic modifications remain critical hurdles in their widespread adoption [4].

4.2.4 *Adult Stem Cells*

ASCs, also called somatic or tissue-specific stem cells, are undifferentiated cells found throughout the body in various tissues and organs, like bone marrow, skin, liver, adipose tissue, dental pulp, or muscles [2]. Unlike ESCs, ASCs are

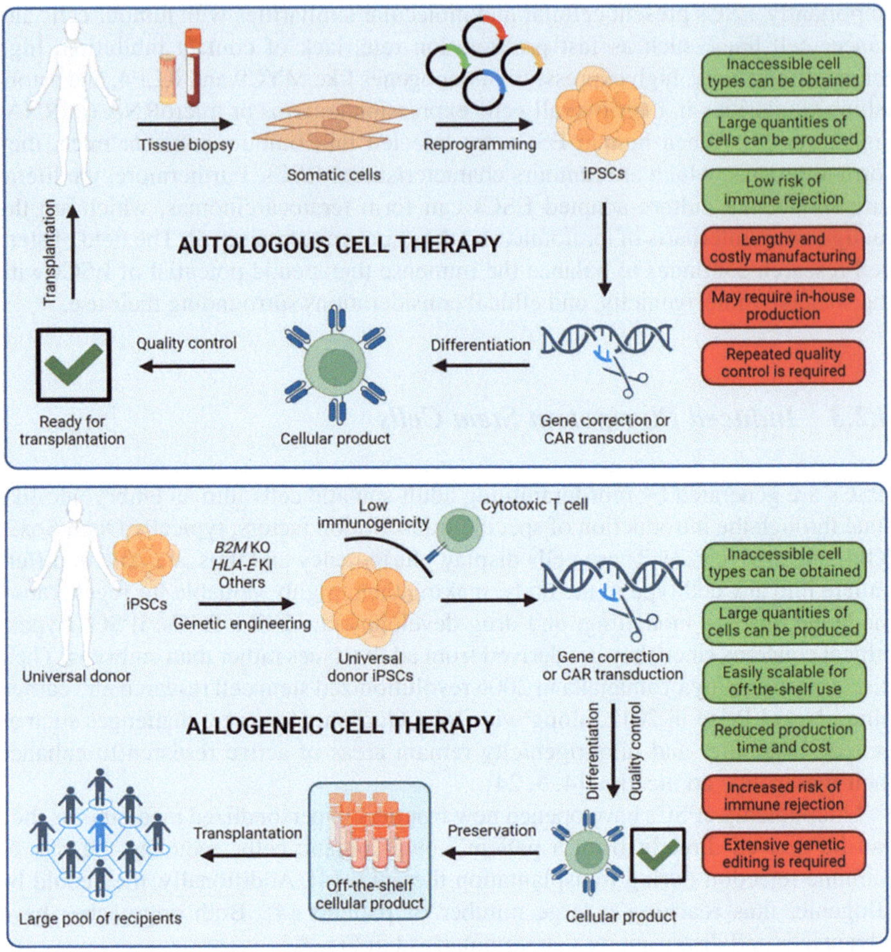

Fig. 4.2 Comparison between autologous and allogeneic iPSC-based cell therapy. In the autologous approach (top box), somatic cells are isolated from the same patient that will receive the cell transplant. The reprogrammed cells can then be genetically engineered for a particular purpose. The resulting cellular product for transplantation must be rigorously evaluated for quality. In the allogeneic approach (bottom box), iPSCs are taken from a biobank, genetically engineered for immune cloaking, and genetically modified for a specific purpose. Cellular products are rigorously assessed for quality, stored and distributed for transplantation into multiple recipients. KO, knockout; KI, knockin; CAR, chimeric antigen receptor. Image adapted from Cerneckis et al. 2024 [4], open access article distributed under the terms and conditions of the Creative Commons Attribution (CC BY) license

multipotent, meaning they have a more limited ability to differentiate into various cell types. They play a crucial role in tissue homeostasis by replenishing damaged or lost cells, as well as contributing to the body's natural healing processes [25]. Advances in TERM therapies have highlighted the potential of ASCs in treating conditions such as leukaemia, spinal cord injuries, and degenerative diseases, although challenges in their extraction, growth, and application remain active areas of research. Importantly, they offer a clear advantage over ESCs in terms of ethical constraints [20].

Of particular relevance are mesenchymal stem cells (MSCs), a type of multipotent ASCs primarily found in bone marrow, from where they were originally isolated and characterized, but are also present in adipose tissue, umbilical cord blood, placenta, and other tissues [26–28]. These cells have the remarkable ability to differentiate into various cell types, including bone, cartilage, muscle, and fat cells, making them a cornerstone of regenerative medicine [26, 27]. MSCs are known for their immunomodulatory properties, which enable them to modulate immune responses and reduce inflammation, providing potential therapeutic benefits for autoimmune diseases and inflammatory conditions [29]. Their ability to home in on injury sites and support tissue repair has led to their use in clinical trials targeting osteoarthritis, cardiovascular diseases, and graft-versus-host disease [29]. Despite their promise, challenges such as ensuring consistent efficacy, scalability, and understanding their full mechanisms of action remain areas of ongoing study.

4.3 The Stem Cell Niche

Stem cell niches are the dynamic in vivo microenvironments inhabited by stem cells where they receive the signals that regulate and determine their fate [6]. Stem cell niches are not merely physical locations where stem cells are found, but should be thought of as highly complex and dynamic microenvironments where a plethora of biochemical and biophysical cues are displayed that together contribute to stem cell fate [6]. All these cues ultimately elicit an intracellular response leading to changes in gene expression, thereby affecting stem cell events like proliferation, migration or differentiation. Therefore, stem cell fate could be directed towards specific lineages or their stemness preserved by manipulating these special microenvironments [30]. But first, it is important to obtain a comprehensive understanding of the different elements present in the niche, how it functions and what mechanisms are employed for regulating stem cell fate (Fig. 4.3).

4.3.1 Microenvironment of the Stem Cell Niche

Stem cells are surrounded by an extracellular matrix (ECM), which is composed of water, proteins and polysaccharides, and provides physical support to cells as well as biochemical and biophysical cues [32, 33]. Stem cells are not the only cellular component found in the microenvironment of the niche, where other differentiated tissue cells and immune cells are also present, thereby contributing to the myriad of signalling factors that influence stem cell function [6]. Signalling factors can be

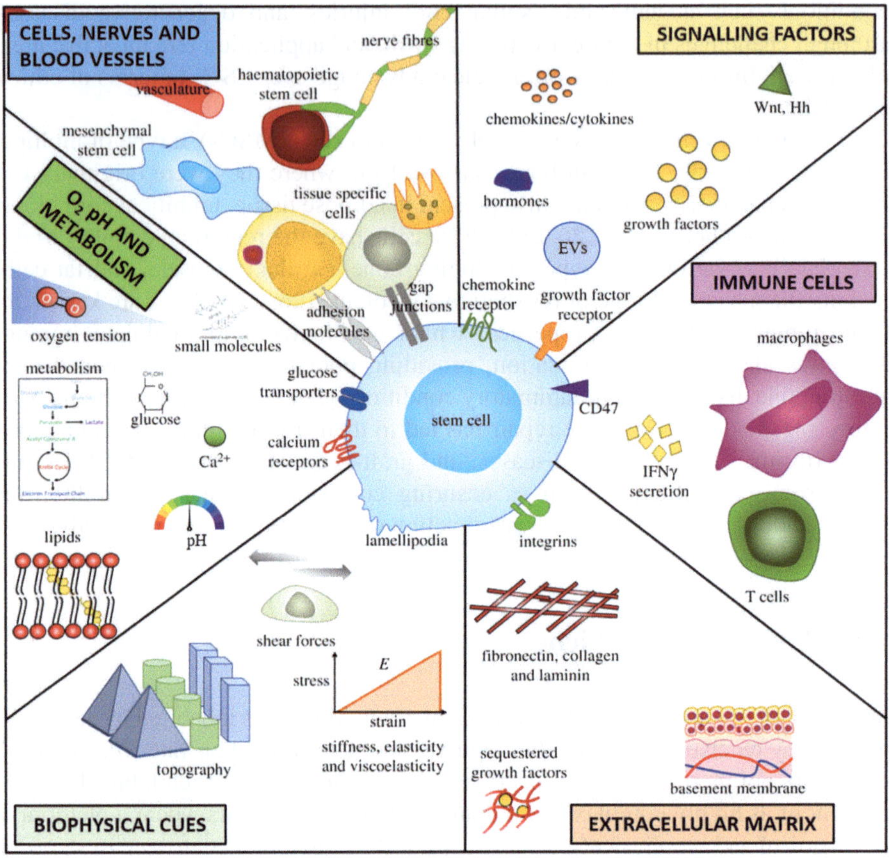

Fig. 4.3 Scheme depicting the main elements present in stem cell niches along with some examples per category. EVs are extracellular vesicles, Wnt are Wingless-INTs, Hh are sonic hedgehog proteins, and IFNγ is interferon-gamma. Image adapted from Donnelly et al. [31], open access article distributed under the terms and conditions of the Creative Commons Attribution (CC BY) license

soluble or immobilised in the matrix and include growth factors, cytokines/chemokines, metabolic signals, hormones, extracellular vesicles, and morphogens [6]. Examples of signalling factors found in numerous niches across different species include fibroblast growth factors (FGFs), bone morphogenetic proteins (BMPs), Wingless-INTs (WNTs), and sonic hedgehog proteins [34]. Another example are extracellular vesicles (EVs), which are used in cell-cell communication and have been found to regulate adult neurogenic niches or modulate the bone marrow niche [35, 36]. It is important to take into account that signalling factors concentrations are spatially and temporally regulated, which adds levels of complexity when engineering stem cell niches in vitro. In vivo, many growth factors and morphogens are found immobilised in the ECM, which binds and releases signalling factors, thereby directly controlling their activity [37, 38]. The ECM actively sequestrates and stores factors, protects them from degradation, and concentrates them in the vicinity of target cells [38]. Moreover, the presence of a peri-vascular niche, where blood vessels interface with the stem cell niche, facilitates the delivery of long-range factors to the niche as well as the recruitment of circulating cells into it [39]. Nerves have also been shown to influence stem cell fate, and therefore, are another element of the stem cell niche to bear in mind [25, 40].

Particular attention should be paid to the niche ECM and the biochemical and biophysical cues it provides [33]. Other chapters of this book offer a detailed account of the ECM, such as Chap. 7 for example, which concerns cell-scaffold interactions. Briefly, two types of ECM exist in our body, namely the basement membrane and the interstitial ECM [33]. Whilst the basement membrane has a sheet-like structure, the interstitial ECM resembles a 3D fibrous mesh [41]. The basement membrane is found in the basal side of endothelial and epithelial cells and is also responsible for the compartmentalization of tissues. The interstitial ECM surrounds stromal cells. Both types of ECM contain the same main components, i.e. water, proteins and polysaccharides, but differ in types and concentrations [33, 41]. As already mentioned, the ECM serves as a storage of signalling factors. Furthermore, it provides a physicochemical environment where factors such as pH, oxygen levels, topography, architecture, mechanical forces, or electrical signals are of vital importance in the regulation of stem cell fate [6, 42, 43]. Cell-matrix interactions, mainly mediated via integrins, facilitate sensing of the ECM composition, architecture and mechanical forces [44]. Furthermore, direct cell-cell interactions are also of relevance and play an important role in the maintenance of the niche [6]. Cell-cell interactions are not only between stem cells but also between stem cells and other differentiated cells of the niche [45].

4.3.2 Engineering the Stem Cell Niche

The stem cell niche ultimately consists of four main components, namely cells, both stem and differentiated cells, an ECM, biochemical and biophysical factors, and a peri-vasculature. Therefore, using tissue engineering principles, the stem cell niche could be engineered by combining a scaffold that acts as the ECM niche, relevant cells and factors, and blood vessels [6]. Of particular importance is the engineering of stem cell niches for maintenance of self-renewal capacity [31]. However, the technical difficulties of achieving a highly biomimetic niche are enormous. Furthermore, we are still understanding and generating knowledge about the complexity and functioning of stem cell niches. Nevertheless, promising attempts to engineer these anatomical structures, or certain aspects of them in order to increase knowledge, have been carried out and are worth bringing to the attention of the reader. For instance, Soleas and co-workers have demonstrated that when human PSC-derived lung progenitor cells are confined in a tube that has a diameter mimicking those seen during lung development, their differentiation is altered towards SOX2 – SOX9+ lung cells, highlighting the importance of architecture in stem cell fate [46]. Another interesting example is the work by Braham and colleagues where the authors aimed at modelling the endosteal and perivascular niches of the bone marrow microenvironment, where multiple myeloma is preferentially found. The authors used 3D bioprinting with calcium phosphate cement and osteogenic multipotent mesenchymal stromal cells (O-MSCs) for the endosteal niche, whilst the perivascular niche was modelled with Matrigel, endothelial progenitor cells and MSCs. Primary CD^{138+} myeloma cells were seeded and cultured in the models containing one or both of the niches and found that their proliferation was significantly increased in the model with combined niches. Furthermore, results revealed that the perivascular niche had a prevalent role in supporting myeloma cells over the endosteal niche [47]. A last example is the work by El Hajj et al. who aimed at understanding the niche interactions between blood and bone in the context of bone regeneration. The authors used a pullulan/dextran/hydroxyapatite hydrogel scaffold that was seeded with MSCs and human umbilical vein endothelial cells (HUVECs), which were co-cultured to allow spheroid formation under both flow and static conditions. Flow culture increased viability, proliferation, mineralization and collagen IV production compared with static conditions. Endothelial cells were seen to cluster at the spheroid core. Lattice Boltzmann simulations suggested that endothelial migration and cell distribution was probably influenced by a strong oxygen gradient [48].

As seen in this section, stem cell niches are truly intriguing anatomical structures, and their understanding is essential to be able to manipulate and control stem cell fate and behaviour for TERM strategies. However, the majority of approaches to engineer these structures, or some aspects of them, have been reductionist in nature, given the highly complex nature of niches. Nevertheless, these approaches are useful to generate knowledge of the factors, cells and matrix aspects that govern stem cell fate and efforts should continue to increase this knowledge.

4.4 Other Cells Used in TERM

Stem cells are not the only cells used in TERM strategies and applications. Undoubtedly, stem cells offer a regenerative edge over the differentiated cells that are found in adult tissues. Nevertheless, since differentiated cells are essential for tissue formation, function and homeostasis, many efforts are underway to include them in TERM strategies. Examples are fibroblasts, endothelial cells, chondrocytes, epithelial cells, keratinocytes, or vascular smooth muscle cells.

Fibroblasts are the most abundant cell type in the connective tissues of our body [49, 50]. They are a heterogeneous cell population that have a mesenchymal origin and are the main producers of the collagen-rich ECM that is typical of these tissues [49, 50]. Apart from various types of collagen, fibroblasts also secrete elastin, laminin, fibronectin or glycosaminoglycans, all essential components of the ECM, which highlights the vital role of fibroblasts in the production, maintenance and remodelling of the cellular microenvironment [50, 51]. Fibroblasts play other major biological roles in wound healing, inflammation and angiogenesis [50, 51]. Furthermore, it has been shown that organs with a high self-renewing capacity necessitate fibroblasts to maintain the stem cell niche. This is the case of the intestine, where intestinal fibroblasts are part of the stem cell niche, where they secrete ECM or growth factors [52]. Another example of fibroblasts participating in the stem cell niche is found in hair production [52]. Given the aforementioned biological roles of fibroblasts, it is not surprising their inclusion in TERM strategies. Important examples are the use of dermal fibroblasts in skin substitutes, with several commercially available products [53, 54], gingival fibroblasts in combination with biodegradable scaffolds for gingiva tissue engineering [7, 55], or corneal fibroblasts for producing tissue-engineered corneas [56]. Beyond their use in TERM therapies for regeneration of tissues, fibroblasts are also used in in vitro models, with cancer associated fibroblasts (CAFs) being a prominent example given their major role in the tumour microenvironment (TME) [57–60].

Endothelial cells (ECs) are also prominently used in TERM strategies. These cells line the lumen of blood vessels, and considering the relevance of establishing a vascular network for the regeneration of multiple tissues, as well as for the in vitro modelling of certain pathological phenomena like cancer cell extravasation or the altered blood-brain barrier in Alzheimer's disease, it is not surprising to find that ECs are extensively used in TERM [8, 61–63]. This book contains a chapter on vascularization in TERM where ECs are particularly featured.

Other differentiated cells used in TERM include chondrocytes, for cartilage tissue engineering approaches [9, 64], epithelial cells, for regeneration of various tissues [10, 65, 66], keratinocytes, for skin tissue engineering [53, 54], or vascular smooth muscle cells for vascular tissue engineering [67].

Before finishing this section, it is worth mentioning the explosion of research in in vitro cancer models that has occurred in the last decade [11, 12, 68–71]. These models aim to imitate the extremely complex TME and continuing research is underway to add complexity and biomimicry to these models [12]. Cancer cell lines as well as patient-derived cancer cells are used in in vitro cancer models, where other cells of the TME are often included, such as the already mentioned CAFs [60].

4.5 Growth Factors

Growth factors are secreted and released into the tissue environment by cells and they act as signalling molecules that bind to specific receptors on the cell surface, triggering a cascade of intracellular events that ultimately influence cell behaviour, like cell proliferation, migration, or differentiation [72, 73]. Growth factors are classified based on their structure, function, and the receptors they bind to, and Table 4.1 offers an overview of the main growth factor families. Table 4.1 highlights the diversity and specialization of growth factors, each fulfilling unique roles in cell signalling, tissue development, and repair. They are essential in various physiological processes, from tissue regeneration and immune response to neuronal

Table 4.1 Summary of growth factor families [74–77]

Family	Key members	Functions	Receptors
Epidermal growth factor (EGF)	EGF TGF-α (transforming growth factor-alpha) Amphiregulin Heparin-binding EGF Betacellulin Epiregulin	Stimulates cell growth, proliferation, and differentiation, especially in epithelial cells	EGF/ErbB receptor family, which includes the EGF receptor (EGFR), ErbB-2, ErbB-3, and ErbB-4
Fibroblast growth factor (FGF)	FGF1 (acidic FGF) FGF2 (basic FGF) Keratinocyte growth factor (KGF or FGF-7) Androgen-induced growth factor (AIGF or FGF-8) Glia activating factor (GAF or FGF-9) Others (FGF3-FGF23)	Involved in embryonic development, wound healing, and tissue maintenance	FGF receptors (FGFRs), which are receptor tyrosine kinases
Transforming growth factor-Beta (TGF-β)	TGF-β1 TGF-β2 TGF-β3 Bone morphogenetic proteins (BMPs)	Regulates immune response, cell proliferation, differentiation, and apoptosis	Binds to TGF-β receptors, which activate SMAD proteins
Platelet-derived growth factor (PDGF)	PDGF-AA PDGF-BB PDGF-CC PDGF-DD	Essential for blood vessel formation, wound healing, and the development of certain tissues	Binds to PDGF receptors (PDGFR-α and PDGFR-β), which activate pathways promoting cell survival and migration
Vascular endothelial growth factor (VEGF)	VEGF-A VEGF-B VEGF-C VEGF-D Placenta growth factor	Plays a crucial role in angiogenesis and increasing vascular permeability	Binds to VEGF receptors (VEGFR-1, VEGFR-2, VEGFR-3) which are expressed on endothelial cells as well as on non-endothelial cells
Insulin-like growth factor (IGF)	IGF-1 IGF-2	Promotes cell growth, development, and survival, particularly in muscle and bone tissues	Binds to IGF receptors (IGF-1R, IGF-2R), which are similar to insulin receptors

(continued)

Table 4.1 (continued)

Family	Key members	Functions	Receptors
Neurotrophins	NGF (nerve growth factor) BDNF (brain-derived neurotrophic factor) NT-3 NT-4/5	Crucial for the development, survival, and function of neurons	Binds to Trk receptors (e.g., TrkA, TrkB) and p75 neurotrophin receptor
Hepatocyte growth factor (HGF)	HGF (also known as scatter factor)	Stimulates liver regeneration, wound healing, and promotes cell motility	Binds to the c-met receptor
Angiopoietins	Ang1 Ang2 Ang3 Ang4	Normal remodelling, maturation, and stabilization of the developing vasculature	TIE tyrosine kinases
Agrin	Agrin	Induces the aggregation of acetylcholine receptors and other synaptic proteins Role in neurogenesis, skeletal muscle development, and mesenchymal-epithelial interaction during organ formation	Agrin receptors and co-receptors, such as MuSK (muscle-specific kinase) or Lrp4
Cytokines with growth factor functions	Interleukins (e.g., IL-6) Colony-stimulating factors (e.g., G-CSF)	Support immune cell growth, haematopoiesis, and inflammatory response	Bind to specific cytokine receptors and often have overlapping roles in immunity and tissue repair

development and cancer biology. Therefore, unsurprisingly, a large number of TERM strategies incorporate growth factors given their ability to stimulate and direct cell behaviour [72, 73].

Growth factors can be incorporated into TERM strategies by adding them to the scaffold component, which can be done by physical encapsulation/immobilization, covalent conjugation, enzymatic conjugation, creation of fusion proteins, or non-covalent interactions [14, 73, 78–83]. Encapsulation of growth factors into microspheres, that could then be incorporated into a scaffold is another strategy [13]. Growth factors can also be encapsulated into nanoparticles, i.e. nanocarriers [13, 14]. Nanocarriers can be made from a variety of components, like natural (proteins and polysaccharides) and synthetic polymers, or lipids, i.e. micelles. Interesting examples are the work by Marchioli et al. [84], Limongi et al. [85], and Herrán et al. [86]. Marchioli and colleagues aimed at developing a 3D scaffold to boost vascularization for extra-hepatic islet transplantation [84]. The developed scaffold was shaped as a 3D ring using polycaprolactone (PCL) coated with heparin to electrostatically bind vascular endothelial growth factor (VEGF). Inside the ring, a hydrogel core made of crosslinked alginate for islets encapsulation was introduced. The authors used a chicken chorioallantoic membrane (CAM) model, widely used for the study of angiogenesis of TERM scaffolds and constructs, where they observed that the immobilized VEGF enhanced angiogenesis not only on the scaffolds surface but also in close proximity of the scaffolds. Furthermore, islets encapsulated in the scaffold responded to glucose stimuli [84]. Limongi and co-workers introduced a micropilllared PCL scaffold for controlled release of brain-derived neurotrophic factor (BDNF). The authors showed that the designed morphology enhanced drug delivery capability with progressive release of BDNF over 21 days in vitro, thereby increasing synaptic density, and neuronal survival in primary neuronal cultures [85]. Lastly, Herrán et al. encapsulated VEGF and glial cell line-derived neurotrophic factor (GDNF) in poly(lactic-co-glycolic acid) nanospheres that when administered into the brain of 6-hydroxydopamine partially lesioned rats provided a continuous and simultaneous VEGF and GDNF release [86].

The inclusion of growth factors in scaffolds and constructs to elicit cell responses has enormous potential. However, certain challenges need to be addressed for their effective use, with the main one being sustained release over a relevant period of time to allow tissue repair and regeneration to effectively occur [13, 14].

4.6 Exosomes

Exosomes are nano-sized (~40–160 nm) endosomal derived membrane biovesicles that are released into surrounding body fluids by virtually all eukaryotic cells and are implicated in cell-cell communication, both at short and long distances [15]. Exosomes form upon fusion of multivesicular bodies and the plasma membrane and were identified in the extracellular compartment in the late 1980s (Table 4.2). Since then, they have received much attention, especially in the last decade, given their

Table 4.2 Description of the main features of exosomes

Feature	Description
Origin	Fusion of multivesicular bodies with plasma membrane
Release mechanism	Constitutive or inducible, depending on the cell type
Size	~40–160 nm
Density	1.13–1.19 g/mL
Markers	Membrane impermeable (PI negative), CD9, CD63, CD81, TSG101, HSP70, HSP90, ALIX, floillin 1
Contents	The cellular origin and physiological state of the cell determines the contents, which include proteins, lipids, different RNA species, DNA, aminoacids, and metabolites
Lipids	Phosphatidylserine, phosphatidic acid, fatty acids, cholesterol, sphingomyelin, prostaglandins, leukotrienes

major role in immune response and signal transduction in both healthy and disease states [15, 87, 88].

4.6.1 Biogenesis and Composition of Exosomes

Explained in a succinct way, the biogenesis of exosomes involves the invagination of late endosomal membranes to yield intraluminal vesicles (ILVs), which form within large multivesicular bodies (MVBs). Late endosomes are generated by inward budding of the MVBs membrane. Upon fusion with the plasma membrane, ILVs are released into the extracellular compartment, and from this point on are called "exosomes" [15, 87].

When observed under transmission electron microscopy in solution, exosomes display a spherical morphology. However, exosomes show a cup-like shape when produced by artificially drying during preparation. Exosomes show heterogeneity on different fronts, like shape, source, content, or functionality.

Exosomes contain a large variety of proteins, RNA species, and lipids that can act as autocrine and/or paracrine factors (Fig. 4.4) [87, 89]. Proteins found in

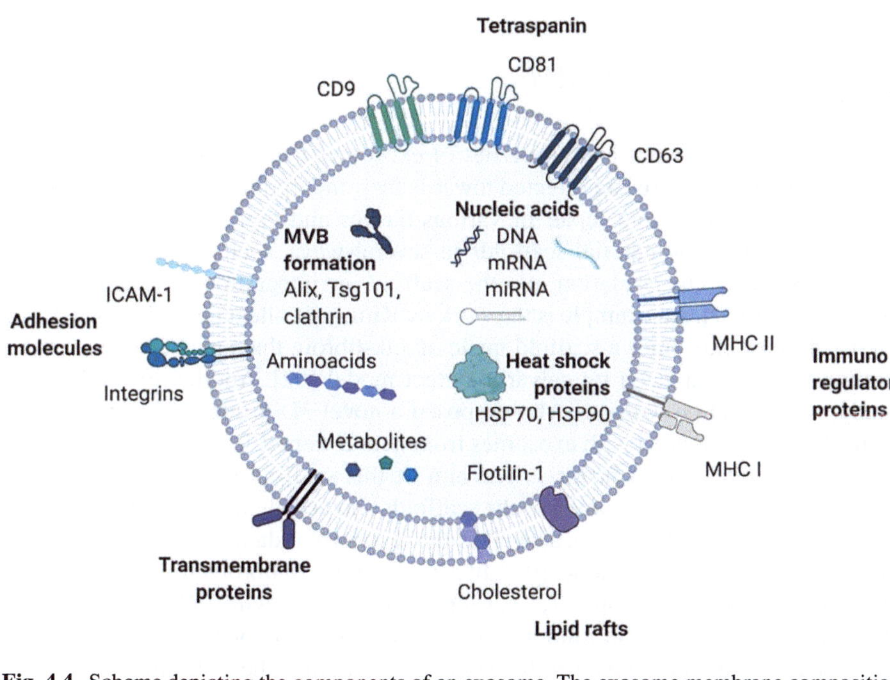

Fig. 4.4 Scheme depicting the components of an exosome. The exosome membrane composition and soluble contents represent the cell of origin from which the exosome is released. Image from Valencia et al. [88], open access article distributed under the terms and conditions of the Creative Commons Attribution (CC BY) license

exosomes can be classified into (1) membrane transport and fusion related proteins, e.g. heat shock proteins (HSPs) including Hsp60, Hsp70, and Hsp90, or flotillin; (2) tetraspanins, which is a large family of small transmembrane proteins found on the surface of almost every eukaryotic cell [90], including CD9, CD63, CD81, CD82, CD106, Tspan8, ICAM (intercellular adhesion molecule)-1; (3) MVBs related proteins, e.g. ALIX, TSG101; and (4) other proteins, like integrins, actin or myosin [87]. Exosomes are enriched in various lipids that together confer these biovesicles with stability and structural rigidity [89]. These lipids include phosphatidylserine, phosphatidic acid, fatty acids, cholesterol, sphingomyelin, prostaglandins, and leukotrienes [89]. Additionally, lipolytic enzymes are also seen in exosomes, which produce units of bioactive lipids [89]. In terms of the RNA species found in exosomes, microRNAs (miRs) are the most abundant. Other RNA species include ribosomal RNA, long non-coding RNA, piwi-interacting RNA, transfer RNA, small nuclear RNA, and small nucleolar RNA [89]. The miRs in exosomes unidirectionally transfer between cells, thereby eliciting transient or persistent phenotypic changes in the recipient cells [89]. Long RNA species, especially long non-coding RNAs (lncRNAs) and circular RNAs (circRNAs) also affect several biological processes that include cancer development [89]. Finally, apart from proteins, lipids and RNA, other components that are found in exosomes cargos are DNA, aminoacids, and metabolites [15, 88].

4.6.2 *Exosomes in TERM*

Given the characteristics and properties of exosomes exposed above, it is not surprising that many efforts are directed towards their incorporation into TERM strategies for regeneration and repair of various tissues and organs. Exosomes can be incorporated into biomaterial scaffolds in several ways, such as grafting onto the scaffold surface, encapsulation into the scaffold, or integration of exosomes with nanoparticles [91]. An example is the work by Kim and colleagues, who coated with MSCs derived exosomes a scaffold made of silk-fibroin that enhanced bone regeneration in a critical-sized rat calvarial defect model [92]. Another example is the study by Liu and co-workers, who proposed a novel 3D-printed collagen/chitosan scaffold that was loaded with exosomes from neural stem cells that had been treated with insulin-like growth factor-1. The aim of this scaffold was to treat traumatic brain injury repair. Implantation of the scaffolds showed significant improvement in motor and cognitive functions, as well as the recovery of damaged nerve tissue, in a rat traumatic brain injury model [93]. In a last example, a fibrous scaffold fabricated by electrospinning of copper 1,4-benzenedicarboxylate with poly(lactic acid-co-glycolic acid) (PLGA) and immobilizing exosomes from bone marrow MSCs was proposed by Xu and colleagues for bone regeneration. The scaffold showed dual release of exosomes and bioactive copper ions, thereby promoting osteogenesis and angiogenesis, both needed for effective bone regeneration [94].

A promising strategy in TERM is the use of hydrogel-exosome systems [95]. Hydrogels are very interesting materials for TERM since they mimic certain aspects of the ECM of tissues, such as their viscoelastic and hydrated nature [96]. Furthermore, they are biocompatible, have adjustable physico-chemical properties and can be fabricated from a variety of polymers, both natural and synthetic [96]. Exosomes present some technical challenges for their use in TERM, including their isolation, loading, and release. To address these challenges, hydrogels are gaining relevance as exosome carriers. An example is the work by Zhang and colleagues who proposed a gelatin-methacryloyl (GelMA) hydrogel that was loaded with extracellular vesicles from tenocytes, the cells that secrete and build the tendon ECM. The GelMA-exosome system was administered in liquid form and subsequently polymerised with UV irradiation. Such system and strategy when applied to injured Achilles tendon tissue allowed the effective coverage of irregular surfaces [97]. With the aim of targeting the complicated diabetic wounds, Hu and co-workers used an extrusion-based cryogenic 3D printing technique to fabricate hydrogel constructs made of decellularized small intestinal submucosa, mesoporous bioactive glass and exosomes from bone marrow derived-MSCs. The fabricated 3D construct provided sustained release of the exosomes. In vivo results suggested that the hydrogel construct accelerated diabetic wound healing through increasing blood flow, stimulating angiogenesis, and promoting granulation tissue formation with deposition of well-organized collagen fibers [98]. A last example is by Wang and colleagues, who used bone marrow-derived MSCs-derived exosomes that were preloaded with curcumin via electroporation, with hydrogels made of decellularized porcine cardiac ECM. The aim was to produce injectable hydrogels with antifibrotic properties for the treatment of myocardial infarction [99].

In conclusion, incorporation of exosomes in TERM scaffolds can be achieved following a variety of strategies, with encapsulation into hydrogels showing promising results. If effective and successful incorporation of exosomes into scaffolds is achieved, as well as their subsequent sustained release, they could potentially substitute the use of therapeutic cells in TERM, thereby bypassing some of the discussed challenges that certain cells pose. Furthermore, exosomes could be designed a-la-carte to incorporate a specific cargo that upon delivery elicits the desired cell response [100]. All of this makes exosomes exciting and promising cell-derived components for improved and effective TERM strategies.

4.7 Cell Sheet Tissue Engineering

As mentioned throughout this book, tissue engineering requires three elements, i.e. scaffolds, cells and biophysical/biochemical cues, for producing tissue and organ analogues. When the scaffold component is merely used as a temporary support for cells, we can talk about scaffold-free tissue engineering, which was proposed by Okano and co-workers as a method that entails the preparation and harvesting of cell sheets for implantation into injured tissues [17, 18]. This is the so-called cell sheet tissue engineering, which has achieved human clinical use for regeneration and repair of various tissues such as cornea, heart, knee, cartilage, or esophagus. Furthering their application, cell sheets have also been used in tissue models [101]. Cell sheets maintain cell-to-cell contact and adhesive proteins belonging to the ECM are found on the basal side of the cells. These adhesive ECM proteins are secreted by the cells themselves as part of their normal cell activity. Cell sheets adhere to the application site thereby integrating with the host tissue [102]. This is a clear advantage compared with the injection of cells that have been harvested by enzymatic treatment, where adhesive ECM proteins are lost and cells do not attach to the target site [17, 102]. Cell sheets offer certain versatility as they can be monolayer or multilayer, composed of one type of cell (monoculture) or several (co-culture), cytokines can be added to them, and can be vascularized by for example adding endothelial cells [17] (Fig. 4.5).

4.7.1 Harvesting of Cell Sheets

Central to cell sheet tissue engineering is the ability of harvesting an intact cell sheet, which means without digesting or damaging the ECM layer deposited by the cells on their basal side. This ECM layer contains cell-ECM adhesion molecules as well as biochemical cues for correct cell function. Several responsive systems have been proposed and are discussed next.

Initially, the use of thermoresponsive surfaces was proposed. Okano's group invented an intelligent polymer surface that upon changing the temperature from 37 °C to 20 °C the polymer changed from hydrophobic, that allows cell attachment and proliferation, to hydrophilic, thereby promoting cell detachment [18, 103] (Fig. 4.6a). The polymer is poly(N-isopropylacrylamide) (PIPAAm) which is placed on tissue culture polystyrene plates via electron beam polymerization [103]. PIPAAm is non-toxic and biocompatible, which has been demonstrated by culturing numerous cell types on it. Its main disadvantage is the slow detachment of the cell sheet, which is due to slow diffusion of water molecules beneath the polymer [101]. Therefore, some strategies have been investigated to overcome this issue, such as placing a highly water-permeable microporous membrane between the PIPAAm surface and the cell sheet, developing comb-type grafted PIPAAm gels, or grafting ultrathin double polymeric nanolayers of PIPAAm and polyacrylamide [101]. Another disadvantage is that placing the PIPAAm layers on plates requires special equipment, or alternatively can be commercially acquired at a high cost [101].

Temperature responsive surfaces of methylcellulose have also been proposed [104–106]. The sol-gel transition of methylcellulose depends on the concentration of methylcellulose and the addition of salts. Chen and colleagues described a method to fabricate cell sheets that consisted of an 8% w/v methylcellulose solution mixed with Na_2SO_4 and phosphate-buffered saline on tissue culture plates at room temperature (~20 °C). The solution turned into a hydrogel at 37 °C (Fig. 4.6b) and was coated with collagen for cell attachment and growth until forming a confluent

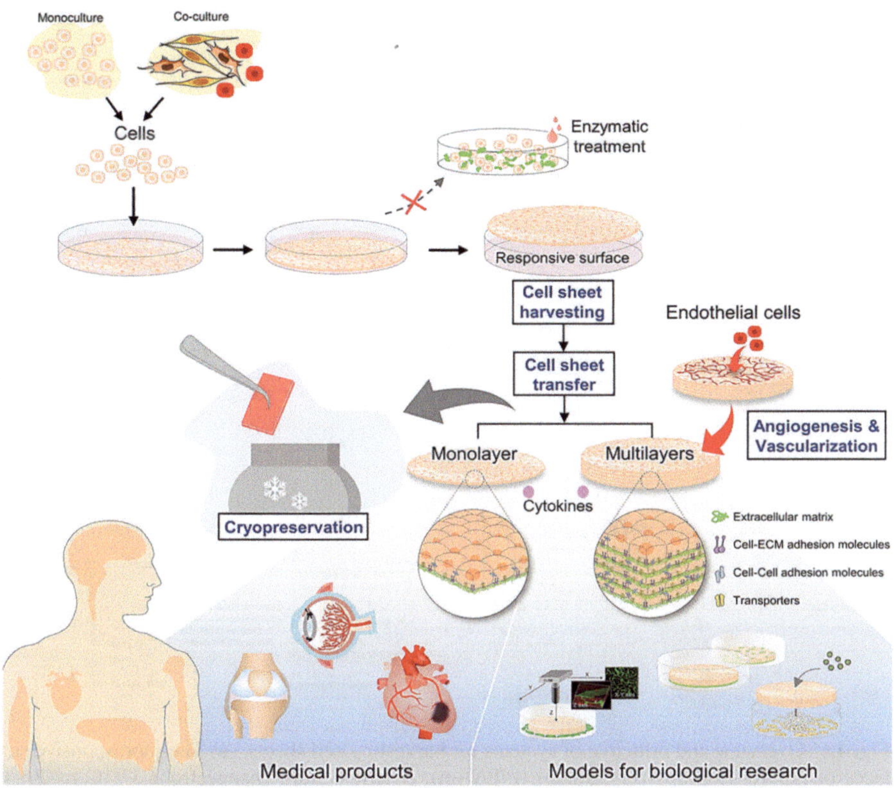

Fig. 4.5 Overview of cell sheet tissue engineering or scaffold-free tissue engineering. Image adapted from Thummarati et al. [101], open access article distributed under the terms and conditions of the Creative Commons Attribution (CC BY) license

monolayer. The cell sheets containing human foreskin fibroblasts were harvested at 20 °C [105]. There is a risk that the produced hydrogel is not homogeneous and therefore, optimisation into the concentrations of both methylcellulose and salts is recommended before culturing the cells [101]. The advantages of this method are that it is inexpensive and does not require special equipment. However, it has been found that methylcellulose surface decreases cell proliferation compared with PIPAAm surfaces [106].

Since highly sensitive cells can be damaged by the temperature change associated with temperature-responsive surfaces discussed above, as well as altering metabolism, cell cycle, gene expression, or cell function, other methods such as ion-induced cell detachment have been proposed [101, 107]. Advantageously, this method does not require specialised equipment or facilities. This system is based on the principle that cell adhesion is promoted by high surface energy. Copolymers comprising nonpolar and polar polymers are deposited on tissue culture plates and their ratios can be modulated to tune the adsorption of ECM proteins [107]

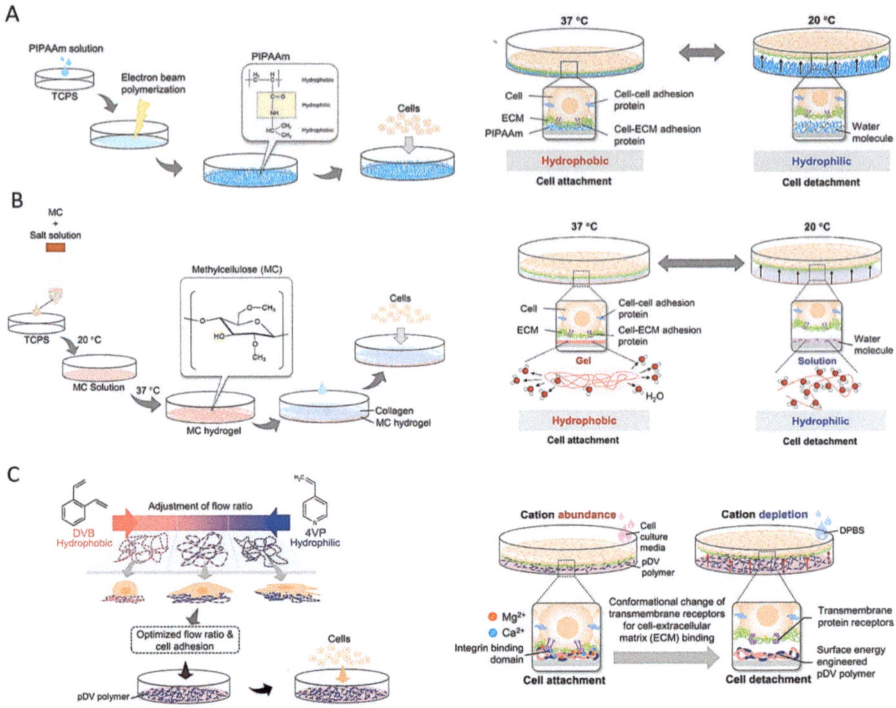

Fig. 4.6 Overview and principle of systems for harvesting cell sheets. (**a**) Thermoresponsive surfaces of poly(N-isopropylacrylamide) (PIPAAm). A temperature change from 37 °C to 20 °C induces cell sheet detachment. (**b**) Thermoresponsive surfaces of methylcellulose (MC) coated with collagen for cell adhesion. A temperature change from 37 °C to 20 °C induces cell sheet detachment. (**c**) Ion-induced cell detachment using copolymers of nonpolar hydrophobic (divinylbenzene, DVB) and polar hydrophilic polymers (4-vinylpyridine, 4VP). Simple addition of DPBS (Dulbecco's phosphate buffered saline) without Ca^{2+} and Mg^{2+} induces cell sheet detachment. Adapted from Thummarati et al. [101], open access article distributed under the terms and conditions of the Creative Commons Attribution (CC BY) license

(Fig. 4.6c). Once the cell sheet is ready for harvesting, depletion of divalent cations, e.g. Ca^{2+} and Mg^{2+}, induces the conformational change of integrins, a key family of cell membrane proteins that facilitate interaction between cells and ECM or substrates [107, 108]. Thus, cell sheets are harvested under isothermal conditions and has been shown to be faster than thermoresponsive surface systems [107].

Other methods also developed for harvesting of cell sheets include electro-responsive surfaces, photo-responsive systems and pH-responsive systems [101].

4.7.2 Applications of Cell Sheet Tissue Engineering

As mentioned earlier, cell sheets have reached patients and there are successful clinical trials using this TERM technology for repair of cartilage [9], cornea [65], heart muscle [109], esophagus epithelium [10], and periodontal ligament [110]. Sato and colleagues reported a combination therapy for osteoarthritis of the knee comprising conventional surgical treatment followed by autologous chondrocyte sheet transplantation. Eight patients were treated and preoperatively and postoperatively assessed reporting that the combination therapy was effective [9]. In a clinical trial with four patients, Nishida et al. concluded that reconstruction of corneal surfaces and vision restoration of bilateral severe disorders of the ocular surface was possible with transplantation of cell sheets of autologous oral mucosal epithelial cells that were used without sutures [65]. Fifteen patients with ischemic cardiomyopathy and twelve patients with dilated cardiomyopathy were treated with cell sheets of cells derived from autologous muscle, without additional interventional treatments. The authors reported promising results regarding safety and functional recovery [109]. Ohki and co-workers reported that endoscopic transplantation of cell sheets of autologous oral mucosal epithelial cells effectively promoted re-epithelialization of the esophagus after esophageal endoscopic submucosal dissection to remove superficial esophageal neoplasms [10]. Three-layered autologous periodontal ligament-derived cell sheets were used to treat patients with chronic periodontitis, reporting no severe adverse reactions and therapeutic effects [110]. Beyond these clinical successes, further research is being carried out into clinical therapies based on cell sheets for regeneration of various tissues. An example is the work by Mito and colleagues who explored the use of bone marrow-derived mesenchymal stem cell sheets for enhancing bone formation [111]. The authors implanted the cell sheets in a bilateral femoral critical bone defect in rats concluding that these cell sheets could be used as a complement to surgery in large weight-bearing bone defects or as a standalone approach that would be followed by internal fixation [111].

Cell sheets are also being used in in vitro tissue models [101]. A prominent example is cancer models. For instance, Lee et al. developed an in vitro cell-sheet epithelial cancer model intended for chemotherapeutic screening [69]. In order to mimic the microenvironment of epithelial cancer, their cell sheet was formed of a layer of keratinocytes, cancer spheroids with or without cancer-associated fibroblasts (CAFs), and a layer of plasma fibrin matrix that contained fibroblasts. The authors used this cell sheet to assess resistance to sorafenib and cisplatin, invasion of cancer-CAF spheroids into the fibrin matrix, expansion of viable cancer cells, and expression of TGF-β1, N-cadherin, and vimentin mRNA [69]. Therefore, this model allowed study of key cancer events as well as effectiveness of chemotherapy treatments. Another interesting example is by Li and Kino-oka who built a tumor cell sheet with human embryonal rhabdomyosarcoma cells and human skeletal muscle myoblasts [112]. Rhabdomyosarcoma is the most common soft tissue sarcoma in children, and this cell sheet mimicked muscle tissue invaded by rhabdomyosarcoma cells. Besides, green fluorescent protein expressing human umbilical vein endothelial cells (GFP-HUVECs) were co-cultured with the cell sheet. Results suggested that disorganized tissue structure, typical of rhabdomyosarcoma, activates endothelial cell migration thereby promoting tumor angiogenesis [112]. Beyond cancer, other authors have explored the use of cells sheets in in vitro modelling. The work by Shahin-Samsabadi and Cappuccitti serves as an example [113]. Cell sheet engineering has technical limitations including creating suitable 3D shapes with effective anchorage. To approach this problem, the authors proposed an innovative and versatile platform for in vitro modelling by way of creating self-assembling 3D tissue constructs. Scaffold-free muscle fibers and lung epithelial cell cultures were used as examples for the study [113].

Cell sheet tissue engineering has proved to be a useful technology for clinical use as well as for creating tissue analogues for further study of disease or tissue development. Their versatility in terms of cell types used, in monoculture or co-culture, number of layers, or addition of a vascular network makes this technology particularly attractive for the advancement of the TERM field. Nevertheless, certain limitations exist, such as obtaining 3D tissue equivalents, the large number of cells needed to achieve clinical success, meaning prolonged culture time, which is not ideal in autologous approaches, and accelerating ECM deposition with biochemical cues such as hypoxia or macromolecular crowding, or biophysical cues such as mechanical stimulation [17]. Further and exciting research is under way to tackle these limitations.

4.8 Conclusions

The paramount cell element of TERM has been comprehensively overviewed in this chapter. Of special relevance are stem cells, given their regenerative potential and their prominent role in tissue formation, homeostasis and healing. Depending on source and potency stem cells have important ethical implications. For instance, the use of ESCs, which are pluripotent and can give rise to cell types from the three germ layers, involves the destruction of embryos and have potential tumorigenicity once implanted. Easier to work with from the ethical point of view are ASCs and differentiated cells, part of adult tissues and as such, easier to obtain and more accessible. The use of a particular cell type should take into account the specific TERM goal and consider ethical implications, source, availability, and resources for growth and maintenance. Furthermore, when working with stem cells, the strategy of recreating their natural niche is also being actively researched since the niche's environment regulates and directs stem cell fate.

As part of their normal activity, cells release growth factors and exosomes into the environment. These are involved in cell communication and can effect a particular response in target cells. Therefore, many efforts are underway to incorporate these cell-derived components to scaffolds and constructs with the technical difficulty of achieving a sustained release during the regeneration process.

Cell sheets for implantation can be produced by using a temporary support that allows their fabrication and facile release. This technology is called cell sheet tissue engineering and is already in clinical use, signalling the enormous potential of the cell element of TERM in regenerative medicine.

Questions

- Explain the stem cell hierarchy.
- Explain the differences between ESCs and iPSCs as well as their respective advantages and disadvantages.
- What is the main feature of ASCs? Name a type.
- What is the stem cell niche, and which are its main components?
- Name five examples of cells, other than stem cells, used in TERM.
- What are growth factors?
- What are exosomes?
- Name strategies for incorporating exosomes into biomaterial scaffolds.
- What is the principle of cell sheet tissue engineering?
- Name at least two systems that have been proposed for harvesting of cell sheets.

References

1. Langer, R., & Vacanti, J. P. (1993). Tissue engineering. *Science, 260*(5110), 920. http://science.sciencemag.org/content/260/5110/920.abstract
2. Tian, Z., Yu, T., Liu, J., Wang, T., & Higuchi, A. (2023). Chapter one - Introduction to stem cells. In A. Higuchi, Y. Zhou, & S.-H. Chiou (Eds.), *Progress in molecular biology and translational science* (pp. 3–32). Academic. https://www.sciencedirect.com/science/article/pii/S1877117323000571
3. Kwon, S. G., Kwon, Y. W., Lee, T. W., Park, G. T., & Kim, J. H. (2018). Recent advances in stem cell therapeutics and tissue engineering strategies. *Biomaterials Research, 22*(1). https://doi.org/10.1186/s40824-018-0148-4
4. Cerneckis, J., Cai, H., & Shi, Y. (2024). Induced pluripotent stem cells (iPSCs): molecular mechanisms of induction and applications. *Signal Transduction and Targeted Therapy, 9*(1), 112.
5. Ben-David, U., & Benvenisty, N. (2011). The tumorigenicity of human embryonic and induced pluripotent stem cells. *Nature Reviews Cancer, 11*(4), 268–277.
6. Scodellaro, C., Pina, R. R., Ferreira, F. C., Sanjuan-Albertc, P., & Fernandes, T. G. (2024). Unlocking the potential of stem cell microenvironments in vitro. *Bioengineering, 11*(3), 289. https://www.mdpi.com/2306-5354/11/3/289
7. Yang, M., Guo, Z., Li, T., Li, J., Chen, L., Wang, J., et al. (2019). Synergetic effect of chemical and topological signals of gingival regeneration scaffold on the behavior of human gingival fibroblasts. *Journal of Biomedial Materials Research Part A, 107*(9), 1875–1885.
8. Rouwkema, J., Rivron, N. C., & van Blitterswijk, C. A. (2008). Vascularization in tissue engineering. *Trends in Biotechnology, 26*(8), 434–441. https://www.sciencedirect.com/science/article/pii/S0167779908001583
9. Sato, M., Yamato, M., Mitani, G., Takagaki, T., Hamahashi, K., Nakamura, Y., et al. (2019). Combined surgery and chondrocyte cell-sheet transplantation improves clinical and structural outcomes in knee osteoarthritis. *npj Regenerative Medicine, 4*, 4.

10. Ohki, T., Yamato, M., Ota, M., Takagi, R., Murakami, D., Kondo, M., et al. (2012 Sep). Prevention of esophageal stricture after endoscopic submucosal dissection using tissue-engineered cell sheets. *Gastroenterology, 143*(3), 582–588.e2.
11. Parkinson, G. T., Salerno, S., Ranji, P., Håkansson, J., Bogestål, Y., Wettergren, Y., et al. (2020). Patient-derived scaffolds as a model of colorectal cancer. *Cancer Medicine, 10*(3), 867–882. https://pubmed.ncbi.nlm.nih.gov/33356003
12. Sontheimer-Phelps, A., Hassell, B. A., & Ingber, D. E. (2019). Modelling cancer in microfluidic human organs-on-chips. *Nature Reviews Cancer, 19*, 65–81.
13. Caballero Aguilar, L. M., Silva, S. M., & Moulton, S. E. (2019). Growth factor delivery: Defining the next generation platforms for tissue engineering. *Journal of Controlled Release, 306*, 40–58.
14. Wang, Z., Wang, Z., Lu, W. W., Zhen, W., Yang, D., & Peng, S. (2017). Novel biomaterial strategies for controlled growth factor delivery for biomedical applications. *NPG Asia Materials, 9*(10), e435.
15. Kalluri, R., & LeBleu, V. S. The biology, function, and biomedical applications of exosomes. *Science, 367*(6478), eaau6977. https://doi.org/10.1126/science.aau6977
16. Ferreira, J. R., Teixeira, G. Q., Santos, S. G., Barbosa, M. A., Almeida-Porada, G., & Gonçalves, R. M. (2018). Mesenchymal stromal cell secretome: Influencing therapeutic potential by cellular pre-conditioning. *Frontiers in Immunology, 9*, 2837.
17. De Pieri, A., Rochev, Y., & Zeugolis, D. I. (2021). Scaffold-free cell-based tissue engineering therapies: Advances, shortfalls and forecast. *npj Regenerative Medicine, 6*(1), 18. https://www.nature.com/articles/s41536-021-00133-3
18. Yang, J., Yamato, M., Shimizu, T., Sekine, H., Ohashi, K., Kanzaki, M., et al. (2007). Reconstruction of functional tissues with cell sheet engineering. *Biomaterials, 28*(34), 5033–5043.
19. Erceg Ivkošić, I., Fureš, R., Ćosić, V., Mikelin, N., Bulić, L., Dobranić, D., et al. (2023). Unlocking the potential of mesenchymal stem cells in Gynecology: Where are we now? *Journal of Personalized Medicine, 13*(8), 1253.
20. Dulak, J., Szade, K., Szade, A., Nowak, W., & Józkowicz, A. (2015). Adult stem cells: Hopes and hypes of regenerative medicine. *Acta Biochimica Polonica, 62*(3), 329–337.
21. Evans, M. J., & Kaufman, M. H. (1981). Establishment in culture of pluripotential cells from mouse embryos. *Nature, 292*(5819), 154–156.
22. Thomson, J. A., Itskovitz-Eldor, J., Shapiro, S. S., Waknitz, M. A., Swiergiel, J. J., Marshall, V. S., et al. (1998). Embryonic stem cell lines derived from human blastocysts. *Science, 282*(5391), 1145. https://doi.org/10.1126/science.282.5391.1145
23. Takahashi, K., & Yamanaka, S. (2006). Induction of pluripotent stem cells from mouse embryonic and adult fibroblast cultures by defined factors. *Cell, 126*(4), 663–676.
24. Yamanaka, S. (2012). Induced pluripotent stem cells: Past, present, and future. *Cell Stem Cell, 10*(6), 678–684.
25. Redondo, P. A., Pavlou, M., Loizidou, M., & Cheema, U. (2017). Elements of the niche for adult stem cell expansion. *Journal of Tissue Engineering*. https://doi.org/10.1177/2041731417725464
26. Pittenger, M. F., Mackay, A. M., Beck, S. C., Jaiswal, R. K., Douglas, R., Mosca, J. D., et al. (1999). Multilineage potential of adult human mesenchymal stem cells. *Science, 284*(5411), 143.
27. Sensebé, L., Krampera, M., Schrezenmeier, H., Bourin, P., & Giordano, R. (2010). Mesenchymal stem cells for clinical application. *Vox Sanguinis, 98*(2), 93–107.
28. Lee, O. K., Kuo, T. K., Chen, W. M., Der, L. K., Hsieh, S. L., & Chen, T. H. (2004). Isolation of multipotent mesenchymal stem cells from umbilical cord blood. *Blood, 103*(5), 1669–1675.
29. Song, N., Scholtemeijer, M., & Shah, K. (2020 Sep). Mesenchymal stem cell immunomodulation: Mechanisms and therapeutic potential. *Trends in Pharmacological Sciences, 41*(9), 653–664.

30. Guller, A., & Igrunkova, A. (2022). Engineered microenvironments for 3D cell culture and regenerative medicine: Challenges, advances, and trends. *Bioengineering, 10*(1), 17. https://www.mdpi.com/2306-5354/10/1/17
31. Donnelly, H., Salmeron-Sanchez, M., & Dalby, M. J. (2018). Designing stem cell niches for differentiation and self-renewal. *Journal of the Royal Society Interface, 15*(145), 20180388. https://doi.org/10.1098/rsif.2018.0388
32. Kular, J. K., Basu, S., & Sharma, R. I. (2014). *The extracellular matrix: Structure, composition, age-related differences, tools for analysis and applications for tissue engineering* (Vol. 5, p. 2041731414557112).
33. Yue, B. (2014). Biology of the extracellular matrix: An overview. *Journal of Glaucoma, 23*(8 Suppl 1), S20–S23. https://pubmed.ncbi.nlm.nih.gov/25275899
34. Schliermann, A., & Nickel, J. (2018). Unraveling the connection between fibroblast growth factor and bone morphogenetic protein Signaling. *International Journal of Molecular Sciences, 19*(10), 3220.
35. Bátiz, L. F., Castro, M. A., Burgos, P. V., Velásquez, Z. D., Muñoz, R. I., Lafourcade, C. A., et al. (2015). Exosomes as novel regulators of adult neurogenic niches. *Frontiers in Cellular Neuroscience, 9*, 501.
36. Niazi, V., Parseh, B., Ahani, M., Karami, F., Gilanchi, S., Atarodi, K., et al. (2020). Communication between stromal and hematopoietic stem cell by exosomes in normal and malignant bone marrow niche. *Biomedicine & Pharmacotherapy, 132*, 110854.
37. Taipale, J., & Keski-Oja, J. (1997). Growth factors in the extracellular matrix. *The FASEB Journal, 11*(1), 51–59.
38. Schultz, G. S., & Wysocki, A. (2009). Interactions between extracellular matrix and growth factors in wound healing. *Wound Repair and Regeneration, 17*(2), 153–162.
39. Putnam, A. J. (2014). The instructive role of the vasculature in stem cell niches. *Biomaterials Science, 2*(11), 1562–1573.
40. Pei, F., Ma, L., Jing, J., Feng, J., Yuan, Y., Guo, T., et al. (2023). Sensory nerve niche regulates mesenchymal stem cell homeostasis via FGF/mTOR/autophagy axis. *Nature Communications, 14*(1), 344.
41. García-Gareta, E. (2023). Protein-based materials for wound healing. In F. R. A. Maia, J. M. Oliveira, & R. L. Reis (Eds.), *Handbook of the extracellular matrix: Biologically-derived materials* (pp. 1–28). Springer International Publishing. https://doi.org/10.1007/978-3-030-92090-6_20-1
42. Vining, K. H., & Mooney, D. J. (2017). Mechanical forces direct stem cell behaviour in development and regeneration. *Nature Reviews. Molecular Cell Biology, 18*(12), 728–742.
43. Thrivikraman, G., Boda, S. K., & Basu, B. (2018). Unraveling the mechanistic effects of electric field stimulation towards directing stem cell fate and function: A tissue engineering perspective. *Biomaterials, 150*, 60–86.
44. Driscoll, T. P., Bidone, T. C., Ahn, S. J., Yu, A., Groisman, A., Voth, G. A., et al. (2021). Integrin-based mechanosensing through conformational deformation. *Biophysical Journal, 120*(20), 4349–4359.
45. Dong, J., Pan, Y.-B., Wu, X.-R., He, L.-N., Liu, X.-D., Feng, D.-F., et al. (2019). A neuronal molecular switch through cell-cell contact that regulates quiescent neural stem cells. *Science Advances, 5*(2).
46. Soleas, J. P., D'Arcangelo, E., Huang, L., Karoubi, G., Nostro, M. C., McGuigan, A. P., et al. (2020). Assembly of lung progenitors into developmentally-inspired geometry drives differentiation via cellular tension. *Biomaterials, 254*, 120128.
47. Braham, M. V. J., Ahlfeld, T., Akkineni, A. R., Minnema, M. C., Dhert, W. J. A., Öner, F. C., et al. (2018). Endosteal and perivascular subniches in a 3D bone marrow model for multiple myeloma. *Tissue Engineering. Part C, Methods, 24*(5), 300–312.
48. El Hajj, S., Ntaté, M. B., Breton, C., Siadous, R., Aid, R., Dupuy, M., et al. (2024). Bone spheroid development under flow conditions with mesenchymal stem cells and human umbil-

ical vein endothelial cells in a 3D porous hydrogel supplemented with hydroxyapatite. *Gels, 10*(10), 666.
49. Nolte, S. V., Xu, W., Rennekampff, H.-O., & Rodemann, H. P. (2008). Diversity of fibroblasts—A review on implications for skin tissue engineering. *Cells Tissues Organs, 187*(3), 165–176. https://doi.org/10.1159/000111805
50. Kendall, R. T., & Feghali-Bostwick, C. A. (2014). Fibroblasts in fibrosis: Novel roles and mediators. *Frontiers in Pharmacology, 5*, 1–13.
51. Costa-Almeida, R., Soares, R., & Granja, P. L. (2018). Fibroblasts as maestros orchestrating tissue regeneration. *Journal of Tissue Engineering and Regenerative Medicine, 12*(1), 240–251.
52. Gomes, R. N., Manuel, F., & Nascimento, D. S. (2021). The bright side of fibroblasts: molecular signature and regenerative cues in major organs. *npj Regenerative Medicine, 6*(1), 43.
53. García-Gareta, E. (2019). Introduction to biomaterials for skin repair and regeneration. In *Biomaterials for skin repair and regeneration* (pp. xiii–xxvii). Woodhead Publishing. https://www.sciencedirect.com/science/article/pii/B978008102546802001X
54. Davison-Kotler, E., Sharma, V., Kang, N. V., & García-Gareta, E. (2018). A universal classification system of skin substitutes inspired by factorial design. *Tissue Engineering Reviews, 24*(4), 279–288. https://doi.org/10.1089/ten.teb.2017.0477
55. Cheung, J. W. C., McCulloch, C. A. G., & Santerre, J. P. (2014). Establishing a gingival fibroblast phenotype in a perfused degradable polyurethane scaffold: Mediation by TGF-β1, FGF-2, β1-integrin, and focal adhesion kinase. *Biomaterials, 35*(38), 10025–10032.
56. Guérin, L.-P., Le-Bel, G., Desjardins, P., Couture, C., Gillard, E., Boisselier, É., et al. (2021). The human tissue-engineered cornea (hTEC): Recent Progress. *International Journal of Molecular Sciences, 22*(3), 1291.
57. Zhang, T., Ren, Y., Yang, P., Wang, J., & Zhou, H. (2022). *Cancer-associated fibroblasts in pancreatic ductal adenocarcinoma. Vol. 13, Cell death and disease*. Springer Nature.
58. Kalluri, R., & Zeisberg, M. (2006). Fibroblasts in cancer. *Nature Reviews. Cancer, 6*(5), 392–401.
59. Shiga, K., Hara, M., Nagasaki, T., Sato, T., Takahashi, H., & Takeyama, H. (2015). Cancer-associated fibroblasts: Their characteristics and their roles in tumor growth. *Cancers, 7*(4), 2443–2458.
60. Jiang, L., Khawaja, H., Tahsin, S., Clarkson, T. A., Miranti, C. K., & Zohar, Y. (2024). Microfluidic-based human prostate-cancer-on-chip. *Frontiers in Bioengineering and Biotechnology, 12*, 1302223.
61. Yang, G., Mahadik, B., Choi, J. Y., & Fisher, J. P. (2020). Vascularization in tissue engineering: Fundamentals and state-of-art. *Progress in Biomedical Engineering, 2*(1), 012002. https://doi.org/10.1088/2516-1091/ab5637
62. Zhang, S., Wan, Z., & Kamm, R. D. (2021). Vascularized organoids on a chip: Strategies for engineering organoids with functional vasculature. *Lab Chip, 21*(3), 473–488. https://doi.org/10.1039/D0LC01186J
63. Ko, E. C., Spitz, S., Pramotton, F. M., Barr, O. M., Xu, C., Pavlou, G., et al. (2023). Accelerating the in vitro emulation of Alzheimer's disease-associated phenotypes using a novel 3D blood-brain barrier neurosphere co-culture model. *Frontiers in Bioengineering and Biotechnology, 11*, 1251195. https://doi.org/10.3389/fbioe.2023.1251195
64. Xu, W., Zhu, J., Hu, J., & Xiao, L. (2022). Engineering the biomechanical microenvironment of chondrocytes towards articular cartilage tissue engineering. *Life Sciences, 309*, 121043.
65. Nishida, K., Yamato, M., Hayashida, Y., Watanabe, K., Yamamoto, K., Adachi, E., et al. (2004). Corneal reconstruction with tissue-engineered cell sheets composed of autologous Oral mucosal epithelium. *The New England Journal of Medicine, 351*(12), 1187–1196.
66. Pang, K., Du, L., & Wu, X. (2010). A rabbit anterior cornea replacement derived from acellular porcine cornea matrix, epithelial cells and keratocytes. *Biomaterials, 31*(28), 7257–7265. https://www.sciencedirect.com/science/article/pii/S0142961210007192

67. Avolio, E., Alvino, V. V., Ghorbel, M. T., & Campagnolo, P. (2017). Perivascular cells and tissue engineering: Current applications and untapped potential. *Pharmacology & Therapeutics, 171*, 83–92.
68. Sitarski, A. M., Fairfield, H., Falank, C., & Reagan, M. R. (2018). *3D tissue engineered in vitro models of cancer in bone. Vol. 4, ACS biomaterials science and engineering* (pp. 324–336). American Chemical Society.
69. Lee, J., Shin, D., & Roh, J.-L. (2018). Development of an in vitro cell-sheet cancer model for chemotherapeutic screening. *Theranostics, 8*(14), 3964–3973.
70. Hernández-Hatibi, S., Guerrero, P. E., García-Aznar, J. M., & García-Gareta, E. (2024). Polydopamine interfacial coating for stable tumor-on-a-Chip models: Application for pancreatic ductal adenocarcinoma. *Biomacromolecules, 25*(8), 5169–5180.
71. Alamán-Díez, P., Borau, C., Enrique Guerrero, P., Amaveda, H., Mora, M., Fraile, J. M., et al. (2023). Collagen-Laponite Nanoclay hydrogels for tumor spheroid growth. *Biomacromolecules, 24*, 2879–2891.
72. Omorphos, N. P., Gao, C., Tan, S. S., & Sangha, M. S. (2021). Understanding angiogenesis and the role of angiogenic growth factors in the vascularisation of engineered tissues. *Molecular Biology Reports, 48*(1), 941–950. https://doi.org/10.1007/s11033-020-06108-9
73. Ren, X., Zhao, M., Lash, B., Martino, M. M., & Julier, Z. (2020). Growth factor engineering strategies for regenerative medicine applications. *Frontiers in Bioengineering and Biotechnology, 21*, 7. https://doi.org/10.3389/fbioe.2019.00469/full
74. Bafico, A., & Aaronson, S. (2003). Classification of growth factors and their receptors. In *Holland-Frei cancer medicine* (6th ed.). BC Decker. https://www.ncbi.nlm.nih.gov/books/NBK12423/
75. Kim, N., Stiegler, A. L., Cameron, T. O., Hallock, P. T., Gomez, A. M., Huang, J. H., et al. (2008). Lrp4 is a receptor for Agrin and forms a complex with MuSK. *Cell, 135*(2), 334–342.
76. De Falco, S. (2012). The discovery of placenta growth factor and its biological activity. *Experimental & Molecular Medicine, 44*(1), 1.
77. Zhang, Y., Xia, M., Jin, K., Wang, S., Wei, H., Fan, C., et al. (2018). Function of the c-met receptor tyrosine kinase in carcinogenesis and associated therapeutic opportunities. *Molecular Cancer, 17*(1), 45.
78. Orlowska, A., Perera, P. T., Al Kobaisi, M. A., Dias, A., Nguyen, H. K. D., Ghanaati, S., et al. (2017). The effect of coatings and nerve growth factor on attachment and differentiation of Pheochromocytoma cells. *Materials, 11*(1), 60.
79. do Amaral, R. J. F. C., Cavanagh, B., O'Brien, F. J., & Kearney, C. J. (2019). Platelet-derived growth factor stabilises vascularisation in collagen–glycosaminoglycan scaffolds in vitro. *Journal of Tissue Engineering and Regenerative Medicine, 13*(2), 261–273. https://doi.org/10.1002/term.2789
80. Lee, K., Olmer, M., Baek, J., D'Lima, D. D., & Lotz, M. K. (2018). Platelet-derived growth factor-coated decellularized meniscus scaffold for integrative healing of meniscus tears. *Acta Biomaterialia, 76*, 126–134. https://www.sciencedirect.com/science/article/pii/S1742706118303672
81. Jin, H., Quesada, C., Aliabouzar, M., Kripfgans, O. D., Franceschi, R. T., Liu, J., et al. (2021). Release of basic fibroblast growth factor from acoustically-responsive scaffolds promotes therapeutic angiogenesis in the hind limb ischemia model. *Journal of Controlled Release, 338*, 773–783. https://linkinghub.elsevier.com/retrieve/pii/S0168365921004922
82. Matsumine, H., & Niimi, Y. (2022). Basic fibroblast growth factor-impregnated collagen gelatin sponge completes formation of dermis-like tissue within 2 weeks: A prospective cohort study. *Regenerative Therapy, 1*(21), 210–215. https://linkinghub.elsevier.com/retrieve/pii/S2352320422000785
83. Lee, J. Y., Lim, H., Ahn, J. W., Jang, D., Lee, S. H., Park, K., et al. (2018). Design of a 3D BMP-2-delivering Tannylated PCL scaffold and its anti-oxidant, anti-inflammatory, and osteogenic effects in vitro. *International Journal of Molecular Sciences, 19*, 3602.

84. Marchioli, G., Luca, A., de Koning, E., Engelse, M., Van Blitterswijk, C. A., Karperien, M., et al. (2016). Hybrid polycaprolactone/alginate scaffolds functionalized with VEGF to promote de novo vessel formation for the transplantation of islets of Langerhans. *Advanced Healthcare Materials, 5*(13), 1606–1616.
85. Limongi, T., Rocchi, A., Cesca, F., Tan, H., Miele, E., Giugni, A., et al. (2018). Delivery of brain-derived neurotrophic factor by 3D biocompatible polymeric scaffolds for neural tissue engineering and neuronal regeneration. *Molecular Neurobiology, 55*(12), 8788–8798.
86. Herran, E., Requejo, C., Ruiz-Ortega, J. A., Aristieta, A., Igartua, M., Bengoetxea, H., et al. (2014). Increased antiparkinson efficacy of the combined administration of VEGF- and GDNF-loaded nanospheres in a partial lesion model of Parkinson's disease. *International Journal of Nanomedicine, 9*, 2677.
87. Dai, J., Su, Y., Zhong, S., Cong, L., Liu, B., Yang, J., et al. (2020). Exosomes: Key players in cancer and potential therapeutic strategy. *Signal Transduction and Targeted Therapy, 5*(1), 145.
88. Valencia, K., & Montuenga, L. M. (2021). Exosomes in liquid biopsy: The nanometric world in the pursuit of precision oncology. *Cancers, 13*(9), 2147. https://www.mdpi.com/2072-6694/13/9/2147
89. Zhang, Y., Liu, Y., Liu, H., & Tang, W. H. (2019). Exosomes: biogenesis, biologic function and clinical potential. *Cell & Bioscience, 9*(1), 19.
90. Kummer, D., Steinbacher, T., Schwietzer, M. F., Thölmann, S., & Ebnet, K. (2020). Tetraspanins: Integrating cell surface receptors to functional microdomains in homeostasis and disease. *Medical Microbiology and Immunology, 209*(4), 397–405.
91. Hu, W., Wang, W., Chen, Z., Chen, Y., & Wang, Z. (2024). Engineered exosomes and composite biomaterials for tissue regeneration. *Theranostics, 14*(5), 2099–2126.
92. Kyung Kim, D., Lee, S., Kim, M., Jeong, Y., & Lee, S. (2021 Feb). Exosome-coated silk fibroin 3D-scaffold for inducing osteogenic differentiation of bone marrow derived mesenchymal stem cells. *Chemical Engineering Journal, 406*, 127080.
93. Liu, X.-Y., Feng, Y.-H., Feng, Q.-B., Zhang, J.-Y., Zhong, L., Liu, P., et al. (2023). Low-temperature 3D-printed collagen/chitosan scaffolds loaded with exosomes derived from neural stem cells pretreated with insulin growth factor-1 enhance neural regeneration after traumatic brain injury. *Neural Regeneration Research, 18*(9), 1990–1998.
94. Xu, C., Kang, Y., Dong, X., Jiang, D., & Qi, M. (2023 Feb). Integration exosomes with MOF-modified multifunctional scaffold for accelerating vascularized bone regeneration. *Chinese Chemical Letters, 34*(2), 107528.
95. Fan, M.-H., Pi, J.-K., Zou, C.-Y., Jiang, Y.-L., Li, Q.-J., Zhang, X.-Z., et al. (2024). Hydrogel-exosome system in tissue engineering: A promising therapeutic strategy. *Bioactive Materials, 38*, 1–30.
96. Redaelli, F., Sorbona, M., & Rossi, F. (2017). Synthesis and processing of hydrogels for medical applications. In G. Perale & J. Hilborn (Eds.), *Bioresorbable polymers for biomedical applications* (pp. 205–228). Elsevier. https://linkinghub.elsevier.com/retrieve/pii/B9780081002629000100
97. Zhang, T., Wu, Y., Li, X., Zhang, A., Liu, H., Shi, M., et al. (2023). Small extracellular vesicles derived from tendon stem cells promote the healing of injured Achilles tendons by regulating miR-145-3p. *Acta Biomaterialia, 172*, 280–296.
98. Hu, Y., Wu, B., Xiong, Y., Tao, R., Panayi, A. C., Chen, L., et al. (2021). Cryogenic 3D printed hydrogel scaffolds loading exosomes accelerate diabetic wound healing. *Chemical Engineering Journal, 426*, 130634.
99. Wang, Y., Wang, J., Liu, C., Li, J., Lu, K., Yu, Q., et al. (2023). Injectable decellularized extracellular matrix hydrogel loaded with exosomes encapsulating curcumin for prevention of cardiac fibrosis after myocardial infarction. *Journal of Materials Science and Technology, 167*, 50–58.
100. Liu, C., & Su, C. (2019). Design strategies and application progress of therapeutic exosomes. *Theranostics, 9*(4), 1015–1028.

101. Thummarati, P., Laiwattanapaisal, W., Nitta, R., Fukuda, M., Hassametto, A., & Kino-oka, M. (2023). Recent advances in cell sheet engineering: From fabrication to clinical translation. *Bioengineering, 10*(2), 211. https://www.mdpi.com/2306-5354/10/2/211
102. Kobayashi, J., Kikuchi, A., Aoyagi, T., & Okano, T. (2019). Cell sheet tissue engineering: Cell sheet preparation, harvesting/manipulation, and transplantation. *Journal of Biomedical Materials Research Part A, 107*(5), 955–967. https://doi.org/10.1002/jbm.a.36627
103. Akiyama, Y., Kikuchi, A., Yamato, M., & Okano, T. (2004). Ultrathin poly(*N*-isopropylacrylamide) grafted layer on polystyrene surfaces for cell adhesion/detachment control. *Langmuir, 20*(13), 5506–5511.
104. Forghani, A., & Devireddy, R. (2018). Methylcellulose based thermally reversible hydrogels. In B. A. Bunnell & J. M. Gimble (Eds.), *Methods in molecular biology* (pp. 41–51). Humana Press.
105. Chen, C.-H., Tsai, C.-C., Chen, W., Mi, F.-L., Liang, H.-F., Chen, S.-C., et al. (2006). Novel living cell sheet harvest system composed of thermoreversible methylcellulose hydrogels. *Biomacromolecules, 7*(3), 736–743.
106. Forghani, A., Kriegh, L., Hogan, K., Chen, C., Brewer, G., Tighe, T. B., et al. (2017). Fabrication and characterization of cell sheets using methylcellulose and PNIPAAm thermoresponsive polymers: A comparison study. *Journal of Biomedial Materials Research Part A, 105*(5), 1346–1354.
107. Baek, J., Cho, Y., Park, H., Choi, G., Lee, J. S., Lee, M., et al. (2020). A surface-tailoring method for rapid non-thermosensitive cell-sheet engineering via functional polymer coatings. *Advanced Materials, 32*(16), 1907225.
108. Barczyk, M., Carracedo, S., & Gullberg, D. (2010). Integrins. *Cell and Tissue Research, 339*(1), 269–280. https://doi.org/10.1007/s00441-009-0834-6
109. Miyagawa, S., Domae, K., Yoshikawa, Y., Fukushima, S., Nakamura, T., Saito, A., et al. (2017). Phase I clinical trial of autologous stem cell–sheet transplantation therapy for treating cardiomyopathy. *Journal of the American Heart Association, 6*(4), e003918.
110. Iwata, T., Yamato, M., Washio, K., Yoshida, T., Tsumanuma, Y., Yamada, A., et al. (2018). Periodontal regeneration with autologous periodontal ligament-derived cell sheets—A safety and efficacy study in ten patients. *Regenerative Therapy, 9*, 38–44.
111. Mito, K., Lachnish, J., Le, W., Chan, C., Chang, Y.-L., & Yao, J. (2024). Scaffold-free bone marrow-derived mesenchymal stem cell sheets enhance bone formation in a weight-bearing rat critical bone defect model. *Tissue Engineering. Part A, 30*(3-4), 107–114.
112. Li, M., & Kino-oka, M. (2017). Degradation of endothelial network in disordered tumor-containing cell sheet. *Journal of Bioscience and Bioengineering, 123*(6), 748–753.
113. Shahin-Shamsabadi, A., & Cappuccitti, J. (2024). Anchored cell sheet engineering: A novel scaffold-free platform for in vitro modeling. *Advanced Functional Materials, 34*(13), 2308552.

Chapter 5
Scaffolds for Tissue Engineering and Regenerative Medicine

Abstract The paradigm of tissue engineering and regenerative medicine (TERM) involves biomaterial scaffolds, cells, and bio-chemical/physical cues to produce tissue equivalents, that could be either implanted or used as tissue models. A biomaterial scaffold that allows cell attachment, infiltration, growth, function, and ultimately remodelling is necessary to produce functional and relevant tissue equivalents for various TERM applications. This chapter focuses on the scaffold component of the TERM paradigm. Scaffolds' requirements and characteristics will be initially discussed, followed by a description of the different biomaterials used in the fabrication of scaffolds, as well as the various fabrication techniques used. The chapter highlights hydrogels as recent scaffolds of choice for bioengineered tissue analogues. The chapter ends with conclusions and questions to reinforce the different concepts and aspects covered in this chapter.

5.1 Introduction

The paradigm of the developing field of tissue engineering and regenerative medicine (TERM) involves the use of biomaterial scaffolds, cells, and bio-chemical/physical cues to produce tissue analogues or equivalents [1]. These bioengineered tissues could be either implanted to regenerate or restore tissue function or be used as models for the study of tissue development and disease processes, among other applications. Of these three elements of thc TERM paradigm, a biomaterial that acts as a scaffold to allow cell attachment, infiltration, growth, function, and ultimately remodelling of its original composition and structure is key to produce functional and relevant tissue analogues [1].

A number of requirements and characteristics are desirable in a scaffold. For instance, if a tissue analogue is intended for implantation in vivo and subsequent regeneration of the target tissue, then bioactivity, which is the ability to interact with living tissue, is a requirement [2]. Similarly, implanted biodegradable tissue analogues should have optimal degradation rates in sync with new tissue formation [3].

© The Author(s), under exclusive license to Springer Nature Switzerland AG 2026

E. García-Gareta, J. M. García-Aznar, *Contemporary Tissue Engineering and Regenerative Medicine*, Series in BioEngineering,
https://doi.org/10.1007/978-3-032-03441-0_5

Another important example is the structural properties of porosity and permeability, which allow cell and blood vessel infiltration in vivo, cell infiltration and nutrients/ growth factors diffusion in vitro, as well as removal or diffusion of metabolites in both scenarios [4].

TERM scientists possess a biomaterial toolbox for designing and fabricating scaffolds. This toolbox includes polymers, both natural and synthetic, ceramics, metals and carbon-based materials. All of these materials have advantages and disadvantages. For example, natural polymers, which are made up of proteins, polysaccharides, or DNA are inherently biocompatible, bioactive, and biodegradable through natural metabolic pathways, as they are composed of the building blocks that constitute living tissues [5–7]. On the other hand, they display fast degradation rates and weak mechanical properties. A possible strategy to overcome these disadvantages is to combine natural polymers with synthetic ones, which possess tuneability of mechanical and degradation properties as well as versatile chemistry [8]. The low bioactivity and biocompatibility of some synthetic polymers can in turn be overcome by mixing them with natural polymers [9, 10]. Therefore, composites, i.e. combinations of two or more materials, are favoured in TERM, as composites can be designed and fabricated to cater for a specific need [11].

Many of the scaffold's properties are dictated by the way it is fabricated, that is the fabrication technique chosen to produce it. There are numerous fabrication technologies, and the choice of one or another depends on various factors, one of which is the desired scaffold's architecture. A typical example is electrospinning, a fibre production fabrication method that renders ultrafine fibres (up to nano-diameters) by charging and ejecting a polymer melt or polymer solution under a high-voltage electric field. Filaments are produced and deposited on a surface collector thereby forming a fibrous scaffold. Electrospun scaffolds possess a characteristic fibrous architecture that resembles that of the natural fibres of the extracellular matrix (ECM) of tissues, the natural scaffold of our body [12]. Some fabrication techniques like additive manufacturing ones can produce custom-fit implants with intricate structures that perfectly fit the defect to be treated [13, 14].

Scaffolds can have different forms and shapes, as well as formulations. These include sponges, fibrous meshes, porous sheets, nano and microspheres, solid simple porous shapes like cylinders, or intricate ones like the custom-fit porous solid implants generated by additive manufacturing, or the very popular hydrogels. Specifically, hydrogels have gained much attention due to their hydrated nature that mimics important features of the natural ECM of body tissues [7, 15–17].

This chapter offers an overview of the scaffold component of the TERM paradigm and discusses scaffolds' requirements and characteristics, biomaterials used in their fabrication, as well as the various fabrication techniques used. The chapter also highlights hydrogels as recent scaffolds of choice for bioengineered tissue analogues. Finally, the chapter offers some conclusions, ending with questions to reinforce the different concepts and aspects covered in this chapter.

5.2 Scaffold Requirements and Characteristics

Scaffolds should meet certain requirements for optimal function, whether intended for in vivo implantation or in vitro experimentation in tissue models. These requirements include biocompatibility, bioactivity, biodegradability, suitable mechanical properties, optimal surface properties, optimal architecture, and for hard tissues, osteoconductivity and osteoinductivity. Besides, for certain tissues such as myocardium or nerves, electrical properties are desired.

Biocompatibility is the ability of a material to be accepted by the surrounding tissues and eventually body without eliciting an adverse immune response [18]. Adverse reactions can be cytotoxicity, mutagenesis, carcinogenesis, immunogenicity, and genotoxicity. Bioactivity is defined as the ability of a material to interact with living tissues [2]. This is of vital important if the scaffold is intended for implantation. Biodegradability would allow the implanted scaffold to eventually disappear leaving behind the formed new tissue [3]. This way, no foreign material would remain in the body once the regeneration or repair process is complete. The key is the synchronizing of the scaffold's degradation rate with that of the new tissue formation.

Body tissues and organs are load-bearing and respond to mechanical stimuli; therefore, it is intuitive to pay attention to the mechanical properties of TERM scaffolds [19]. These should have optimal mechanical properties to for example, be able to resist the typical forces experienced by the target tissue. The scaffold surface will initially be responsible for interacting with body fluids, cells and host tissues and therefore, its properties are fundamental to ensure the desired response [20]. The scaffold's architecture, from its macrostructure to its internal one, is of vital importance to ensure mechanical coupling, nutrients and oxygen flow, or cell and blood vessel infiltration, among other responses [4, 20]. In some tissues like nerves or the myocardium, self-generated cyclic electric waves that travel to peripheral regions of the body are part of their tissue functions and therefore, electrically conductive scaffolds are required to adequately restore these tissues function [21].

Finally, in the case of hard tissue regeneration, e.g. skeletal and alveolar bone, osteoconductivity and osteoinductivity are highly sought [22]. Osteoconductivity is the ability of a material to allow bone growth on its surface. This ability is important for implant/host bone fixation and integration. Osteoconduction can be added to bone scaffold by means of surface treatments for example [22]. Osteoinductivity is highly desired as it is the ability to induce the differentiation of local stem cells into the osteogenic lineage, thereby inducing the formation of bone tissue [23]. Only a handful of materials possess this ability, mostly ceramics, and therefore, the engineering of osteoinductive scaffolds is an exciting and unmet challenge [23].

In order to fulfil the scaffold requirements just mentioned, important scaffold characteristics to design, optimise and ultimately control are macrostructure, porosity, permeability and pore size, mechanical properties, electrical properties, surface properties and degradation.

5.2.1 Macrostructure

Scaffolds can have different shapes and macrostructures, e.g. sponges, thin films, porous sheets, injectable gels, hydrogels, meshes, etc. [16, 24–28]. The shape and macrostructure is dictated by the intended scaffold's application, and depending on those chosen, the fabrication technique employed in the scaffold's production will be key to achieve the desired macrostructure.

5.2.2 Porosity, Permeability and Pore Size

Porosity and permeability are major architectural features of scaffolds to allow cell infiltration, vascularization, innervation, fluid flow, nutrients and oxygen diffusion, or removal of metabolites [4, 29, 30]. Furthermore, porosity increases surface area for cell-scaffold interaction. Critically porosity should be interconnected to avoid dead ends or cul-de-sacs that would lead to concentration of metabolites, cell death and necrosis.

Porous scaffolds display pores in their structure that can have various sizes, i.e. pore size range, or a single pore size, as well as different morphologies [31–33]. Most TERM scaffolds have different pore morphologies depending on the fabrication method used in their production. Pore width and curvature influence tissue growth and morphology, where an increased growth rate has been observed with a higher curvature [34]. Similarly, increased cell proliferation was seen at the short edges of rectangular pores compared with the long edges [35]. It has also been shown that tissue formation favours concave surfaces as they allow cell alignment and greater density of actin and myosin fibres that advances cell migration [35].

In the in vivo scenario, cells are in contact with a layer of tissue ECM that possess different pore sizes, therefore displaying a pore size range represents better the native ambient of cells [30]. In terms of sizes, according to the literature, the ideal pore size range for skin regeneration is 20–125 μm [36], whilst the ideal pore sizes reported for bone regeneration are usually larger, i.e. 200–350 μm was optimal for osteoblast proliferation [34].

The degree of porosity and pore size range are often limited by the resulting mechanical properties, since the higher the porosity and the larger the pore sizes, the lower the mechanical properties of a given scaffold [2]. The work by Torres-Sanchez et al. illustrates this point [37]. In this work, the authors studied the effect of porosity and pore size on the elastic modulus and compressive strength in porous Titanium scaffolds, where the Young's moduli and compressive strengths correlated with porosity: scaffolds with a nominal porosity of 50% obtained Young's modulus in the range 15.46–18.12 GPa and a compressive strength at yield point of 222–274 MPa, whilst scaffolds with a nominal porosity of 70% had values of 5.29–6.72 GPa for the Young's modulus and 67–159 MPa for the compressive strength [37].

5.2.3 Mechanical Properties

Body tissues and organs are mechanical, where mechanotransduction is the response of tissue-specific cells to mechanical stimuli, and it is the process by which external loading is converted to biochemical or biophysical signals that culminate in a cellular response [38] (Fig. 5.1). For example, an increase in mechanical load due to physical activity can lead to growth of skeletal and muscular tissues [38].

One of the key challenges in the design of TERM scaffolds is to obtain a similarity in the mechanical properties between the scaffold and the biological tissue [3]. This similarity is crucial for a successful tissue-engineered strategy. The right choice of materials, architectural features and fabrication techniques is crucial to engineer the mechanical properties of scaffolds [19]. These properties include stiffness, strength or elasticity. These mechanical properties vary substantially between different tissues. For instance, cortical bone tensile properties are in the range of 89–114 MPa for the tensile strength and 3.9–11.7 GPa for the elastic modulus [3]. In contrast, the tensile strength of skin is in the range of 7–100 MPa and its elastic modulus is 20–150 MPa [3]. It is also important to consider that soft tissues like skin or cartilage have viscoelastic properties, and therefore, scaffolds produced for tissue engineering of these tissues should have their viscoelastic properties investigated and tuned by using rheology [31]. Similarly, hydrogels, which are viscoelastic in nature, like the ECM, also should have their rheological properties characterized [16, 17]. Finally, some tissues, e.g. skin, present some interesting mechanical

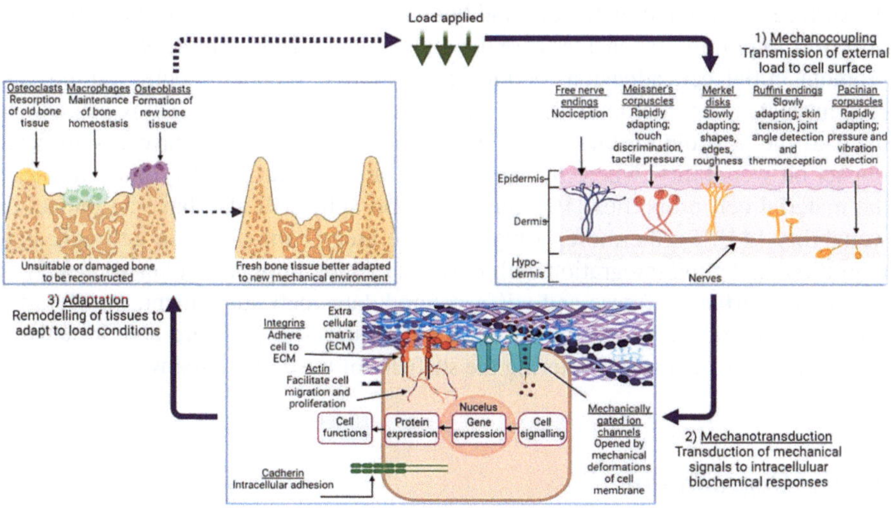

Fig. 5.1 Mechanotransduction: bone remodelling in response to external load applied to skin. (1) Types of mechanoreceptors present in human dermis. (2) Mechanotransduction pathway because of ECM deformation. (3) Bone remodelling in response to this load and the cells involved in the process. Image from Johnston and Callanan [38], open access article distributed under the terms and conditions of the Creative Commons Attribution (CC BY) license

properties like auxeticity, i.e. a negative Poisson's ratio, and therefore, auxetic material are being proposed and investigated as scaffolds in auxetic tissues like skin [39].

5.2.4 Electrical Properties

Electrically conductive materials have gained much attention as components of conductive scaffolds for regeneration of nerves, skeletal muscle or myocardium [40]. Conductive scaffolds allow to locally deliver electrical stimulus, that could be precisely controlled in terms of level and duration, whilst providing a template for tissue repair [21]. Also important are piezoelectric materials, which can convert mechanical load into an electrical signal, for applications such as bone regeneration, since bone has piezoelectric properties [41]. An example is the work by Wang and colleagues who developed a composite scaffold of polycaprolactone and whitlockite, a natural magnesium-containing calcium phosphate with piezoelectric properties [42]. The scaffold had a sustained release of Mg^{2+} for enhanced regeneration of neuro-vascularized bone tissue [42].

5.2.5 Surface Properties

Surface chemistry is of vital importance since the molecular groups presenting on the surface will interact with cells and body fluids such as blood or interstitial fluid [20]. It is important to consider here the inclusion of adhesive molecules that might promote initial cell adhesion, or molecular motifs, such as specific amino acid sequences that can interact with cell surface receptors thereby inducing cell attachment to the scaffold surface [43]. The surface chemistry will have a significant impact on wettability, where hydrophobicity versus hydrophilicity plays a role in the material cell properties [10]. Surface topography will also have an impact on wettability [44]. Furthermore, a high surface area is important in cell adhesion, migration, and good integration with host tissue upon implant implantation [45].

Finally, surface softness and stiffness modulates cell attachment, proliferation and differentiation [44]. At the tissue-implant interface, cells are able to actively modify the surface, thereby altering the stiffness of the cell microenvironment [46].

5.2.6 Degradation

The issue of scaffolds degradation is highly complex and ultimately depends on factors such as scaffold composition, surface chemistry and topography, porosity and pore size range, and overall macrostructure [3, 47]. Ideally, the optimal degradation rate should be in sync with formation of new tissue in vivo, or ECM deposition and

Table 5.1 Definitions of biodegradable, bioresorbable, bioerodible, and bioabsorbable according to Vert et al. [48]

Category	Definition
Biodegradable	Solid polymeric materials and devices which break down due to macromolecular degradation with dispersion in vivo but no proof for the elimination from the body
Bioresorbable	Solid polymeric materials and devices which show bulk degradation and further resorb in vivo
Bioerodible	Solid polymeric materials or devices, which show surface degradation and further, resorb in vivo
Bioabsorbable	Solid polymeric materials or devices, which can dissolve in body fluids without any polymer chain cleavage or molecular mass decrease

remodelling in vitro [3]. Fast degradation rates do not support tissue formation in vivo. Therefore, the degradation rates of scaffolds must be carefully adjusted and engineered. Degradation of scaffolds can happen through various mechanisms and different terms exist to classify the degradation of scaffolds, namely biodegradable, bioresorbable, bioerodible and bioabsorbable [48] (Table 5.1). Regardless, it is worth stressing that the scaffold disappearance rate must coincide with the rate of new tissue or matrix formation.

5.3 Biomaterial Toolbox

The biomaterial toolbox for designing and fabricating scaffolds includes four different families of materials, namely polymers, ceramics, metals, and carbon-based materials, all of which present advantages and disadvantages. Therefore, it is common practice to combine different materials to overcome the disadvantages of the individual biomaterials.

5.3.1 *Polymers: Natural and Synthetic*

The most popular and vast family of biomaterials in this toolbox are polymers. These can be natural or synthetic. Furthermore, natural polymers can be classified into protein-based, polysaccharide-based, or DNA-based [5–7]. Natural polymers are naturally occurring materials that either constitute biological tissues or are secreted by biological species. The most used natural polymer in TERM is collagen, which is the main constituent of the ECM of vertebrates [49]. Other protein-based polymers widely used in TERM are gelatin, which is a hydrolysed form of collagen, fibrin, the first biomatrix that our body produces after an injury, or silk fibroin, a fibrous protein produced by numerous insects like the domestic silk moth Bombyx mori [5]. Polysaccharides are important natural polymers in TERM and examples include alginates, hyaluronan, glycosaminoglycans (GAGs), chitosan, or cellulose [6]. The negatively charged DNA molecule has a polar sugar-phosphate backbone that functions as a strong hydrophilic electrolyte allowing the formation of bonds with water molecules thus rendering hydrogel structures [7].

Synthetic polymers are chemically versatile and as such, a vast array of synthetic polymers exist [8]. It is worth highlighting polyesters since they are biodegradable, and their degradation products are processed through natural metabolic pathways and therefore widely used in TERM scaffolds [8, 33, 47]. Other synthetic polymers worth mentioning are polyethylene glycol (PEG), polyvinyl alcohol (PVA), or polyurethane (PU) [8].

Natural polymers are obvious biomaterials of choice since they are bioactive, biocompatible, and biodegradable [50]. These polymers are made of the building blocks of biological tissues and as such, when used in scaffolds, cells find themselves in a natural and biomimetic environment. Furthermore, cells can remodel these natural scaffolds and replace them for a tissue-like environment. Disadvantages to consider with natural polymers are their poor mechanical properties and fast degradation rates, which need tailoring through approaches such as combination with other materials, or chemical modification [50]. On the other hand, synthetic polymers present superior mechanical properties compared with natural polymers, which can also be adjusted and tailored to the specific need [8]. Some synthetic polymers are biodegradable, such as the mentioned polyesters [8]. However, synthetic polymers do not possess the good biological properties mentioned for natural ones and therefore, tend to be combined with natural materials or modified through surface chemistry to improve their cell properties [9, 51].

5.3.2 Focus on Natural Polymers: Collagen Scaffolds

The dominance of collagen as biomaterial of choice in TERM is understandable since it is the principal constituent of vertebrates' tissues [49]. Collagen is actually a protein super-family of 28 different types, of which fibrillar collagens are the most abundant [52]. These include collagen types I, II, III and V, and together with the basal-lamina collagen IV comprise over 90% of the total collagen found in our bodies, 90% alone in the form of collagen I [52]. Therefore, collagen has been used in scaffolds for both soft and hard tissues such as bone, skin, cartilage, cardiovascular, cornea, liver, tendon, ligament, skeletal muscle, periodontal, or nerve [11, 49, 53]. To give an idea of the importance and pioneering work using collagen in TERM, the first biomaterial scaffolds to reach the clinic for repair of tissues like cartilage or skin were made of collagen [54].

Commonly, collagen is combined with other materials and/or therapeutic agents. For example, collagen/chitosan scaffolds were 3D-printed at low-temperature and loaded with exosomes derived from neural stem cells (NSCs) that had been pretreated with insulin growth factor-1. These therapeutic composite scaffolds showed that they could increase the recruitment of NSCs from the host and improve neural regeneration, therefore improving the chances of neurological functional recovery after suffering a traumatic brain injury [55]. Another example is the combination of collagen and methacrylated gelatin (GelMA) to fabricate convex scaffold implants via 3D printing for corneal regeneration, that when implanted in a rabbit model fitted better with the adjacent host tissue and had stronger cell adhesion than flat implants, leading to accelerated corneal epithelialization [56].

5.3.3 Focus on Natural Polymers: Decellularized Matrix Scaffolds

Within the TERM field there is much focus on the native ECM of tissues since it is the natural scaffold that cells in our bodies find themselves embedded in. This natural scaffold provides the necessary bio-physical and bio-chemical cues for cell and tissue function [57]. Therefore, much research is concerned with using this natural scaffold as an actual one for TERM applications. This involves a process of obtaining fresh tissues and organs and decellularizing them to yield the 3D matrix that supports our tissues [58]. However, this decellularization process is technically complex and the main disadvantage is that ECM components and properties are lost in this process [58]. Nevertheless, this is a promising approach to obtain biomimetic natural scaffolds, which are essentially made of collagen, GAGs, as well as other minor protein and sugar components. An example is decellularized human amniotic membrane, which has been shown to clinically heal chronic wounds [59].

5.3.4 Focus on Synthetic Polymers: Polyester Scaffolds

Polyesters are synthetic polymers that contain the ester functional group in every repeating unit of their polymeric chain. Prominent examples in TERM are poly-lactic acid (PLA), poly-glycolic acid (PGA), poly-lactic-co-glycolic acid (PLGA) or poly-Ɛ-caprolactone (PCL) [8, 33, 47] (Table 5.2). The main advantages of these polymers are that they are biocompatible and biodegradable through chemical hydrolysis reaction of their ester bonds, rendering products such as lactic acid, glycolic acid, caproic acid, butyric acid, succinic acid or valeric acid, which are all processed through natural metabolic pathways [8, 33, 47]. However, some of these degradation products are known to be proinflammatory and therefore, degradation rates should be adjusted to avoid burst release of these products [8]. Finally, these materials are approved by the USA Food and Drug Administration (FDA) and therefore, their regulatory process towards the clinic is easier than for other new or non-approved polymers [8].

Nevertheless, the main disadvantage of polyesters is their poor cell properties compared with their natural counterparts. Therefore, surface treatments or

Table 5.2 Polyesters used in TERM

Polyester	Chemical structure	Properties	Degradation products
Poly-lactic acid (PLA)	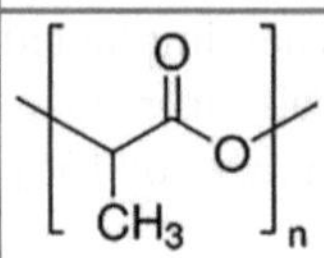	Biocompatible Biodegradable Hydrophobic Chiral monomer Structurally stable Melting temperature range: 180–220 °C FDA approved	Lactic acid
Poly-glycolic acid (PGA)	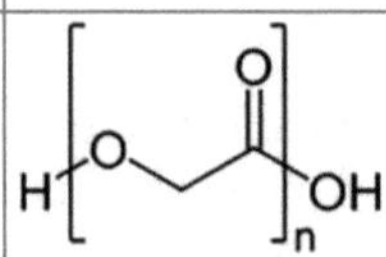	Biocompatible Biodegradable More hydrophilic than PLA High tensile strength Melting point >200 °C Insoluble in most organic solvents FDA approved	Glycolic acid
Poly-lactic-co-glycolic acid (PLGA)	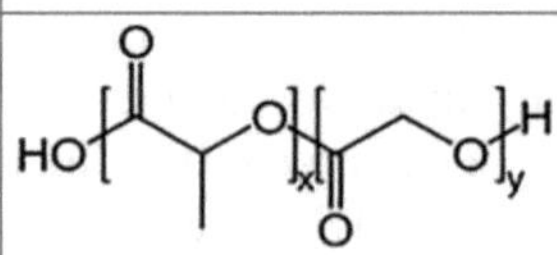	Biocompatible Biodegradable Degradation rate tailored by optimizing the monomer ratios FDA approved	Lactic acid Glycolic acid
Poly-Ɛ-caprolactone (PCL)	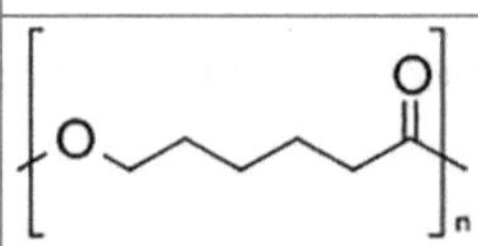	Biocompatible Biodegradable Slower degradation rate than other polyesters Hydrophobic Semicrystalline Low meting point (~60 °C) Good elastic properties FDA approved	Caproic acid Butyric acid Succinic acid Valeric acid

Table adapted from Kohli et al. [8]

combination with other materials are strategies of choice to improve their overall cell properties. Examples are PCL scaffolds produced by 3D printing and coated with different types of ceramics for accelerated reconstruction of calvaria bone defects [60], or nanofibrous scaffolds of PCL and chitosan loaded with melatonin and recombinant human nerve growth factor to regenerate peripheral nerves [61].

5.3.5 *Ceramics*

Ceramics are a family of materials with very compelling properties for TERM applications. These materials are bioactive, biocompatible, biodegradable, chemically versatile, and for bone tissue engineering applications, they offer osteoconductivity and depending on their chemistry and surface properties, osteoinductivity [23]. However, ceramics present brittleness, low fracture strength and their degradation rates are difficult to predict. Nevertheless, these materials can be fabricated in different shapes and morphologies, are popular in composites, either as part of the 3D bulk structure or in coatings, are compatible with additive manufacturing techniques, and can be produced to nano-sizes [27, 28, 51, 62].

Different types of ceramics exist. Calcium phosphates are an important type and of biological relevance since they resemble the chemistry of the biomineral found in bone, which is a calcium phosphate apatite [63]. Examples of calcium phosphates are hydroxyapatite or β-tricalcium phosphate. Particularly, a wealth of research exists on the use of hydroxyapatite for bone scaffolds, either on its own, or most commonly in combination with polymers or metals. Bioactive glasses and glass ceramics are another type of ceramics with the important feature of being able to release therapeutic ions in physiological conditions [64]. Examples are silicate-based bioactive glasses. Finally, there are other ceramics such as calcium silicates or alumina ceramic, which have also been investigated for TERM scaffolds.

The main applications of ceramics are obviously in bone regeneration, both skeletal and dental, skin wound healing and peripheral nerve regeneration [23, 62, 65].

5.3.6 *Focus on Ceramics: Bioactive Glass Scaffolds*

Bioactive glasses were discovered in 1969 by Professor Larry Hench and they can be silicate-based, phosphate-based, and borate-based. These materials present chemical versatility as they can be doped with biologically active ions such as Ag^{+}, Li^{+}, Co^{2+}, Ca^{2+}, Cu^{2+}, Zn^{2+}, Sr^{2+}, Fe^{2+}, Mg^{2+}, Ga^{3+}, or B^{3+}, in order to add or stimulate physiological processes like osteogenesis or angiogenesis, or material properties like bioactivity or antibacterial effects [64, 66, 67].

Bioactive glasses are widely known for their bone repair and regeneration properties, and a wealth of literature exists on their use for bone regeneration purposes. Importantly, they can be combined with other materials into effective composites to address the pressing clinical need of regenerating bone. Examples are combination with PLGA and fibrin, PLA, gelatin and chitosan, or carbon nano-fibres [68–71].

Furthermore, in physiological conditions bioactive glasses release ionic products that play critical roles in accelerating wound healing. Since silicate-based are the oldest bioactive glasses known, these were the first to be explored for skin wound healing. 45S5 Bioglass ionic extracts stimulate fibroblast proliferation and migration and enhance collagen type I secretion. Thus, researchers are introducing this material into skin substitutes. Although the use of bioactive glasses for skin repair is relatively recent, it has good potential for promoting wound healing of challenging wounds, where the controlled release of therapeutic ions will be key in their effectiveness [65].

5.3.7 Metals

Metals have excellent mechanical properties and are ideal for load-bearing orthopaedic applications [23, 26]. Furthermore, fabricated porous metal implants provide bone ingrowth that leads to early fixation of the implant [26]. Disadvantages of metals include lack of tissue adherence, which may result in implant loosening, and a risk of toxicity due to accumulation of metal ions from corrosion. Significant research explores how to increase the bioactivity of these materials. Surface properties of metal implants are very important for fast and stable integration with the host bone tissue and therefore, this research mainly focuses on surface modification and functionalization [72].

Titanium and its alloys are the most used metals in biomedical applications because of their biocompatibility, strength, lightness and high resistance to corrosion. Their biocompatibility comes from a thin titanium dioxide (TiO_2) layer formed on the surface, as titanium is a very reactive element even at room temperature [23]. Porous titanium scaffolds have been fabricated to achieve mechanical properties comparable to those of bone. Although the majority of metals are not degradable, magnesium and its alloys actually are as they can be absorbed or discharged by the human body. Additive manufacturing techniques can be employed to fabricate porous magnesium-based scaffolds with tailored porosity and pore sizes [73].

5.3.8 Carbon-Based Materials

Carbon-based nanomaterials comprise carbon elements and mainly include graphene, fullerene, carbon dots, and carbon nanotubes in the order of dimensional quantum properties [74–76]. Carbon nanomaterials are bioactive, have a large specific surface area/volume ratio, and display an intrinsic antibacterial activity. Particularly, they show great potential for bone tissue engineering applications because of their good mechanical properties, and they have been widely investigated either as a matrix or as reinforcing material in polymeric nano-composites. Importantly, the homogeneous dispersion of carbon nanomaterials in the matrix is key on their performance and therefore, the optimization of their dispersion is critical for developing carbon-based nanomaterial scaffolds. Furthermore, since carbon-based nanomaterials have a dosage response in cells, it is of vital importance to find out the optimal concentration to avoid cytotoxicity and maximize the scaffold's efficiency [74–76].

5.4 Scaffold Fabrication Techniques

Various scaffold fabrication techniques are available in TERM, from classical ones like solvent casting and particulate leaching or freeze-drying, to electrodynamic techniques like electrospinning and melt electrowriting, to the trending additive manufacturing techniques and bioprinting.

5.4.1 Solvent Casting and Particulate Leaching

In the solvent casting and particulate leaching (SCPL) technique a polymer is dissolved in a suitable organic solvent, and then particles with specific dimensions, mainly insoluble salts, are added and mixed with the polymer solution. The resulting mixture is added to a mould to shape it into a specific geometry. The organic solvent is evaporated giving place to a salt-polymer composite, which is washed to dissolve the salt particles leaving behind a 3D porous structure. Advantages of the SCPL technique is its relative simplicity, does not need expensive equipment, and allows control of porosity, pore size and interconnectivity by selecting the appropriate polymer and porogen as well as their relative amounts [77, 78]. Disadvantages of this technique include residual solvent and particles in the scaffolds, and lack of sufficient mechanical integrity for load-bearing TERM applications.

5.4.2 Freeze-Drying

In the freeze drying or lyophilisation method, polymeric solutions are dried following three steps: (1) preparation of the polymeric solution, (2) casting or moulding of the solution, and (3) freezing and drying at low pressure, where ice is removed by direct sublimation. One of the advantages of freeze drying is that scaffolds with approximately 90% interconnected porosity can be produced. Interestingly, the pore size can be controlled by freeze rate, polymer concentration, and temperature [78, 79]. An example on the use of this fabrication method is the work by Dragusin and colleagues where porous hydrogel scaffolds of methacrylamide-modified gelatin and 2-hydroxyethyl methacrylate (HEMA) were produced by freeze-drying. Five different prototypes were produced by varying the concentrations of the two components. The authors reported that the content of HEMA modulated the scaffolds architecture, where the higher the HEMA content the more complex and dense internal porosity. It was also found that the covalently bound gelatin improved the biocompatibility of the resulting hydrogel scaffolds [80].

Perhaps the biggest challenge with using the method is the production of scaffolds with complex hierarchical structures, like for example scaffolds with an internal vasculature system. Furthermore, the scaffolds produced with this method have insufficient mechanical integrity, which precludes their use in load-bearing applications.

5.4.3 Electrospinning

Electrospinning is a technique capable of fabricating nano- and microfibrous scaffolds for TERM. Typical ultrafine filaments are produced by applying a high-voltage electric field to a polymer solution or melt, and are deposited on a surface collector, until a fibrous scaffold is rendered (Fig. 5.2a). Parameters such as polymer concentration, polymer solution viscosity, voltage, or spinneret-to-collector distance can be adjusted to produce scaffolds with different fibre diameter and architectures, including oriented fibres [3] (Fig. 5.2c).

Various synthetic and natural biomaterials are compatible with this fabrication technique, such as PLGA, PLA, PCL, poly(propylene carbonate) (PPC), poly(hydroxybutyrate) (PHB), collagen, chitosan, and silk fibroin. Therefore, scaffolds with nano- and microfibers with different porosities and surface properties can be designed by selecting the appropriate material or combination of materials. For instance, Gautam and colleagues produced gelatin-PCL electrospun scaffolds and their surface was treated to produce a mineral coating of nano-hydroxyapatite in order to produce a potential scaffold for bone tissue engineering [83].

5.4.4 Melt Electrowriting

Melt electrowriting (MEW) is a relatively new fabrication technique that was first introduced in 2011 [84, 85]. MEW combines electrospinning with additive manufacturing and uses high voltage to precisely deposit molten polymer into microfibers, which are deposited on top of each other with extraordinary resolution (Fig. 5.2b, d). Therefore, MEW allows building complex and multi-scale structures. An example is mechanical metamaterials such as those with an auxetic behaviour, which upon application of tensile load they expand in transverse directions. Auxetic materials are of interest since some biological tissues such as skin present this behaviour. MEW is an efficient technique to produce auxetic scaffolds [39].

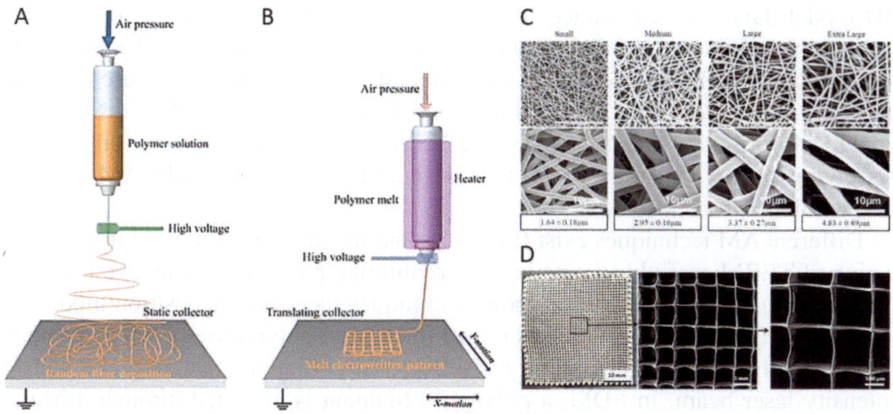

Fig. 5.2 Schematic representation of the difference between electrospinning (**a**) and melt electrowriting (**b**) techniques. (**c**) Electrospun PCL scaffolds incorporating decellularized bovine aorta ECM with increasing fibre diameter. (**d**) MEW PCL scaffolds sheets (40 mm × 40 mm × ~2 mm) fabricated to have a stacked (~50 layers) 90° cross-hatched microarchitecture, with a fibre diameter of 50 µm and spacing of 1 mm. Images from Afghah et al. [81], Reid et al. [82] and Ren et al. [9]. Open access articles distributed under the terms and conditions of the Creative Commons Attribution (CC BY) license

In terms of the materials used for MEW, PCL is considered the "gold standard" polymer for MEW as it produces the highest printing quality [86]. However, as mentioned earlier, the cell properties of PCL need to be improved since PCL is hydrophobic and lacks bioactive cues. Surface treatments are the strategy of choice for improving PCL's bioactive properties. For example, MEW-fabricated box-structured PCL scaffolds with fiber diameters of 15 µm and fiber spacings of 200 µm were sputter-coated with a nanolayer of gold (10–80 nm) to enhance the conductivity of these scaffolds for nerve regeneration [87].

Beyond PCL, other materials that have been used for MEW are water-soluble polymers like poly(2-oxazoline)s, photocurable polymers, or polyesters other than PCL, such as poly(hydroxymethylglycolide-*co*-ε-caprolactone) [86].

5.4.5 Additive Manufacturing Techniques

Additive manufacturing (AM) techniques, popularly known as "3D printing" are gaining much attention due to their advantages compared with other traditional fabrication methods. Some of these advantages include a high degree of freedom in design, low cost, and high efficiency. Furthermore, for biomedical applications, these techniques also offer the advantage of fabricating patient-specific implants that perfectly fit the defect to be treated [13].

AM is defined as "the process of joining materials to make objects from 3D model data, usually layer upon layer" (ASTM, 2012). To produce a TERM scaffold,

3D model data can be obtained from medical imaging techniques, e.g. computer tomography (CT) or magnetic resonance imaging (MRI), and treated by computer-aided design (CAD) and computer-aided manufacturing (CAM) software that converts the 3D model into a standard tessellation language (STL) file. This STL file contains information about the 3D object's surface geometry and is sliced into layers giving rise to a slice file (SLI) that is digitally uploaded into the AM machine that will build the scaffold [14].

Different AM techniques exist (Fig. 5.3), and the most popular used in the fabrication of TERM scaffolds are perhaps stereolithography (SLA), selective laser sintering (SLS), and melt extrusion-based techniques like fused deposition modelling (FDM) [14]. SLA utilises uv light or laser in order to polymerise layers of a photosensitive polymer. SLS is based on the selective sintering of a material with a high-intensity laser beam. In FDM, a polymeric filament is extruded through a heated nozzle, which is continually deposited on a mobile platform. A wealth of literature exists on the use of AM techniques for manufacturing of TERM scaffolds [14, 32, 60, 88, 89], and these techniques are positioning themselves as fabrication methods of choice in TERM.

5.4.6 Bioprinting

Bioprinting, also known as bio-additive manufacturing or biofabrication, is a manufacturing technique that allows control of the spatial distribution of cells within bioinks to print 3D tissue-engineered constructs. Different 3D bioprinting techniques have been developed, including extrusion-based inkjet, laser-assisted stereolithography (SLA), and digital light processing (DLP) techniques [90]. Of key importance in bioprinting is the choice of bioink. Bioinks are natural or synthetic polymers that are biocompatible to accommodate living cells and possess adequate rheological properties for printing. Hydrogels are the main materials used as bioinks as they can fulfil these requirements. In terms of the specific materials, polysaccharides such as alginates, agarose, or hyaluronic acid, and proteins such as gelatin and collagen are popular choices in bioinks. Synthetic polymers like PEG or Pluronic are also widely used in bioinks for bioprinting [91].

As succinctly put by Liu and colleagues, ultimately, the "goal of bioprinting is to produce stable, anatomically-shaped, human-scale functional organs or tissue substitutes that can be implanted" [90]. However, the biofabrication of cell-laden volumetric constructs is still very challenging due to the limited range of materials available that display printability with a high shape fidelity as well as support the various cellular functions. Nevertheless, the future of bioprinting is indeed bright as it could even offer the possibility of in situ bioprinting, where the construct is directly printed onto or into the tissue or organ to be repaired, that actually acts in itself as a biological bioreactor [92, 93]. An example is the work by Urciuolo and co-workers who showed that cell-laden photosensitive polymer hydrogels could be

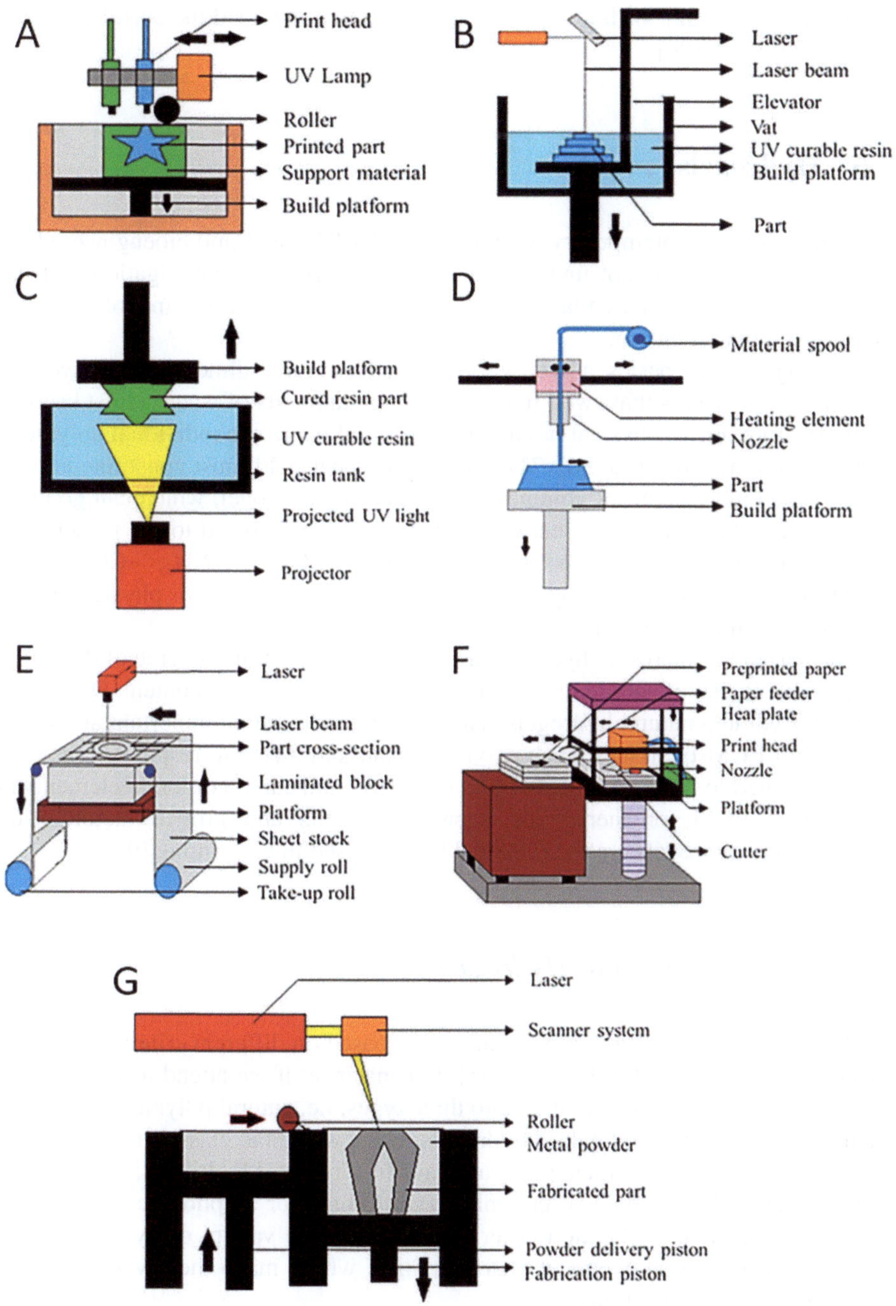

Fig. 5.3 Scheme of popular additive manufacturing techniques: (**a**) PolyJet printing, (**b**) laser-assisted stereolithography (SLA), (**c**) digital light processing (DLP), (**d**) fused deposition modeling (FDM), (**e**) laminated object manufacturing (LOM), (**f**) selective deposition modeling (SDM), and (**g**) selective laser sintering (SLS). Image from Wu et al. [13], open access article distributed under the terms and conditions of the Creative Commons Attribution (CC BY) license

bioprinted across and within tissues of live mice that included the dermis, skeletal muscle and brain [94].

5.5 Hydrogels

The explosion of contemporary research in the TERM field into bioengineered tissue models is perhaps not understood without the parallel investigation that has taken and currently takes place into hydrogels. As such, this kind of scaffolds deserve singular attention.

Starting with the basics, a hydrogel is defined as a three-dimensional network of hydrophilic polymers that swell in and hold large amounts of water whilst keeping their structure due to physical or chemical crosslinking of the individual polymeric chains of the network [95, 96]. The amount of water held must constitute at least 10% of the total weight, or volume, of the resulting hydrogel, which can go up to 95% of the total weight. The network's hydrophilicity is owed to the presence of hydrophilic molecular groups such as –OH, –COOH, $-NH_2$, $-CONH_2$, or SO_3H. These materials undergo a sol-gel phase transition due to physical and/or mechanical stimuli [95, 96].

The particular interest in hydrogels is because they display a degree of flexibility comparable to that of natural tissue due to their significant water content. Due to this intrinsic hydrated nature, hydrogels can be designed to mimic important aspects of the natural ECM of tissues, such as viscoelastic and swelling response. They are also fabricated to be biocompatible and bioactive, and can incorporate elements of the natural ECM, or therapeutic agents like cells and growth factors [96]. Furthermore, hydrogels are the principal bioinks used in bioprinting [91].

5.5.1 Classification of Hydrogels

Hydrogels can be classified into various types based on different criteria or factors, as visually summarized in Fig. 5.4 [95]. For instance, if we attend to the polymer source, hydrogels can be classified into three types, i.e. natural polymer-based, synthetic polymer-based, and hybrid hydrogels. If we wished to classify them in terms of their electrical charge, then we would have four types of hydrogels, namely non-ionic, ionic (either anionic or cationic), zwitterionic, or amphoteric electrolytes. Regardless of the classification criteria chosen, a wide variety of hydrogels exist with different physico-chemical characteristics, which make them versatile scaffolds for TERM applications.

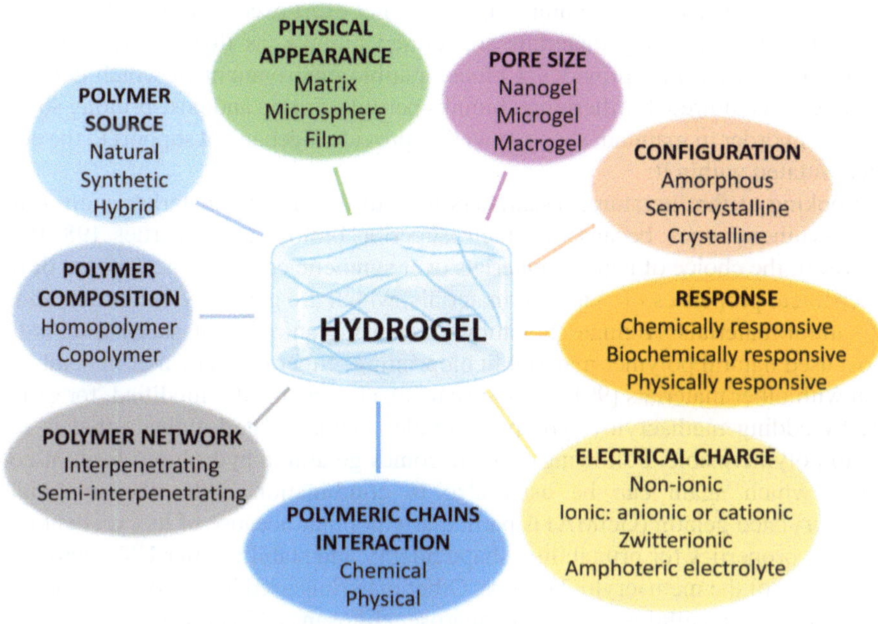

Fig. 5.4 Classification of hydrogels attending to different criteria

5.5.2 Gelation

Hydrogels undergo a sol-gel transition through gelation. Gelation is the formation of a 3D network by physical or chemical crosslinking [97]. Therefore, hydrogel gelation mechanisms can be either physical or chemical. In terms of physical gelation mechanisms, these are classified into strong and weak, thereby yielding strong physical gels and weak physical gels, respectively. Strong physical bonds between polymer chains are formed in strong physical gels, whilst weak gels have reversible links that are formed by non-permanent unions between the polymeric chains. On the other hand, chemical gelation implies covalent bonds, thereby resulting in a strong gel [97]. Hydrogel gelation mechanisms and kinetics are highly complex matters and objects of much study in various disciplines such as materials science, chemistry or physics.

5.5.3 Hydrogels as Bioinks

Hydrogels have a central role in bioprinting, which as explained earlier in this chapter consists of including biological components such as cells in AM. The addition of components like cells greatly complicates the printing process as well as the nature of the printing materials. Importantly, the printing process must be cell-compatible, occur within a narrow temperature range, and not exert excessive shear forces onto the cells. These printing materials are termed bioinks, as they incorporate cells within a substrate that provides a biocompatible environment in which cells are encased and cultured. Furthermore, bioinks permit printing and post-printing solidification in order to maintain shape as well as provide mechanical support to the cells encapsulated within it.

Looking at the literature, researchers tend to use natural materials more often than synthetic ones because of their superior biological properties [98–100]. However, the choice of natural materials over synthetic comes at the expense of the mechanical properties of synthetic materials, which are generally better than those of natural materials. Alginate, a naturally occurring anionic polysaccharide, is the most used natural polymer material in bioprinting, either on its own or in combination with other materials [99]. Alginate can also be chemically modified, for example by adding methacryloyl groups to render methacrylated alginate that is UV photo-polymerizable. Following alginate comes gelatin, a hydrolysed form of collagen, which again can be used also in combination with other materials. Methacrylated gelatin (GelMA) is particularly popular because of its excellent rheological properties for printability, shape fidelity and stability after UV photopolymerization of the methacryloyl groups. Other materials used in hydrogel bioinks are hyaluronic acid, cellulose, collagen, agarose, chitosan, DNA, polyethylene glycol, or polyesters [98, 99]. However, the search for the ideal hydrogel bioink in terms of improved printability and cell viability continues [101].

5.6 Conclusions

The paradigm of TERM involves biomaterial scaffolds, cells, and bio-chemical/physical cues to produce tissue equivalents that could be implanted or used as tissue models. A biomaterial scaffold that allows cell attachment, infiltration, growth, function, and ultimately remodelling is needed to produce functional and relevant tissue equivalents. A number of requirements and characteristics are desirable in a scaffold, such as biocompatibility, bioactivity, biodegradability, or optimal mechanical features. In order to fulfil the scaffold requirements, important scaffold characteristics to design, optimise and ultimately control are structural, mechanical, electrical, surface or degradation properties.

The biomaterial toolbox for designing and fabricating scaffolds includes natural and synthetic polymers, ceramics, metals and carbon-based materials. Since all of these materials present advantages and disadvantages, composites are favoured in TERM, as they can be produced to cater for a specific need. Many of the scaffold's properties are dictated by the fabrication method employed to produce it. There are numerous fabrication technologies, and the choice of one or another depends on various factors, one of which is the desired scaffold's architecture. Some popular fabrication techniques are freeze-drying, electrospinning, melt electrowriting, AM and bioprinting. Scaffolds can have different forms and shapes, as well as formulations. Particularly, hydrogels have gained much attention due to their hydrated nature and good properties as bioinks in bioprinting.

The chapter highlights the intricate relationship between composition, physical and chemical properties, and fabrication technology in order to design and produce ideal scaffolds for various TERM applications.

Questions

- Name four scaffold requirements and explain them.
- Why is porosity important in TERM scaffolds?
- Which is the most used natural polymer in TERM scaffolds and why?
- What is a polyester and what advantages and disadvantages do they possess when used in TERM scaffolds?
- Name the advantages that ceramics offer when used in TERM scaffolds.
- What are the differences between electrospinning and melt electrowriting?
- Name advantages of AM techniques compared to traditional fabrication methods.
- What is a hydrogel and why are they so popular in TERM?
- How are hydrogels classified?
- What is a bioink?

References

1. Langer, R., & Vacanti, J. P. (1993). Tissue engineering. *Science, 260*(5110), 920. http://science.sciencemag.org/content/260/5110/920.abstract
2. Stratton, S., Shelke, N. B., Hoshino, K., Rudraiah, S., & Kumbar, S. G. (2016). Bioactive polymeric scaffolds for tissue engineering. *Materials., 1*, 93–108.
3. Kirillova, A., Yeazel, T. R., Asheghali, D., Petersen, S. R., Dort, S., Gall, K., et al. (2021). Fabrication of biomedical scaffolds using biodegradable polymers. *Chemical Reviews, 121*(18), 11238–11304. https://doi.org/10.1021/acs.chemrev.0c01200
4. Karageorgiou, V., & Kaplan, D. (2005). Porosity of 3D biomaterial scaffolds and osteogenesis. *Biomaterials, 26*(27), 5474–5491.
5. García-Gareta, E. (2023). Protein-based materials for wound healing. In F. R. A. Maia, J. M. Oliveira, & R. L. Reis (Eds.), *Handbook of the extracellular matrix: Biologically-derived materials* (pp. 1–28). Springer International Publishing. https://doi.org/10.1007/978-3-030-92090-6_20-1
6. Jin, M., Shi, J., Zhu, W., Yao, H., & Wang, D.-A. (2020). Polysaccharide-based biomaterials in tissue engineering: A review. *Tissue Engineering Reviews (Part B), 27*(6), 604–626. https://doi.org/10.1089/ten.teb.2020.0208
7. Morya, V., Walia, S., Mandal, B. B., Ghoroi, C., & Bhatia, D. (2020). Functional DNA based hydrogels: Development, properties and biological applications. *ACS Biomaterials Science and Engineering, 6*, 6021–6035.
8. Kohli, N., Sharma, V., Brown, S. J., & Garcia-Gareta, E. (2019). Synthetic polymers for skin biomaterials. *Biomaterials for Skin Repair and Regeneration, 156*, 125–149. https://www.sciencedirect.com/science/article/pii/B9780081025468000054
9. Ren, J., Kohli, N., Sharma, V., Shakouri, T., Keskin-Erdogan, Z., Saifzadeh, S., et al. (2021). Poly-ε-caprolactone/fibrin-alginate scaffold: A new pro-angiogenic composite biomaterial for the treatment of bone defects. *Polymers (Basel), 13*(19), 3399.
10. Sawadkar, P., Mohanakrishnan, J., Rajasekar, P., Rahmani, B., Kohli, N., Bozec, L., et al. (2020). A synergistic relationship between polycaprolactone and natural polymers enhances

the physical properties and biological activity of scaffolds. *ACS Applied Materials & Interfaces, 12*(12), 13587–13597.
11. Poh, P. S. P., Woodruff, M. A., & García-Gareta, E. (2020). Polymer-based composites for musculoskeletal regenerative medicine. In N. E. Vrana, H. Knopf-Marques, & J. Barthes (Eds.), *Biomaterials for organ and tissue Regeneration* (pp. 33–82). Woodhead Publishing. https://www.sciencedirect.com/science/article/pii/B9780081029060000039
12. Buttafoco, L., Kolkman, N. G., Engbers-Buijtenhuijs, P., Poot, A. A., Dijkstra, P. J., Vermes, I., et al. (2006). Electrospinning of collagen and elastin for tissue engineering applications. *Biomaterials, 27*(5), 724–734.
13. Wu, Y., Lu, Y., Zhao, M., Bosiakov, S., & Li, L. (2022). *A critical review of additive manufacturing techniques and associated biomaterials used in bone tissue engineering* (Vol. 14, p. 2117). Polymers.
14. Mota, C., Puppi, D., Chiellini, F., & Chiellini, E. (2015). Additive manufacturing techniques for the production of tissue engineering constructs. *Journal of Tissue Engineering and Regenerative Medicine, 9*(3), 174–190. https://doi.org/10.1002/term.1635
15. Tian, D. M., Wan, H. H., Chen, J. R., Bin, Y. Y., He, Y., Liu, Y., et al. (2022). In-situ formed elastin-based hydrogels enhance wound healing via promoting innate immune cells recruitment and angiogenesis. *Materials Today Bio, 15*, 100300.
16. Alamán-Díez, P., Borau, C., Enrique Guerrero, P., Amaveda, H., Mora, M., Fraile, J. M., et al. (2023). Collagen-laponite nanoclay hydrogels for tumor spheroid growth. *Biomacromolecules, 24*(6), 2879–2891.
17. Pele, K. G., Amaveda, H., Mora, M., Marcuello, C., Lostao, A., Alamán-Díez, P., et al. (2023). Hydrocolloids of egg white and Gelatin as a platform for hydrogel-based tissue engineering. *Gels, 9*(6), 505.
18. Huzum, B., Puha, B., Necoara, R., Gheorghevici, S., Puha, G., Filip, A., et al. (2021). Biocompatibility assessment of biomaterials used in orthopedic devices an overview (review). *Experimental and Therapeutic Medicine, 22*(5), 1315.
19. Sultana, N. (2018). Mechanical and biological properties of scaffold materials. In Y. Deng & J. Kuiper (Eds.), *Functional 3D tissue engineering scaffolds* (pp. 1–21). Woodhead Publishing. https://www.sciencedirect.com/science/article/pii/B978008100979600001X
20. Idaszek, J., Kijeńska, E., Łojkowski, M., & Swieszkowski, W. (2016). How important are scaffolds and their surface properties in regenerative medicine. *Applied Surface Science, 388*, 762–774. https://www.sciencedirect.com/science/article/pii/S0169433216304743
21. Ul Haq, A., Montaina, L., Pescosolido, F., Carotenuto, F., Trovalusci, F., De Matteis, F., et al. (2023). Electrically conductive scaffolds mimicking the hierarchical structure of cardiac myofibers. *Scientific Reports, 13*(1), 2863. https://doi.org/10.1038/s41598-023-29780-w
22. Kazimierczak, P., & Przekora, A. (2020). Osteoconductive and osteoinductive surface modifications of biomaterials for bone regeneration: A concise review. *Coatings, 10*, 971.
23. García-Gareta, E., Coathup, M. J., & Blunn, G. W. (2015). Osteoinduction of bone grafting materials for bone repair and regeneration. *Bone, 81*, 112–121.
24. Owji, N., Mandakhbayar, N., Cha, J. R., Padalhin, A. R., Erdogan, Z. K., Aldaadaa, A., et al. (2022). Inclusion of calcium phosphate does not further improve in vitro and in vivo osteogenesis in a novel, highly biocompatible, mechanically stable and 3D printable polymer. *Scientific Reports, 12*(1), 16977.
25. Sharma, V., Kohli, N., Moulding, D., Afolabi, H., Hook, L., Mason, C., et al. (2017). Design of a novel two-component hybrid dermal scaffold for the treatment of pressure sores. *Macromolecular Bioscience, 17*(11), 1700185.
26. García-Gareta, E., Hua, J., & Blunn, G. W. (2015). Osseointegration of acellular and cellularized osteoconductive scaffolds: Is tissue engineering using mesenchymal stem cells necessary for implant fixation? *Journal of Biomedical Materials Research Part A, 103*(3), 1067–1076.
27. Poh, P. S. P., Hutmacher, D. W., Stevens, M. M., & Woodruff, M. A. (2013). Fabrication and in vitro characterization of bioactive glass composite scaffolds for bone regeneration. *Biofabrication, 5*(4), 045005.

28. Jana, S., Datta, P., Das, H., Jaiswal, S., Ghosh, P. R., Lahiri, D., et al. (2022). Copper and cobalt doped bioactive glass-fish dermal collagen electrospun mat triggers key events of diabetic wound healing in full-thickness skin defect model. *Journal of the Mechanical Behavior of Biomedical Materials, 1*, 134.
29. Lutzweiler, G., Halili, A. N., & Vrana, N. E. (2020). The overview of porous, bioactive scaffolds as instructive biomaterials for tissue regeneration and their clinical translation. *Pharmaceutics, 12*, 602.
30. Loh, Q. L., & Choong, C. (2013). Three-dimensional scaffolds for tissue engineering applications: Role of porosity and pore size. *Tissue Engineering Reviews (Part B), 19*(6), 485–502. https://doi.org/10.1089/ten.teb.2012.0437
31. Sharma, V., Patel, N., Kohli, N., Ravindran, N., Hook, L., Mason, C., et al. (2016). Viscoelastic, physical, and bio-degradable properties of dermal scaffolds and related cell behaviour. *Biomedical Materials, 11*(5), 055001.
32. Owji, N., Aldaadaa, A., Cha, J. R., Shakouri, T., García-Gareta, E., Kim, H. W., et al. (2019). Synthesis, characterization, and 3D Printing of an isosorbide-based, light-curable degradable polymer for potential application in maxillofacial reconstruction. *ACS Biomaterials Science & Engineering, 11*, 2578–2587.
33. Alamán-Díez, P., García-Gareta, E., Napal, P. F., Arruebo, M., & Pérez, M. Á. (2022). In vitro hydrolytic degradation of polyester-based scaffolds under static and dynamic conditions in a customized perfusion bioreactor. *Materials, 15*(7), 2572. https://www.mdpi.com/1996-1944/15/7/2572
34. Abbasi, N., Hamlet, S., Love, R. M., & Nguyen, N.-T. (2020). Porous scaffolds for bone regeneration. *Journal of Science: Advanced Materials and Devices, 5*(1), 1–9. https://www.sciencedirect.com/science/article/pii/S2468217920300071
35. Knychala, J., Bouropoulos, N., Catt, C. J., Katsamenis, O. L., Please, C. P., & Sengers, B. G. (2013). Pore geometry regulates early stage human bone marrow cell tissue formation and organisation. *Annals of Biomedical Engineering, 41*(5), 917–930. https://doi.org/10.1007/s10439-013-0748-z
36. Ikada, Y. (2006). *Tissue engineering: Fundamentals and applications* (1st ed.). Elsevier.
37. Torres-Sanchez, C., Al Mushref, F. R. A., Norrito, M., Yendall, K., Liu, Y., & Conway, P. P. (2017). The effect of pore size and porosity on mechanical properties and biological response of porous titanium scaffolds. *Materials Science and Engineering C, 77*, 219–228. https://www.sciencedirect.com/science/article/pii/S092849311632793X
38. Johnston, A., & Callanan, A. (2023). Recent methods for modifying mechanical properties of tissue-engineered scaffolds for clinical applications. *Biomimetics, 8*(2), 205.
39. Lecina-Tejero, Ó., Pérez, M. Á., García-Gareta, E., & Borau, C. (2023). The rise of mechanical metamaterials: Auxetic constructs for skin wound healing. *Journal of Tissue Engineering.* https://doi.org/10.1177/20417314231177838
40. Ghasemi-Mobarakeh, L., Prabhakaran, M. P., Morshed, M., Nasr-Esfahani, M. H., Baharvand, H., Kiani, S., et al. (2011). Application of conductive polymers, scaffolds and electrical stimulation for nerve tissue engineering. *Tissue Engineering and Regenerative Medicine, 5*(4), e17–e35. https://doi.org/10.1002/term.383
41. Carter, A., Popowski, K., Cheng, K., Greenbaum, A., Ligler, F. S., & Moatti, A. (2021). Enhancement of bone Regeneration Through the Converse piezoelectric effect, a novel approach for applying mechanical stimulation. *Bioelectricity, 3*(4), 255–271. https://doi.org/10.1089/bioe.2021.0019
42. Wang, L., Pang, Y., Tang, Y., Wang, X., Zhang, D., Zhang, X., et al. (2023). A biomimetic piezoelectric scaffold with sustained Mg2+ release promotes neurogenic and angiogenic differentiation for enhanced bone regeneration. *Bioactive Materials, 25*, 399–414. https://www.sciencedirect.com/science/article/pii/S2452199X22004686
43. García-Gareta, E., Levin, A., & Hook, L. (2019). Engineering the migration and attachment behaviour of primary dermal fibroblasts. *Biotechnology and Bioengineering., 116*, 1102–1115.

44. Majhy, B., Priyadarshini, P., & Sen, A. K. (2021). Effect of surface energy and roughness on cell adhesion and growth—Facile surface modification for enhanced cell culture. *RSC Advances, 11*(25), 15467–15476. https://doi.org/10.1039/D1RA02402G
45. Lloyd, A. (2003). Engineering surfaces to enhance cell adhesion: Surface engineering. *Mater Today, 6*(11), 19. https://www.sciencedirect.com/science/article/pii/S1369702103011209
46. El-Mohri, H., Wu, Y., Mohanty, S., & Ghosh, G. (2017). Impact of matrix stiffness on fibroblast function. *Materials Science and Engineering: C, 1*(74), 146–151.
47. Woodruff, M. A., & Hutmacher, D. W. (2010). The return of a forgotten polymer—Polycaprolactone in the 21st century. *Progress in Polymer Science, 35*(10), 1217–1256. https://linkinghub.elsevier.com/retrieve/pii/S0079670010000419
48. Vert, M., Li, S. M., Spenlehauer, G., & Guerin, P. (1992). Bioresorbability and biocompatibility of aliphatic polyesters. *Journal of Materials Science: Materials in Medicine, 3*(6), 432–446. https://doi.org/10.1007/BF00701240
49. García-Gareta, E. (2019). *Collagen, from tissue culture to biomaterials, tissue engineering, and beyond* (1st ed.). Cambridge Scholars Publishing.
50. Chen, M., Jiang, R., Deng, N., Zhao, X., Li, X., & Guo, C. (2022). Natural polymer-based scaffolds for soft tissue repair. *Frontiers in Bioengineering and Biotechnology, 10*, 954699. https://doi.org/10.3389/fbioe.2022.954699
51. Ren, J., Blackwood, K. A., Doustgani, A., Poh, P. P., Steck, R., Stevens, M. M., et al. (2014). Melt-electrospun polycaprolactone strontium-substituted bioactive glass scaffolds for bone regeneration. *Journal of Biomedical Materials Research Part A, 102*(9), 3140–3153.
52. Ricard-Blum, S. (2011). The collagen family. *Cold Spring Harbor Perspectives in Biology, 3*(1), 1–19.
53. Davison-Kotler, E., Marshall, W. S., & García-Gareta, E. (2019). Sources of collagen for biomaterials in skin wound healing. *Bioengineering, 6*(3), 56.
54. Zheng, M., Wang, X., Chen, Y., Yue, O., Bai, Z., Cui, B., et al. (2023). A review of recent progress on collagen-based biomaterials. *Advanced Healthcare Materials, 12*(16), 2202042. https://doi.org/10.1002/adhm.202202042
55. Liu, X. Y., Feng, Y. H., Feng, Q. B., Zhang, J. Y., Zhong, L., Liu, P., et al. (2023). Low-temperature 3D-printed collagen/chitosan scaffolds loaded with exosomes derived from neural stem cells pretreated with insulin growth factor-1 enhance neural regeneration after traumatic brain injury. *Neural Regeneration Research, 18*(9), 1990–1998.
56. Xu, Y., Liu, J., Song, W., Wang, Q., Sun, X., Zhao, Q., et al. (2023). Biomimetic convex implant for corneal regeneration through 3D printing. *Advanced Science, 10*(11), 2205878. https://doi.org/10.1002/advs.202205878
57. Kular, J. K., Basu, S., & Sharma, R. I. (2014). The extracellular matrix: Structure, composition, age-related differences, tools for analysis and applications for tissue engineering. *Journal of Tissue Engineering, 5*, 2041731414557112.
58. García-Gareta, E., Abduldaiem, Y., Sawadkar, P., Kyriakidis, C., Lali, F., & Greco, K. V. (2020). Decellularised scaffolds: Just a framework? Current knowledge and future directions. *Journal of Tissue Engineering., 11*, 2041731420942903.
59. Kyriakidis, C., Lali, F., Greco, K. V., & García-Gareta, E. (2021). Chronic leg ulcers: Are tissue engineering and biomaterials science the solution? *Bioengineering, 8*(5), 62. https://www.mdpi.com/2306-5354/8/5/62
60. Fazeli, N., Arefian, E., Irani, S., Ardeshirylajimi, A., & Seyedjafari, E. (2023). Accelerated reconstruction of rat calvaria bone defect using 3D-printed scaffolds coated with hydroxyapatite/bioglass. *Scientific Reports, 13*(1), 12145. https://doi.org/10.1038/s41598-023-38146-1
61. Zhang, J., Ge, H., Li, J., Chen, L., Wang, J., Cheng, B., et al. (2023). Effective regeneration of rat sciatic nerve using nanofibrous scaffolds containing rat ADSCs with controlled release of rhNGF and melatonin molecules for the treatment of peripheral injury model. *Regenerative Therapy, 24*, 180–189. https://www.sciencedirect.com/science/article/pii/S2352320423000573

62. Kohli, N., & Garcia-Gareta, E. (2020). Bioactive nanoceramics. In *Nanomaterials for theranostics and tissue engineering* (pp. 233–257). Elsevier.
63. Habraken, W., Habibovic, P., Epple, M., & Bohner, M. (2016). Calcium phosphates in biomedical applications: Materials for the future? *Materials Today, 19*(2), 69–87. https://doi.org/10.1016/j.mattod.2015.10.008
64. Hench, L. L. (2006). The story of bioglass. *Journal of Materials Science: Materials in Medicine, 17*(11), 967–978. https://doi.org/10.1007/s10856-006-0432-z
65. Li, H., Wu, Z., Zhou, Y., & Chang, J. (2019). Bioglass for skin regeneration. In E. García-Gareta (Ed.), *Biomaterials for skin repair and regeneration* (pp. 225–250). Woodhead Publishing. http://www.sciencedirect.com/science/article/pii/B978008102546800008X
66. Kaur, G., Pandey, O. P., Singh, K., Homa, D., Scott, B., & Pickrell, G. (2014). A review of bioactive glasses: Their structure, properties, fabrication and apatite formation. *Journal of Biomedical Materials Research Part A, 102*(1), 254–274. https://doi.org/10.1002/jbm.a.34690
67. Hoppe, A., Güldal, N. S., & Boccaccini, A. R. (2011). A review of the biological response to ionic dissolution products from bioactive glasses and glass-ceramics. *Biomaterials, 32*(11), 2757–2774. https://www.sciencedirect.com/science/article/pii/S0142961211000056?via%3Dihub
68. Sharifi, E., Sadati, S. A., Yousefiasl, S., Sartorius, R., Zafari, M., Rezakhani, L., et al. (2022). Cell loaded hydrogel containing ag-doped bioactive glass–ceramic nanoparticles as skin substitute: Antibacterial properties, immune response, and scarless cutaneous wound regeneration. *Bioengineering & Translational Medicine, 7*(3), e10386. https://doi.org/10.1002/btm2.10386
69. Cichos, S., Schätzlein, E., Wiesmann-Imilowski, N., Blaeser, A., Henrich, D., Frank, J., et al. (2023). A new 3D-printed polylactic acid-bioglass composite for bone tissue engineering induces angiogenesis in vitro and in ovo. *International Journal of Bioprinting, 9*(5), 751.
70. Jeyachandran, D., Murshed, M., Haglund, L., & Cerruti, M. A. (2023). Bioglass-poly(lactic-co-glycolic acid) scaffold@fibrin hydrogel construct to support endochondral bone formation. *Advanced Healthcare Materials, 12*(25), 2300211. https://doi.org/10.1002/adhm.202300211
71. Abd El-Aziz, A. M., Serag, E., Kenawy, M. Y., El-Maghraby, A., & Kandil, S. H. (2023). Hydrothermally reinforcing hydroxyaptatite and bioactive glass on carbon nanofiber scaffold for bone tissue engineering. *Frontiers in Bioengineering and Biotechnology, 11*, 1170097. https://doi.org/10.3389/fbioe.2023.1170097
72. Garcia-Gareta, E., Hua, J., Orera, A., Kohli, N., Knowles, J. C., & Blunn, G. W. (2018). Biomimetic surface functionalization of clinically relevant metals used as orthopaedic and dental implants. *Biomedical Materials, 13*(1), 015008.
73. Antoniac, I., Manescu, V., Paltanea, G., Antoniac, A., Nemoianu, I. V., Petrescu, M. I., et al. (2022). Additive manufactured magnesium-based scaffolds for tissue engineering. *Materials, 15*, 8693.
74. Dubey, N., Bentini, R., Islam, I., Cao, T., Castro Neto, A. H., & Rosa, V. (2015). Graphene: A versatile carbon-based material for bone tissue engineering. *Stem Cells International, 2015*, 804213. https://doi.org/10.1155/2015/804213
75. Chen, Y., & Li, X. (2022). The utilization of carbon-based nanomaterials in bone tissue regeneration and engineering: Respective featured applications and future prospects. *Medicine in Novel Technology and Devices, 16*, 100168. https://www.sciencedirect.com/science/article/pii/S2590093522000558
76. Eivazzadeh-Keihan, R., Maleki, A., de la Guardia, M., Bani, M. S., Chenab, K. K., Pashazadeh-Panahi, P., et al. (2019). Carbon based nanomaterials for tissue engineering of bone: Building new bone on small black scaffolds: A review. *Journal of Advanced Research, 18*, 185–201. https://www.sciencedirect.com/science/article/pii/S2090123219300773
77. Sola, A., Bertacchini, J., D'Avella, D., Anselmi, L., Maraldi, T., Marmiroli, S., et al. (2019). Development of solvent-casting particulate leaching (SCPL) polymer scaffolds as improved three-dimensional supports to mimic the bone marrow niche. *Materials Science*

and Engineering C, 96, 153–165. https://www.sciencedirect.com/science/article/pii/S0928493117340699
78. Thavornyutikarn, B., Chantarapanich, N., Sitthiseripratip, K., Thouas, G. A., & Chen, Q. (2014). Bone tissue engineering scaffolding: Computer-aided scaffolding techniques. *Progress in Biomaterials, 3*(2), 61–102. https://doi.org/10.1007/s40204-014-0026-7
79. Lu, T., Li, Y., & Chen, T. (2013). Techniques for fabrication and construction of three-dimensional scaffolds for tissue engineering. *International Journal of Nanomedicine, 8*, 337–350.
80. Dragusin, D. M., Van Vlierberghe, S., Dubruel, P., Dierick, M., Van Hoorebeke, L., Declercq, H. A., et al. (2012). Novel gelatin-PHEMA porous scaffolds for tissue engineering applications. *Soft Matter, 8*(37), 9589–9602.
81. Afghah, F., Dikyol, C., Altunbek, M., & Koc, B. (2019). Biomimicry in bio-manufacturing: Developments in melt electrospinning writing technology towards hybrid biomanufacturing. *Applied Sciences, 9*(17), 3540.
82. Reid, J. A., McDonald, A., & Callanan, A. (2020). Modulating electrospun polycaprolactone scaffold morphology and composition to alter endothelial cell proliferation and angiogenic gene response. *PLoS One, 15*(10), e0240332. https://doi.org/10.1371/journal.pone.0240332
83. Gautam, S., Sharma, C., Purohit, S. D., Singh, H., Dinda, A. K., Potdar, P. D., et al. (2021). Gelatin-polycaprolactone-nanohydroxyapatite electrospun nanocomposite scaffold for bone tissue engineering. *Materials Science and Engineering: C, 119*, 111588. https://www.sciencedirect.com/science/article/pii/S0928493120335062
84. Brown, T. D., Dalton, P. D., & Hutmacher, D. W. (2011). Direct writing by way of melt electrospinning. *Advanced Materials, 23*(47), 5651–5657. https://doi.org/10.1002/adma.201103482
85. Dalton, P. D. (2017). Melt electrowriting with additive manufacturing principles. *Current Opinion in Biomedical Engineering, 2*, 49–57. https://www.sciencedirect.com/science/article/pii/S2468451117300156
86. Kade, J. C., & Dalton, P. D. (2021). Polymers for melt electrowriting. *Advanced Healthcare Materials, 10*(1), 2001232. https://doi.org/10.1002/adhm.202001232
87. Wang, Y., Zhang, Y., Zhang, Z., Su, Y., Wang, Z., Dong, M., et al. (2020). An injectable high-conductive bimaterial scaffold for neural stimulation. *Colloids Surfaces B Biointerfaces, 195*, 111210. https://www.sciencedirect.com/science/article/pii/S092777652030566X
88. Fazeli, N., Arefian, E., Irani, S., Ardeshirylajimi, A., & Seyedjafari, E. (2021). 3D-printed PCL scaffolds coated with nanobioceramics enhance osteogenic differentiation of stem cells. *ACS Omega, 6*(51), 35284–35296. https://doi.org/10.1021/acsomega.1c04015
89. Zarei, M., Shabani Dargah, M., Hasanzadeh Azar, M., Alizadeh, R., Mahdavi, F. S., Sayedain, S. S., et al. (2023). Enhanced bone tissue regeneration using a 3D-printed poly(lactic acid)/Ti6Al4V composite scaffold with plasma treatment modification. *Scientific Reports, 13*(1), 3139. https://doi.org/10.1038/s41598-023-30300-z
90. Liu, S., Cheng, L., Liu, Y., Zhang, H., Song, Y., Park, J.-H., et al. (2023). 3D bioprinting tissue analogs: Current development and translational implications. *Journal of Tissue Engineering, 1*(14), 20417314231187110. https://doi.org/10.1177/20417314231187113
91. Gopinathan, J., & Noh, I. (2018). Recent trends in bioinks for 3D printing. *Biomaterials Research, 22*(1), 11. https://doi.org/10.1186/s40824-018-0122-1
92. Xie, M., Shi, Y., Zhang, C., Ge, M., Zhang, J., Chen, Z., et al. (2022). In situ 3D bioprinting with bioconcrete bioink. *Nature Communications, 13*(1), 3597. https://doi.org/10.1038/s41467-022-30997-y
93. MacAdam, A., Chaudry, E., McTiernan, C. D., Cortes, D., Suuronen, E. J., & Alarcon, E. I. (2022). Development of in situ bioprinting: A mini review. *Frontiers in Bioengineering and Biotechnology, 10*, 940896. https://doi.org/10.3389/fbioe.2022.940896
94. Urciuolo, A., Poli, I., Brandolino, L., Raffa, P., Scattolini, V., Laterza, C., et al. (2020). Intravital three-dimensional bioprinting. *Nature Biomedical Engineering, 4*(9), 901–915. https://doi.org/10.1038/s41551-020-0568-z

95. Ahmed, E. M. (2015). Hydrogel: Preparation, characterization, and applications: A review. *Journal of Advanced Research, 6*(2), 105–121. http://www.sciencedirect.com/science/article/pii/S2090123213000969
96. He, W., Reaume, M., Hennenfent, M., Lee, B. P., & Rajachar, R. (2020). Biomimetic hydrogels with spatial- and temporal-controlled chemical cues for tissue engineering. *Biomaterials Science, 8*, 3248–3269. https://doi.org/10.1039/D0BM00263A
97. Redaelli, F., Sorbona, M., & Rossi, F. (2017). Synthesis and processing of hydrogels for medical applications. In G. Perale & J. Hilborn (Eds.), *Bioresorbable polymers for biomedical applications* (pp. 205–228). Woodhead Publishing. https://www.sciencedirect.com/science/article/pii/B9780081002629000100
98. Dell, A. C., Wagner, G., Own, J., & Geibel, J. P. (2022). 3D bioprinting using hydrogels: Cell inks and tissue engineering applications. *Pharmaceutics, 14*, 2596.
99. Mancha Sánchez, E., Gómez-Blanco, J. C., López Nieto, E., Casado, J. G., Macías-García, A., Díaz Díez, M. A., et al. (2020). Hydrogels for bioprinting: A systematic review of hydrogels synthesis, bioprinting parameters, and bioprinted structures behavior. *Frontiers in Bioengineering and Biotechnology*, 8. https://doi.org/10.3389/fbioe.2020.00776
100. Unagolla, J. M., & Jayasuriya, A. C. (2020). Hydrogel-based 3D bioprinting: A comprehensive review on cell-laden hydrogels, bioink formulations, and future perspectives. *Applied Materials Today, 18*, 100479. https://www.sciencedirect.com/science/article/pii/S2352940719305980
101. Liu, S., Kilian, D., Ahlfeld, T., Hu, Q., & Gelinsky, M. (2023). Egg white improves the biological properties of an alginate-methylcellulose bioink for 3D bioprinting of volumetric bone constructs. *Biofabrication, 15*(2), 025013.

Chapter 6
Nanomaterials for Tissue Engineering and Regenerative Medicine

Abstract Many of the synergistic interactions between the three elements of the tissue engineering and regenerative medicine (TERM) paradigm occur at the nanoscale. Thus, nanotechnology becomes an ally of TERM to design effective strategies and therapies. The focus of this chapter is on the use and incorporation of nanomaterials in TERM strategies. These nanomaterials include nanoparticles, nanofibrous scaffolds and nanopatterned surfaces. The alliance between nanotechnology and bioprinting is of particular interest. The chapter ends with relevant conclusions as well as questions to underpin the knowledge acquired through the reading of this chapter.

6.1 Introduction

The tissue engineering and regenerative medicine (TERM) paradigm involves the combination of three elements, namely scaffolds, cells, and bio-chemical/physical cues, to produce tissue and organ analogues either for implantation or as models of tissue development and disease processes [1]. In recent decades, significant advances are being made in the field of TERM through notable progress in the development and understanding of the three TERM elements, as well as their synergistic interactions. Many of these interactions occur at the nanoscale, and examples are protein adsorption on biomaterial surfaces, essential for subsequent cell-scaffold interaction, cell adhesion to biomaterials, or interaction of cell surface receptors with relevant growth factors. Therefore, nanotechnology is certainly an ally of TERM in order to devise innovative and precise strategies for tissue regeneration.

Nanotechnology enables the development of nanomaterials that can be incorporated into TERM strategies. These include nanoparticles, nanofibers, and nanopatterned surfaces. Nanoparticles can be fabricated from various materials like ceramics, metals and polymers, and are being researched and proposed for a variety of applications in TERM [2, 3]. Nanofibers are biologically significant as the most relevant molecules that comprise the extracellular matrix (ECM) of tissues arrange

© The Author(s), under exclusive license to Springer Nature Switzerland AG 2026

E. García-Gareta, J. M. García-Aznar, *Contemporary Tissue Engineering and Regenerative Medicine*, Series in BioEngineering,
https://doi.org/10.1007/978-3-032-03441-0_6

in interwoven, nanoscaled fibrous structures. Since in vivo cells interact with these nanoscaled fibers, scaffolds that feature nanofibers are the focus of much study [4, 5]. Nanopatterned surfaces are very important to not only understand cell/scaffold surface interactions but also to control them, so functional TERM scaffolds and models can be built [6, 7]. Nanopatterned surfaces can be used to for example direct cell fate by controlling cell shape [8].

Particularly interesting is the combination of nanotechnology and 3D bioprinting, which have become an innovative tool for TERM applications [9]. As seen in the previous chapter, the choice of a biomaterial ink to achieve a suitable bioink is critical, with hydrogels dominating this space. Nevertheless, hydrogels present some important drawbacks such as weak mechanical properties or lack of structural integrity of the bioprinted scaffolds. Nanotechnology can be used to reinforce and customize hydrogel networks to advance the bioprinting technology for TERM strategies [10]. Specifically, nanomaterial reinforcement to yield a nanocomposite is the most used and advanced approach [10].

The aim of this chapter is to review the current nanomaterials used in TERM, their specific applications and the different fabrication approaches used to produce nanoparticles, nanofibers, and nanopatterned surfaces. The alliance between nanotechnology and bioprinting will also be discussed. The chapter will end with conclusions in this area of TERM as well as questions related to the content of this chapter.

6.2 Nanoparticles

Nanoparticles range in size from about 10 to 1000 nm and can be prepared either in solid or colloidal form [2] (Fig. 6.1). Importantly, they can be made from a wide range of materials, including ceramics, polymers, both natural and synthetic, metals and carbon-based materials [2, 3] (Fig. 6.1). The composition and distinctive advantages of nanoparticles make them highly preferred candidates in the TERM field [2, 3]. Advantages include high penetration ability and high surface area with adjustable surface properties [2]. Nanoparticles are widely used for imaging, enhancing mechanical strength, as supplements in bioinks, and as carriers for antimicrobial and bioactive agents [2].

6.2.1 Nanoceramics

Nanoceramics are perhaps the most utilized and studied family of nanoparticles and are made of ceramic materials, which are defined as inorganic, non-metallic, heat-resistant solids comprised of metallic and non-metallic elements [11]. Nanoceramics present good biocompatibility at low concentrations, can direct the differentiation of progenitor cells towards a specific phenotype, increase mechanical integrity, and

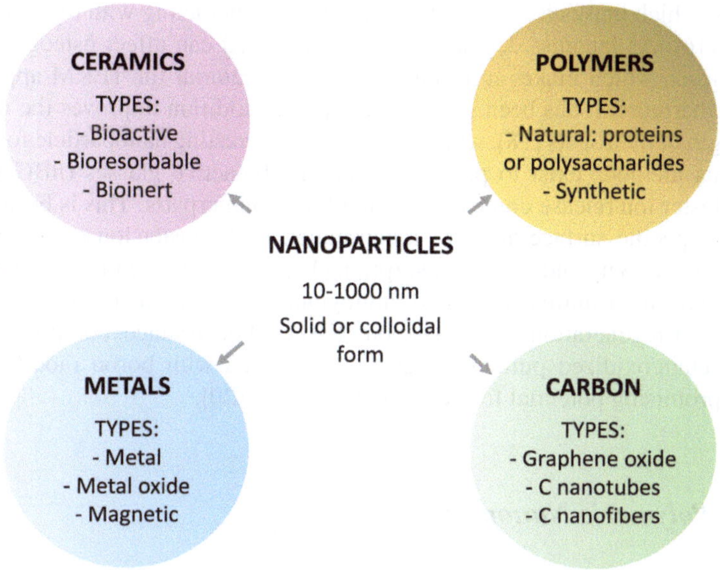

Fig. 6.1 Visual scheme of the different types of nanoparticles, based on composition, as well as the different types within each category

improve the degradation of the resulting scaffolds [10, 11]. Nanoceramics are generally classified in bioactive, e.g. bioactive glass nanoceramics, bioresorbable, e.g. calcium phosphates (CaPs) and laponite, and bioinert nanoceramics, such as titanium, alumina and zirconia-based materials [11].

CaPs are popular bioresorbable nanoceramics since they exhibit chemical compositions that closely resemble those of hard tissues such as bone, both skeletal and alveolar, or dentin [12]. Specifically, hydroxyapatite, a type of CaP, is one of the main components of the bone matrix along with collagen [12]. Therefore, it is widely used in scaffolds intended for bone regeneration [12, 13]. Besides, articular cartilage tissue presents a calcified zone [14]. Nanohydroxyapatite (nHAp) displays enhanced functional properties when compared with microhydroxyapatite because of surface reactivity, homogenous ultrafine structure and increased dissolution rate leading to improved bioactivity [11]. Tricalcium phosphate (TCPs) nanoceramics show similar bioactivity to nHAp. An example of research with CaP nanoparticles is the work by Gautam and co-workers, who aimed to fabricate an electrospun nanocomposite scaffold for bone tissue engineering using gelatin, polycaprolactone (PCL), and nHAp. X-ray diffraction (XRD), Fourier Transform Infrared (FTIR) spectroscopy, and thermogravimetric (TG) analysis confirmed the deposition of nHAp on the scaffold surface. Good viability and significant proliferation of human osteoblasts on the scaffold was demonstrated [15].

The silica-based, synthetic, disc-shaped, bioactive nanoceramic laponite presents the chemical formula of $Na^{+}_{0.7}[(Si_8Mg_{5.5}Li_{0.3})O_{20}(OH)_4]^{-}_{0.7}$ [16]. This laponite nanomaterial adopts a negative charge due to the release of sodium in aqueous solutions, which makes it very hydrophilic thereby interacting with biological components [16]. As laponite degrades, the ions released can affect osteogenesis and angiogenesis, which makes it an interesting nanomaterial for TERM applications [17]. Furthermore, it has been shown that laponite addition improves the mechanical strength of hydrogels [18], which makes it an interesting nanoparticle for TERM.

Various methods exist to produce nanoscale bioactive glasses (nBGs), which provide faster ion release compared to their bulk counterparts. This is because their enhanced specific surface area. As a consequence, the faster ion release provides enhanced bioactivity and protein adsorption [19]. The excellent bioactivity of nBGs coupled with their antibacterial and angiogenic properties make them promising materials for regeneration of various body tissues. For instance, composites of cellulose acetate/oxidized pullulan/gelatin incorporated with boron modified nBGs showed promising potential for dentin regeneration [20].

6.2.2 Polymeric Nanoparticles

Polymeric nanoparticles are classified attending to different criteria like composition, structure, or manufacturing method [2]. In terms of composition, various biodegradable and biocompatible natural and synthetic polymers have been used in the design and fabrication of polymeric nanoparticles [2]. Polymers used include proteins (e.g. gelatin, elastin, silk, albumin), polysaccharides (e.g. dextran, alginate, chitosan, hyaluronic acid, heparin, starch), and a vast choice of synthetic polymers including polyesters, polyamides, polyurethanes, or polyacrylates. These polymers have also been used in combination with specific functional moieties or other materials. Polymeric nanoparticles can come in different morphologies, which will influence their macroscopic behaviour. These morphologies include nanospheres, nanogels, polymeric micelles or nanocapsules. Regarding manufacturing methods, emulsification, self-assembly, nanoprecipitation, salting-out, dialysis, supercritical fluid technology, and microfluidic systems have been proposed and utilized [2].

An interesting example concerning nanoparticles made of the serum protein albumin and the anticoagulant polysaccharide heparin was provided by Zhang and colleagues, who prepared these nanoparticles for the encapsulation of the chemoattractant stromal cell-derived factor-1α (SDF-1α). This factor intervenes in the homing of bone marrow resident mesenchymal stem cells (BM-MSCs), and the authors showed a dose-dependent effect of SDF-1α over BM-MSCs migration [21]. With the purpose of encapsulating dexamethasone for its intracellular release for bone regeneration, Santo and co-workers designed a micelle from gelatin grafted with lactic acid oligomers that was pH responsive. These micelles were incorporated into gelatin hydrogels and seeded with rat BM-MSCs to create a tissue engineered construct for regeneration of bone that showed efficient formation of mineralized bone tissue up to 4 weeks after implantation of the construct in rat ulna [22]. A last example about a wound healing strategy is the work by Kheradvar et al., who designed a vitamin E delivery system based on starch nanoparticles incorporated in a nanofibrous scaffold of silk fibroin/PVA/aloe vera fabricated by electrospinning. Vitamin E has interesting features for wound healing such as antioxidant, anti-inflammatory and scar-prevention properties. Results showed increased cell viability and cell-matrix interactions due to the presence of the vitamin E-loaded starch nanoparticles and aloe vera [23].

6.2.3 Metallic Nanoparticles

Metallic nanoparticles can be classified into metal, metal oxide, and magnetic nanoparticles [3, 24].

Metal nanoparticles possess unique antibacterial properties, catalytic activity, mechanical strength, and electrical conductivity, making them particularly suitable for TERM applications [3]. For example, gold nanoparticles have excellent biocompatibility and ease of synthesis, whilst silver nanoparticles are the most widely used metallic nanoparticles due to their antimicrobial properties [2, 25]. Metal oxide nanoparticles are versatile and used for a variety of purposes. For instance, they are used as contrast agents in imaging applications, in biosensors, or drug delivery and catalysis applications [3]. The most important advantage of these metallic nanoparticles is their flexibility of structure, size, and surface chemistry [3]. Magnetic nanomaterials exhibit significant magnetism, influenced by applied magnetic fields, and are used in various applications such as targeted drug delivery, tissue growth enhancement, and reducing implant infections, among others [3]. Magnetic nanomaterials with superparamagnetic properties, e.g. magnetite (Fe_3O_4), spinel ferrites (AFe_2O_4), or maghemite (γ-Fe_2O_3), combine the characteristics of nanomaterials with magnetic responsiveness and superparamagnetic properties. These materials can localize in magnetic fields, absorb electromagnetic waves, and produce heat in alternating magnetic fields, making them suitable for delivering drugs to specific targets and for hyperthermia treatments to kill or inhibit cancer cells [3, 24].

The work by Ahmed and colleagues serves as an example on the use of metallic nanoparticles in TERM. In this study, the authors explored the potential of an injectable hydrogel, rich in liver ECM and enhanced with silver nanoparticles (AgNPs), for the treatment of acute liver failure, which results from severe liver damage and leads to a rapid decline in normal liver function. Intraperitoneal administration of liver ECM hydrogel, a collagen–AgNP mixture, or a liver ECM hydrogel–AgNP mixture was done in an acute liver failure model. Treatment with the liver ECM hydrogel–AgNP mixture significantly improved liver injury and promoted regeneration, demonstrating promising therapeutic and regenerative potential for treating liver injury [26].

A last example is the study by Xia and co-workers, where they used γFe_2O_3 and αFe_2O_3 superparamagnetic nanoparticles to develop novel CaP cement scaffolds that were seeded with human dental pulp stem cells (hDPSCs) for bone tissue engineering. Results showed that the osteogenic differentiation of hDPSCs was significantly enhanced by the incorporation of the superparamagnetic nanoparticles into the scaffold, which was attributed to the scaffold surface nanotopography as well as the internalization of nanoparticles released from the scaffold by the cells. The authors concluded that the developed scaffold incorporating superparamagnetic nanoparticles showed promise for applications where substantial enhancement of bone regeneration is needed [27].

6.2.4 Carbon-Based Nanoparticles

The revolutionary nanomaterial graphene, which consists of carbon atoms arranged in a honeycomb-like 2-dimensional sheet, shows multiple applications in the energy, microelectronics and biotechnology sectors [28, 29]. Properties of this wonder material include striking electrical conductivity, large surface area, and excellent mechanical strength [28]. For biomedical applications, graphene is barely biocompatible; however, its oxide formations display acceptable cytocompatibility at low concentrations [10]. Carbon nanotubes (CNTs) and carbon nanofibers (CNFs) show similar properties to those of graphene. Whilst CNTs are essentially cylindrical folded sheets of graphene monolayers, CNFs display a less canonical and more erratic stacking pattern [30]. Nevertheless, their particular geometry is useful in certain TERM applications, as geometry is important in functionalizing medical devices [30, 31].

An example of carbon-based nanoparticles in TERM is the work by Kumar and colleagues on the development of polyethyleneimine (PEI)/graphene oxide (GO)/polycaprolactone (PCL) composite scaffolds for bone tissue engineering. The composite scaffolds encouraged adhesion and proliferation of human mesenchymal stem cells and induced osteogenic differentiation of these cells. Furthermore, the authors reported that the composite scaffolds exhibited intrinsic bactericidal activity [32]. Another example is the study by Liang et al., who reported that nHAp/collagen/poly(lactic-co-glycolic acid) (PLGA)/GO composite scaffolds could significantly stimulate adhesion and proliferation of MC3T3-E1 osteoprogenitor cells. In terms of material properties, addition of 1.5 wt% GO increased the mechanical strength of the scaffold. Furthermore, the wettability of the scaffolds also improved due to the presence of more hydroxyl functional groups in the GO [33].

6.3 Nanofibers

Nanofibers are prominent nanomaterials employed in scaffolds for TERM applications [4, 5]. This is due to their resemblance with the natural nanofibers that comprise the ECM [5]. ECM molecules typically exhibit intricate fibrous arrangements on the nanoscale, facilitating cell adhesion and bioactivity [5]. Consequently, the development of scaffolds mirroring ECM architecture has emerged as an active domain within TERM. Fibrous scaffolds can be fabricated using polymer self-assembly, phase separation, and, particularly, electrospinning [5, 34].

6.3.1 Fabrication of Nanofibrous Scaffolds by Polymer Self-assembly

Self-assembly is considered a powerful tool for obtaining supramolecular materials. Polymer self-assembly typically involves peptides associating intermolecularly, thereby forming organized, well-defined, and stable structures through non-covalent forces like van der Waals interactions, electrostatic forces, hydrogen bonds, and π–π stacking [5, 35]. The combination of these weak forces determines the structural conformation and stability of the assembly, significantly influencing its interaction with other molecules, cells, and tissues [35]. Designing peptides with chemical and structural compatibility is crucial for self-assembly, whilst electrolytes in the solution can drive peptide self-assembly, which results from reduced electrostatic repulsion between similarly charged peptide monomers due to electrolyte ions. This reduction enables peptides to come closer together, enhancing hydrophobic interactions and forming stable nanofibers. The further organization of nanofibers into 3D structures leads to hydrogel formation, where various factors, including the amino acid sequence and charge, peptide solution concentration, electrolyte ion concentration and type, pH, and temperature, influence the mechanical properties of the resulting hydrogel [5, 35]. Self-assembling peptide hydrogels serve as an ideal platform for TERM due to their ease of use, non-immunogenicity, non-toxicity and biodegradability. Importantly, their degradation products can be metabolized by cells [5, 35]. However, self-assembled nanofibers require complex processing, have low productivity, high cost, and inferior mechanical properties compared with nanofibers made using other techniques. An example is the work by Rufaihah and colleagues, who reported that nanofiber scaffolds fabricated from synthetic glycosaminoglycan (GAG) mimetic peptides were capable of inducing neovascularization in cardiovascular tissue after myocardial infarction. These scaffolds did not contain any growth factors or therapeutic cells and exerted their neovascularization capabilities on their own [36].

6.3.2 Fabrication of Nanofibrous Scaffolds by Phase Separation

This nanofiber fabrication method involves polymer dissolution, gelation, solvent extraction from the gel using water, and freeze-drying [5]. Initially, a polymer is dissolved in a solvent at a high temperature. Then, through cooling or non-solvent exchange, a liquid–liquid or solid–liquid phase separation of the polymer is induced, whereby the polymer becomes thermodynamically insoluble leading to the formation of polymer-rich phases within the solvent. After the solvent is extracted, the scaffold is frozen to preserve its structure and freeze-dried to produce highly porous fibrous polymeric scaffolds with micro-spherical pores [5]. Various factors, including the type of polymer and solvent, gelation temperature and time, or thermal treatment, are known to affect the morphology of nanofibers [5]. Nevertheless, the phase separation method offers simplicity, low cost, and ease of fabrication. Besides, utilizing specific and computer-aided designed moulds, this method allows for the fabrication of 3D scaffolds with precise dimensions and complex structures. On the other hand, the method presents long processing times, laboratory-scale nanofiber production, nanofiber structure instability, difficulty in controlling porosity, and limitation to polymers that can undergo phase separation [5].

A relevant example is the work by Zhao and co-workers, where they used a dual thermally induced phase separation technique to fabricate a biomimetic nanofibrous poly(l-lactide) scaffold which was decorated by a chitosan nanofiber network inside the macropores. With the first phase separation, a nanofibrous poly(l-lactide) scaffold with interconnected macropores was obtained. With the second phase separation, chitosan nanofibers of approximately 500 nm in diameter were incorporated into the macropores. The degradation rate of the scaffold was slowed down by the chitosan nanofibers. Moreover, when mesenchymal stem cells were cultured on the composite scaffolds, the cells would rather attach and stretch along the chitosan nanofibers, thereby showing the advantage of adding chitosan nanofibers to the scaffold [37].

6.3.3 Fabrication of Nanofibrous Scaffolds by Electrospinning

Electrospinning is, without a doubt, the most employed technique in the fabrication of nanofibrous scaffolds [5, 34, 38]. This technique offers high porosity, a large surface area, and an adjustable pore size distribution in nanofibrous scaffolds [34].

As already explained in the previous chapter, this fabrication technique applies a high-voltage electric field to a polymer solution or melt, whereby ultrafine filaments are produced and deposited on a surface collector, until eventually a fibrous scaffold is obtained [5, 34]. More in detail, an electric field is generated between a counter electrode and a positively charged spinneret that is filled with a polymer solution or melt. Since the electrostatic charge is larger than the surface tension of the solution, a polymer jet is formed at the metal needle that moves through the charged spinneret to the counter electrode. During this process, the solvent evaporates thereby forming continuous polymer fibers with diameters in the nano or microscale [5, 34]. Various parameters influence the diameter and morphology of the fabricated electrospun nanofibers (Table 6.1). These parameters include properties of the polymeric solution or melt such as concentration, viscosity, polymer chains entanglement, conductivity, or surface tension. There are also parameters related to the electrospinning process, including the applied voltage, the existing distance between the

Table 6.1 Electrospinning parameters that affect fiber morphology. Table compiled from the works by Zhang et al. [34] and Abu Owida et al. [38]

Category	Parameter	Effect on fiber morphology
Properties of the polymer solution or melt	Concentration	Fiber diameter increases with increasing concentration
	Viscosity	Increasing viscosity results in increased fiber diameter
	Entanglement of the polymer chains	Associated with polymer molecular weight and solution concentration An optimal level of chain entanglement promotes a stable jet during solvent evaporation leading to fibrous structures. If it is not optimal, beads, or not-uniform fibers are formed
	Conductivity	In general, higher conductivities result in smaller fibers Increasing conductivity facilitates the production of bead-free fibers
	Surface tension	As the surface tension increases, the quantity of beads increases
Parameters related to the electrospinning process	Applied voltage	Inconclusive results regarding the relationship between applied voltage and fiber diameter
	Distance between the nozzle and the surface collector	A minimum distance is needed to acquire dry fibers Beading is observed at either too short or too long distances
	Flow rate of the solution	Fibers with smaller diameters are produced at lower flow rates High flow rates may result in fibers that are not dried when they arrive at the surface collector
Environmental conditions	Temperature	As the temperature increases, the viscosity of the solution decreases, resulting in fibers with smaller diameters
	Humidity	Increasing humidity causes the development of circular pores in the fibers

nozzle and the surface collector, or the flow rate of the polymer solution, and environmental conditions like temperature or humidity [34]. For instance, the viscosity of the solution decreases with increasing temperature, which results in fibers with a smaller diameter. A wide range of materials that includes synthetic and natural polymers, and composite systems are compatible with the electrospinning fabrication technique. The literature shows that the size and the morphology of the electrospun nanofibers that result from these materials are controlled by tuning the material viscosity and conductivity, operational parameters like voltage or needle-collector distance, and environmental factors like humidity, and temperature, among various parameters [34, 38] (Table 6.1).

A wealth of literature exists on electrospun nanofibers for TERM applications [34, 38–40]. An example is the work by Malekpour et al. in periodontal tissue engineering [40]. Since statins positively affect the proliferation and regeneration of periodontal tissues, simvastatin-loaded poly (lactic-co-glycolic acid) (SIM-PLGA-NF) nanofibers were prepared by electrospinning. The authors determined the optimal conditions for preparing SIM-PLGA-NF using a 2^3 factorial design. The optimized nanofibers with a diameter of 640.2 ± 32.5 nm and a simvastatin entrapment efficiency of 99.6 ± 1.5% were surface-modified with a 1% w/v hyaluronic acid solution (1%HA-SIM-PLGA-NF) to enhance their compatibility with fibroblasts. The 1%HA-SIM-PLGA-NF exhibited a uniform, bead-free, interwoven morphology similar to the ECM. The mechanical performance and simvastatin release profile of SIM-PLGA-NFs were significantly improved after coating with hyaluronic acid. In vitro cell work indicated that fibroblast cell adhesion, proliferation, and differentiation were positively enhanced on the surface of 1%HA-SIM-PLGA-NF. The authors concluded that 1%HA-SIM-PLGA-NFs have potential as a scaffold for periodontal tissue engineering [40].

Another example is the work by Davoudabadi and co-workers where in order to create a healing-accelerating antibacterial wound dressing, various electrospun polyurethane (PU)-based nanofibers containing *Calendula officinalis* and Propolis ethanolic extracts were developed [39]. The minimum inhibitory concentrations against Methicillin-resistant *Staphylococcus aureus* (MRSA) were measured to determine the appropriate concentrations of the extracts. The produced nanofibers—polyurethane (PU), polyurethane/Calendula officinalis (PU/CO), polyurethane/Propolis (PU/PR), and polyurethane/Calendula officinalis/Propolis (PU/CO/PR)—were evaluated in vivo in terms of their antibacterial properties and wound-healing efficacy. The electrospun PU/CO/PR nanofiber exhibited the highest levels of antibacterial, antioxidant, and cell proliferation efficiency. In vivo testing and histological analysis demonstrated significant improvement in the complete, well-organized wound-healing process in MRSA-infected wounds treated with PU/CO/PR. These results suggested that PU/CO/PR electrospun nanofibers can serve as an effective wound dressing, significantly enhancing wound healing with notable antibacterial, antioxidant, and cell proliferation properties [39]. This work by Davoudabadi and colleagues is an interesting example of the combination of different materials to render nanofibrous scaffolds for wound dressing applications, where the research on electrospun nanofibers containing plant extracts with wound healing activity, such as *Centella asiatica*, green tea, aloe vera, chamomile, grape seed, or *Calendula officinalis,* is very active [38].

6.4 Nanopatterned Surfaces

As we will see in the following chapters of this book, cell-scaffold interactions at the surface level are key to many biological phenomena related to tissue regeneration [6]. Therefore, knowledge, understanding and control of these interactions is vital in order to design functional tissue substitutes either for implantation or as models [41]. It is important to consider that these interactions occur at the nanoscale and therefore, research into nanoscaled surface features is continually active since in our body cells interact and recognize topographical features and changes that occur in the ECM, of which two types exist, namely the basement membrane and the interstitial ECM [6, 7, 41, 42]. As an example, the basement membrane of endothelium or epithelium displays nanopores, nanofibers, and nanoridges [43]. In the interstitial ECM, which is a porous and interconnecting three-dimensional network, collagen fibers contain tropocollagen that arranges into microfibrils with a periodicity of 66 nm [44].

The so-called nanopatterned surfaces allow manipulation of cell behaviour in a controlled manner [6, 7, 42]. In the field of TERM, there are basic nanopatterned geometries that are often used to investigate cell attachment, proliferation, differentiation, and phenotype in vitro. These geometries include nanogrooves, nanopillars and nanopits [42]. Fabrication techniques of these surfaces include photolithography, electron beam lithography, soft lithography and microcontact printing, hot embossing, or microfluidics [6]. Next, we will discuss some examples of this important area of TERM research.

In this early work by Yim et al. [8], the behaviour of bovine pulmonary artery smooth muscle cells (SMC), including their morphology, proliferation, and motility, was investigated on poly(methyl methacrylate) (PMMA) and poly(dimethylsiloxane) (PDMS) surfaces with nanopatterned gratings of 350 nm linewidth, 700 nm pitch, and 350 nm depth. Results showed that over 90% of the cells aligned with the gratings and were significantly more elongated compared to SMC cultured on non-patterned surfaces. The nuclei were also elongated and aligned. Cell proliferation was notably reduced on the nanopatterned surfaces. The polarization of microtubule organizing centers (MTOC), which play a role in cell migration, in SMC cultured on nanopatterned surfaces showed a preference for the axis of cell alignment in an in vitro wound healing assay. Conversely, the MTOC of SMC on non-patterned surfaces tended to polarize towards the wound edge. The authors proposed that this nanoimprinting technology offered a valuable platform for studying cell-substrate interactions and for the development of medical devices featuring nanoscale characteristics [8].

Since surface physical features of biomaterials are known to influence macrophage cell shape and phenotype, thereby offering opportunities to design novel 'immune-instructive' topographies to modulate foreign body response, the study by Li and co-workers utilized nanopatterned PDMS substrates with well-defined nanopillars and nanopits to evaluate the response of RAW264.7 macrophages to different feature sizes [45] (Fig. 6.2). Macrophages on small nanopillars (SNPLs; 450 nm diameter, 300 nm average edge-to-edge spacing) exhibited larger, more spread-out morphologies (Fig. 6.3). Increasing the interpillar distance to 800 nm in large nanopillars (LNPLs) resulted in macrophages displaying morphologies similar to those on flat control surfaces (Fig. 6.3). Macrophages on nanopits (NPTs; 150 nm depth, 800 nm average edge-to-edge spacing) showed significant elongation (Fig. 6.3). This elongation and spreading led to the expression of anti-inflammatory/

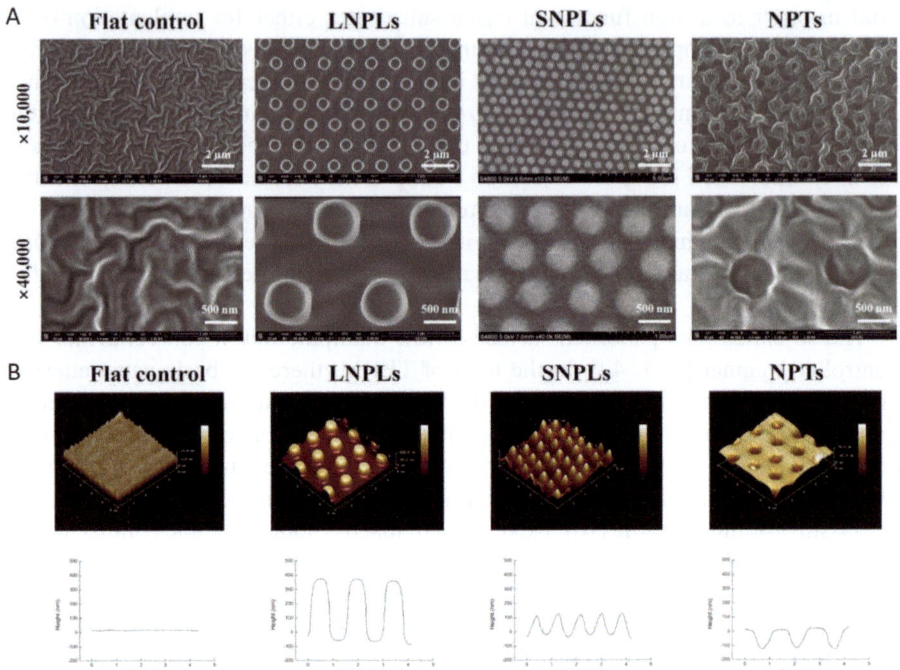

Fig. 6.2 Surface characterization of the nanopatterned and flat substrates, where LNPLs is large nanopillars, SNPLs is small nanopillars and NPTs is nanopits. (**a**) Scanning electros microscopy (SEM) images of the surfaces. (**b**) Atomic force microscopy (AFM) images of the surfaces with 3D images (top) and height profiles (bottom). Image from Li et al. [45]. Open access article distributed under the terms and conditions of the Creative Commons Attribution (CC BY) license

pro-healing (M2) phenotypic markers and reduced expression of inflammatory cytokines. Both SNPLs and NPTs, with high availability of fibronectin integrin binding regions, facilitated integrin β1 expression and enhanced focal adhesion formation. Increased integrin β1 expression in macrophages on SNPLs and NPTs was necessary for activating the PI3K/Akt pathway, promoting cell spreading, and negatively regulating NF-κB activation, as indicated by similar rounded cell shapes and higher NF-κB levels after PI3K inhibition. These findings suggested that changes in macrophage cell shape due to nanotopographical features may provide essential cues for directing macrophage phenotype [45].

Pluripotent stem cells (PSCs) are capable of self-renewal and eventually develop into all cells of the adult body [46]. The regulation of human PSCs behaviour has primarily been studied through biochemical factors. However, biophysical signals in the stem cell microenvironment, including nanotopographical cues, can significantly regulate adult stem cell behaviours such as self-renewal and differentiation [47]. In their study, Chen and colleagues utilized a newly developed, large-scale nanofabrication technique based on reactive-ion etching to create random nanoscale structures on glass surfaces with high precision and reproducibility [47]. The

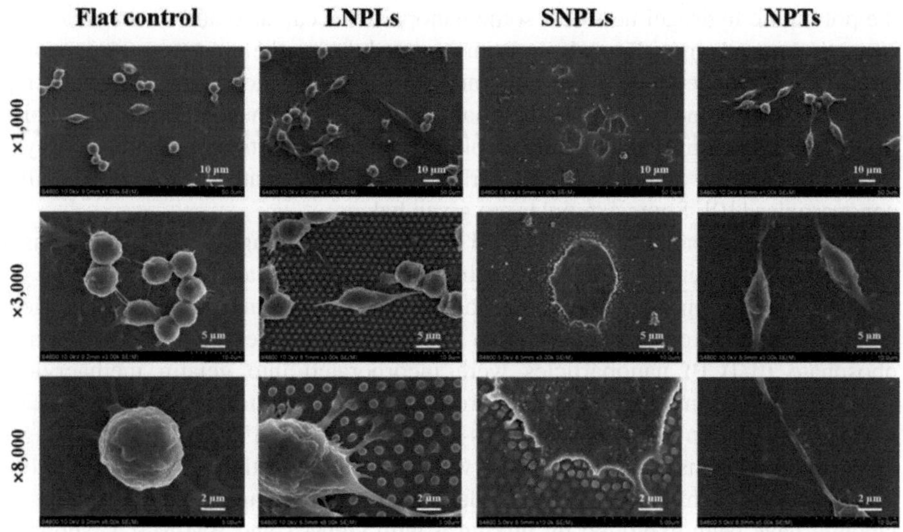

Fig. 6.3 SEM images of RAW264.7 macrophages on the nanopatterned and flat surfaces after 1 day of incubation. LNPLs is large nanopillars, SNPLs is small nanopillars and NPTs is nanopits. Image from Li et al. [45]. Open access article distributed under the terms and conditions of the Creative Commons Attribution (CC BY) license

authors found that human PSCs are responsive to nanotopographical cues, and this sensitivity can be harnessed to enhance directed neuronal differentiation. Specifically, they demonstrated that nanoengineered topographic substrates can promote early neuroepithelial conversion and motor neuron (MN) progenitor differentiation of human PSCs. Additionally, they investigated how human PSCs detect substrate nanotopography and transmit this biophysical signal through a regulatory network involving cell adhesion, the actomyosin cytoskeleton, and Hippo/YAP signalling to facilitate neuroepithelial induction. The study showed an efficient method for large-scale production of MNs from human PSCs, which is valuable for TERM applications [47].

6.5 Nanomaterials in Bioinks for 3D Bioprinting

As mentioned in the introduction of this chapter, nanotechnology and bioprinting can be combined to produce superior bioinks [9]. Nanomaterial reinforcement of hydrogels is a common strategy since the high surface area, unique shape and size of nanomaterials can significantly influence the properties of the hydrogel network, which allows tailoring of its functionalities [10]. Nanoparticles are of particular interest for optimizing the material properties of hydrogels by enhancing surface interactions between the nanoparticles and the hydrogel polymer network. Other than the enhanced mechanical strength that the addition of nanoparticles confers to the polymeric hydrogel network, some nanoparticles can also add further features. Nevertheless, adverse effects have been reported due to the presence of nanoparticles, such as reduced biocompatibility and slower degradation rates of the resulting scaffolds [10]. Some of the nanomaterials that are used in nanocomposites for bioprinting are bioactive and bioresorbable ceramic nanoparticles, natural and synthetic polymeric nanoparticles, carbon-based nanomaterials, and metallic nanoparticles [10]. Some examples are given below.

Nanohydroxyapatite (nHAp) is used for bone and cartilage 3D bioprinting, as it appropriately disperses in aqueous solutions and has a relatively low cost. As an example, it was combined with the polysaccharides gellan and alginate to yield bioinks for chondrogenesis [48]. Also for cartilage regeneration, beta-tricalcium phosphate (β-TCP) nanoparticles were incorporated into gelatin methacryloyl (GelMA) and alginate with bone marrow mesenchymal stem cells (BM-MSCs), where upregulation of chondrogenesis-related factors and presence of cartilage-associated collagens was observed, as well as greater printing shape fidelity [49]. Since bone is a highly vascularized tissue, the potential pro-angiogenic properties of laponite have been exploited by mixing it with GelMA and BM-MSCs, showing that the material supported osteogenesis and angiogenesis of the construct [17]. Another study investigated strontium-doped GelMA nanocomposite bioinks with human mesenchymal stromal cells (hMSCs) for bone regeneration [50].

The excellent electroconductivity and mechanical properties of graphene can affect the phenotype of progenitor cells and aid in the optimization of the bioprinting process [51]. CNTs and CNFs have similar properties to those of graphene. Furthermore, their shape and architecture are useful in various TERM applications, including bioprinting. An interesting study by Izadifar and colleagues explored the bioprinting of a nanoreinforced hybrid cardiac patch for myocardial tissue engineering. The authors employed a hybrid 3D printing/3D bioprinting fabrication technique with methacrylated collagen, alginate and CNTs. The aim was to fabricate a multilayered composite, which avoided direct contact of the cells with the CNTs to avoid cytotoxicity issues, with enhanced mechanical and electrical properties of the final composite [52].

Finally, Zengin and co-workers developed a novel type of nanocomposite bioink that included thiolated mesoporous silica nanoparticles (MSNs) as active fillers within norbornene-functionalized alginate hydrogels [53]. The MSNs rapidly formed covalent crosslinks within the hydrogels when exposed to UV light. By varying the concentration of MSNs, the mechanical properties of the gels were adjusted, ranging from 9.3 to 19.7 kPa. The effectiveness of MSNs in covalently crosslinking polymer networks was notably affected by the polymer architecture and the number of functional groups present. By modifying the outer surface of MSNs with matrix metalloproteinase (MMP) sensitive peptides, they developed bioinks that respond to proteinase K and MMP-9 enzymes, making them biodegradable. Incorporating cysteine-modified RGD peptides further enhanced cell-matrix interactions and reduced gelation time, facilitating bioprinting. Encapsulated MG63 cells remained viable and proliferated within the bioinks. The authors concluded that the developed nanocomposite bioinks showed promise for creating bioprinted constructs with adjustable mechanical and degradation properties [53].

6.6 Conclusions

Many of the synergistic interactions between scaffolds, cells and bio-chemical/physical cues occur at the nanoscale, like protein adsorption on the scaffold surface, cell adhesion to scaffolds, or interaction of cell surface receptors with growth factors of interest. This makes nanotechnology an important ally of TERM to devise innovative, effective, and precise strategies for tissue regeneration. Nanotechnology enables the development of nanomaterials that include nanoparticles, nanofibers, and nanopatterned surfaces. Nanoparticles are fabricated from ceramics, polymers, metals or carbon-based materials for a variety of applications in TERM. Nanofibrous scaffolds are biologically significant as the molecules of the ECM arrange in interwoven, nanoscaled fibrous structures. Electrospun nanofibrous scaffolds are particularly relevant. Nanopatterned surfaces are very important to understand and control cell/scaffold surface interactions, so functional TERM scaffolds and models can be developed.

The combination of nanotechnology and 3D bioprinting has become an innovative tool to reinforce and customize hydrogel networks to advance the bioprinting technology for TERM applications. Nanomaterial reinforcement to yield a nanocomposite is the most used and advanced approach.

This chapter offers an overview of the current nanomaterials used in TERM, their applications and fabrication approaches used to produce nanoparticles, nanofibers, and nanopatterned surfaces, with a discussion of the alliance between nanotechnology and bioprinting.

Questions

- Name the three types of nanomaterials used in TERM.
- What type of materials are used in the fabrication and development of nanoparticles?
- How are nanoceramics usually classified?
- Name three examples of nanoceramics.
- How can metallic nanoparticles be classified into?
- Name three types of carbon-based nanoparticles
- Name three techniques for fabricating nanofibrous scaffolds.
- What parameters influence the diameter and morphology of the fabricated electrospun nanofibers?
- Name up to three basic nanopatterned geometries often used to investigate cell attachment, proliferation, differentiation, and phenotype in vitro.
- What are the advantages of using nanomaterial reinforcement of hydrogels for 3D bioprinting?

References

1. Langer, R., & Vacanti, J. P. (1993). Tissue engineering. *Science, 260*, 920. https://doi.org/10.1126/science.8493529
2. Fathi-Achachelouei, M., Knopf-Marques, H., Ribeiro da Silva, C. E., Barthès, J., Bat, E., Tezcaner, A., & Vrana, N. E. (2019). Use of nanoparticles in tissue engineering and regenerative medicine. *Frontiers in Bioengineering and Biotechnology, 7*, 113. https://doi.org/10.3389/fbioe.2019.00113
3. Zheng, X., Zhang, P., Fu, Z., Meng, S., Dai, L., & Yang, H. (2021). Applications of nanomaterials in tissue engineering. *RSC Advances, 11*, 19041–19058. https://doi.org/10.1039/D1RA01849C
4. Sharma, R., Kumar, S., Bhawna, A., Gupta, N., Dheer, P., Jain, P., & Singh, V. K. (2022). An insight of nanomaterials in tissue engineering from fabrication to applications. *Tissue Engineering and Regenerative Medicine, 19*, 927–960. https://doi.org/10.1007/s13770-022-00459-z
5. Nemati, S., Kim, S., Shin, Y. M., & Shin, H. (2019). Current progress in application of polymeric nanofibers to tissue engineering. *Nano Convergence, 6*, 36. https://doi.org/10.1186/s40580-019-0209-y
6. Ermis, M., Antmen, E., & Hasirci, V. (2018). Micro and Nanofabrication methods to control cell-substrate interactions and cell behavior: A review from the tissue engineering perspective. *Bioactive Materials, 3*, 355–369. https://doi.org/10.1016/j.bioactmat.2018.05.005
7. Barlian, A., & Vanya, K. (2022). Nanotopography in directing osteogenic differentiation of mesenchymal stem cells: Potency and future perspective. *Future Science OA, 8*, FSO765. https://doi.org/10.2144/fsoa-2021-0097
8. Yim, E. K. F., Reano, R. M., Pang, S. W., Yee, A. F., Chen, C. S., & Leong, K. W. (2005). Nanopattern-induced changes in morphology and motility of smooth muscle cells. *Biomaterials, 26*, 5405–5413. https://doi.org/10.1016/j.biomaterials.2005.01.058
9. Cai, Y., Chang, S. Y., Gan, S. W., Ma, S., Lu, W. F., & Yen, C.-C. (2022). Nanocomposite bioinks for 3D bioprinting. *Acta Biomaterialia, 151*, 45–69. https://doi.org/10.1016/j.actbio.2022.08.014
10. Loukelis, K., Helal, Z. A., Mikos, A. G., & Chatzinikolaidou, M. (2023). Nanocomposite bioprinting for tissue engineering applications. *Gels, 9*, 103. https://doi.org/10.3390/gels9020103
11. Kohli, N., & Garcia-Gareta, E. (2020). Bioactive nanoceramics. In *Nanomaterials for theranostics and tissue engineering* (pp. 233–257). Elsevier. https://doi.org/10.1016/B978-0-12-817838-6.00009-7
12. LeGeros, R. Z. (2008). Calcium phosphate-based osteoinductive materials. *Chemical Reviews, 108*, 4742–4753. https://doi.org/10.1021/cr800427g
13. García-Gareta, E., Coathup, M. J., & Blunn, G. W. (2015). Osteoinduction of bone grafting materials for bone repair and regeneration. *Bone, 81*, 112–127. https://doi.org/10.1016/j.bone.2015.07.007
14. Poh, P. S. P., Woodruff, M. A., & García-Gareta, E. (2020). Polymer-based composites for musculoskeletal regenerative medicine. In N. E. Vrana, H. Knopf-Marques, & J. Barthes (Eds.), *Biomaterials for organ and tissue regeneration* (pp. 33–82). Woodhead Publishing. https://doi.org/10.1016/B978-0-08-102906-0.00003-9
15. Gautam, S., Sharma, C., Purohit, S. D., Singh, H., Dinda, A. K., Potdar, P. D., Chou, C.-F., & Mishra, N. C. (2021). Gelatin-polycaprolactone-nanohydroxyapatite electrospun nanocomposite scaffold for bone tissue engineering. *Materials Science and Engineering: C, 119*, 111588. https://doi.org/10.1016/j.msec.2020.111588
16. Stealey, S. T., Gaharwar, A. K., & Zustiak, S. P. (2023). Laponite-based nanocomposite hydrogels for drug delivery applications. *Pharmaceuticals, 16*, 821. https://doi.org/10.3390/ph16060821
17. Cidonio, G., Alcala-Orozco, C. R., Lim, K. S., Glinka, M., Mutreja, I., Kim, Y.-H., Dawson, J. I., Woodfield, T. B. F., & Oreffo, R. O. C. (2019). Osteogenic and angiogenic tissue formation

in high fidelity nanocomposite Laponite-gelatin bioinks. *Biofabrication, 11*, 035027. https://doi.org/10.1088/1758-5090/ab19fd
18. Alamán-Díez, P., Borau, C., Enrique Guerrero, P., Amaveda, H., Mora, M., Fraile, J. M., García-Gareta, E., García-Aznar, J. M., & Pérez, M. A. (2023). Collagen-laponite nanoclay hydrogels for tumor spheroid growth. *Biomacromolecules, 24*, 2879–2891.
19. Boccaccini, A. R., Erol, M., Stark, W. J., Mohn, D., Hong, Z., & Mano, J. F. (2010). Polymer/bioactive glass nanocomposites for biomedical applications: A review. *Composites Science and Technology, 70*, 1764–1776. https://doi.org/10.1016/j.compscitech.2010.06.002
20. Moonesi Rad, R., Pazarçeviren, E., Ece Akgün, E., Evis, Z., Keskin, D., Şahin, S., & Tezcaner, A. (2019). In vitro performance of a nanobiocomposite scaffold containing boron-modified bioactive glass nanoparticles for dentin regeneration. *Journal of Biomaterials Applications, 33*, 834–853. https://doi.org/10.1177/0885328218812487
21. Zhang, H., Yu, S., Zhao, X., Mao, Z., & Gao, C. (2018). Stromal cell-derived factor-1α-encapsulated albumin/heparin nanoparticles for induced stem cell migration and intervertebral disc regeneration in vivo. *Acta Biomaterialia, 72*, 217–227. https://doi.org/10.1016/j.actbio.2018.03.032
22. Santo, V. E., Ratanavaraporn, J., Sato, K., Gomes, M. E., Mano, J. F., Reis, R. L., & Tabata, Y. (2015). Cell engineering by the internalization of bioinstructive micelles for enhanced bone regeneration. *Nanomedicine, 10*, 1707–1721. https://doi.org/10.2217/nnm.15.11
23. Kheradvar, S. A., Nourmohammadi, J., Tabesh, H., & Bagheri, B. (2018). Starch nanoparticle as a vitamin E-TPGS carrier loaded in silk fibroin-poly(vinyl alcohol)-Aloe vera nanofibrous dressing. *Colloids and Surfaces. B, Biointerfaces, 166*, 9–16. https://doi.org/10.1016/j.colsurfb.2018.03.004
24. Dasari, A., Xue, J., & Deb, S. (2022). Magnetic nanoparticles in bone tissue engineering. *Nanomaterials, 12*, 757. https://doi.org/10.3390/nano12050757
25. García-Gareta, E., Davidson, C., Levin, A., Coathup, M. J., & Blunn, G. W. (2016). Biofilm formation in total hip arthroplasty: Prevention and treatment. *RSC Advances, 6*, 80244. https://doi.org/10.1039/c6ra09583f
26. Ahmed, E., Saleh, T., Yu, L., Song, S.-H., Park, K.-M., Kwak, H.-H., & Woo, H.-M. (2020). Decellularized ECM-rich hydrogel–silver nanoparticles mixture as a potential treatment for acute liver failure model. *Journal of Biomedical Materials Research Part A, 108*, 2351–2367. https://doi.org/10.1002/jbm.a.36988
27. Xia, Y., Chen, H., Zhang, F., Wang, L., Chen, B., Reynolds, M. A., Ma, J., Schneider, A., Gu, N., & Xu, H. H. K. (2018). Injectable calcium phosphate scaffold with iron oxide nanoparticles to enhance osteogenesis via dental pulp stem cells. *Artificial Cells, Nanomedicine, and Biotechnology, 46*, 423–433. https://doi.org/10.1080/21691401.2018.1428813
28. Randviir, E. P., Brownson, D. A. C., & Banks, C. E. (2014). A decade of graphene research: Production, applications and outlook. *Materials Today, 17*, 426–432. https://doi.org/10.1016/j.mattod.2014.06.001
29. Dubey, N., Bentini, R., Islam, I., Cao, T., Castro Neto, A. H., & Rosa, V. (2015). Graphene: A versatile carbon-based material for bone tissue engineering. *Stem Cells International, 2015*, 804213. https://doi.org/10.1155/2015/804213
30. Chen, Y., & Li, X. (2022). The utilization of carbon-based nanomaterials in bone tissue regeneration and engineering: Respective featured applications and future prospects. *Medicine in Novel Technology and Devices, 16*, 100168. https://doi.org/10.1016/j.medntd.2022.100168
31. Eivazzadeh-Keihan, R., Maleki, A., de la Guardia, M., Bani, M. S., Chenab, K. K., Pashazadeh-Panahi, P., Baradaran, B., Mokhtarzadeh, A., & Hamblin, M. R. (2019). Carbon based nanomaterials for tissue engineering of bone: Building new bone on small black scaffolds: A review. *Journal of Advanced Research, 18*, 185–201. https://doi.org/10.1016/j.jare.2019.03.011
32. Kumar, S., Raj, S., Sarkar, K., & Chatterjee, K. (2016). Engineering a multi-biofunctional composite using poly(ethylenimine) decorated graphene oxide for bone tissue regeneration. *Nanoscale, 8*, 6820–6836. https://doi.org/10.1039/C5NR06906H

33. Liang, C., Luo, Y., Yang, G., Xia, D., Liu, L., Zhang, X., & Wang, H. (2018). Graphene oxide hybridized nHAC/PLGA scaffolds facilitate the proliferation of MC3T3-E1 cells. *Nanoscale Research Letters, 13*, 15. https://doi.org/10.1186/s11671-018-2432-6
34. Owida, H. A., Al-Nabulsi, J. I., Alnaimat, F., Al-Ayyad, M., Turab, N. M., Al Sharah, A., & Shakur, M. (2022). Recent applications of electrospun nanofibrous scaffold in tissue engineering. *Applied Bionics and Biomechanics*. https://doi.org/10.1155/2022/1953861
35. Eskandari, S., Guerin, T., Toth, I., & Stephenson, R. J. (2017). Recent advances in self-assembled peptides: Implications for targeted drug delivery and vaccine engineering. *Advanced Drug Delivery Reviews, 110-111*, 169–187. https://doi.org/10.1016/j.addr.2016.06.013
36. Rufaihah, A. J., Yasa, I. C., Ramanujam, V. S., Arularasu, S. C., Kofidis, T., Guler, M. O., & Tekinay, A. B. (2017). Angiogenic peptide nanofibers repair cardiac tissue defect after myocardial infarction. *Acta Biomaterialia, 58*, 102–112. https://doi.org/10.1016/j.actbio.2017.06.009
37. Zhao, J., Han, W., Tu, M., Huan, S., Zeng, R., Wu, H., Cha, Z., & Zhou, C. (2012). Preparation and properties of biomimetic porous nanofibrous poly(l-lactide) scaffold with chitosan nanofiber network by a dual thermally induced phase separation technique. *Materials Science and Engineering: C, 32*, 1496–1502. https://doi.org/10.1016/j.msec.2012.04.031
38. Zhang, W., Ronca, S., & Mele, E. (2017). Electrospun nanofibres containing antimicrobial plant extracts. *Nanomaterials, 7*, 42. https://doi.org/10.3390/nano7020042
39. Davoudabadi, M., Fahimirad, S., Ganji, A., & Abtahi, H. (2023). Wound healing and antibacterial capability of electrospun polyurethane nanofibers incorporating Calendula officinalis and Propolis extracts. *Journal of Biomaterials Science, 34*, 1491–1516. https://doi.org/10.1080/09205063.2023.2170138
40. Malekpour, Z., Akbari, V., Varshosaz, J., & Taheri, A. (2021). Preparation and characterization of poly (lactic-co-glycolic acid) nanofibers containing simvastatin coated with hyaluronic acid for using in periodontal tissue engineering. *Biotechnology Progress, 37*. https://doi.org/10.1002/btpr.3195
41. Jeong, E. J., Lee, J. W., Yeon, S.-J., Lee, J.-W., Kwark, Y.-J., Rhee, S.-H., Park, W. H., Kim, S. H., & Lee, K. Y. (2012). Fabrication of nanopatterned surfaces for tissue engineering. In *2012 international conference on biomedical engineering and biotechnology* (pp. 1030–1033). IEEE. https://doi.org/10.1109/iCBEB.2012.220
42. Cao, S., & Yuan, Q. (2022). An update of nanotopographical surfaces in modulating stem cell fate: A narrative review. *Biomaterials Translational, 3*, 55–64. https://doi.org/10.12336/biomatertransl.2022.01.006
43. Abrams, G. A., Goodman, S. L., Nealey, P. F., Franco, M., & Murphy, C. J. (2000). Nanoscale topography of the basement membrane underlying the corneal epithelium of the rhesus macaque. *Cell and Tissue Research, 299*, 39–46. https://doi.org/10.1007/s004419900074
44. Curtis, A., & Wilkinson, C. (2001). Nantotechniques and approaches in biotechnology. *Trends in Biotechnology, 19*, 97–101. https://doi.org/10.1016/S0167-7799(00)01536-5
45. Li, K., Lv, L., Shao, D., Xie, Y., Cao, Y., & Zheng, X. (2022). Engineering nanopatterned structures to orchestrate macrophage phenotype by cell shape. *Journal of Functional Biomaterials, 13*, 31. https://doi.org/10.3390/jfb13010031
46. Andrews, P. W. (2021). Human pluripotent stem cells: tools for regenerative medicine. *Biomaterials Translational, 2*, 294–300. https://doi.org/10.12336/biomatertransl.2021.04.004
47. Chen, W., Han, S., Qian, W., Weng, S., Yang, H., Sun, Y., Villa-Diaz, L. G., Krebsbach, P. H., & Fu, J. (2018). Nanotopography regulates motor neuron differentiation of human pluripotent stem cells. *Nanoscale, 10*, 3556–3565. https://doi.org/10.1039/C7NR05430K
48. Kesti, M., Eberhardt, C., Pagliccia, G., Kenkel, D., Grande, D., Boss, A., & Zenobi-Wong, M. (2015). Bioprinting complex cartilaginous structures with clinically compliant biomaterials. *Advanced Functional Materials, 25*, 7406–7417. https://doi.org/10.1002/adfm.201503423
49. Kosik-Kozioł, A., Costantini, M., Mróz, A., Idaszek, J., Heljak, M., Jaroszewicz, J., Kijeńska, E., Szöke, K., Frerker, N., Barbetta, A., Brinchmann, J. E., & Święszkowski, W. (2019). 3D bioprinted hydrogel model incorporating β-tricalcium phosphate for calcified cartilage tissue engineering. *Biofabrication, 11*, 035016. https://doi.org/10.1088/1758-5090/ab15cb

50. Alcala-Orozco, C. R., Mutreja, I., Cui, X., Kumar, D., Hooper, G. J., Lim, K. S., & Woodfield, T. B. F. (2020). Design and characterisation of multi-functional strontium-gelatin nanocomposite bioinks with improved print fidelity and osteogenic capacity. *Bioprinting, 18*, e00073. https://doi.org/10.1016/j.bprint.2019.e00073
51. Huang, C.-T., Kumar Shrestha, L., Ariga, K., & Hsu, S. (2017). A graphene–polyurethane composite hydrogel as a potential bioink for 3D bioprinting and differentiation of neural stem cells. *Journal of Materials Chemistry B, 5*, 8854–8864. https://doi.org/10.1039/C7TB01594A
52. Izadifar, M., Chapman, D., Babyn, P., Chen, X., & Kelly, M. E. (2017). UV-assisted 3D bioprinting of nanoreinforced hybrid cardiac patch for myocardial tissue engineering. *Tissue Engineering, Part C: Methods, 24*, 74–88. https://doi.org/10.1089/ten.tec.2017.0346
53. Zengin, A., Teixeira, F. C., Feliciano, T., Habibovic, P., Mota, C. D., Baker, M. B., & van Rijt, S. (2023). Matrix metalloproteinase degradable, in situ photocrosslinked nanocomposite bioinks for bioprinting applications. *Biomaterials Advances, 154*, 213647. https://doi.org/10.1016/j.bioadv.2023.213647

Chapter 7
Cell–Scaffold Interactions

Abstract Scaffolds and cells play a central role in the tissue engineering and regenerative medicine (TERM) field, where scaffolds provide physical scaffolding, and biochemical and biophysical cues for regulation of cell function or fate. Scaffolds and cells interact at the surface level through important cell structures like the cell membrane or the cytoskeleton, which are used in vivo for interacting with the dynamic extracellular matrix (ECM). The first event of cell-scaffold interaction is cell adhesion, which is highly dependent on surface properties such as chemistry, wettability, topography or stiffness. Following cell adhesion, other cellular events ensue such as spreading, proliferation or differentiation. Ultimately, TERM scaffolds present three-dimensionality and cells respond to architectural traits like porosity, pore size, pore interconnectivity, or permeability. This chapter focuses on cell—scaffold interactions at the surface level, as well as cell responses to 3D features of scaffolds. The chapter also reviews the cell membrane, the cytoskeleton and the ECM. Finally, conclusions and questions to reinforce the different concepts and aspects covered in this chapter are presented.

7.1 Introduction

Scaffolds and cells are two of the major players of the tissue engineering and regenerative medicine (TERM) paradigm [1]. Scaffolds provide a physical template for cells to infiltrate, proliferate and carry out their associated functions in order to regenerate lost or damaged tissues. This physical scaffolding is equally important in bioengineered tissue models for the study of various tissue phenomena. Scaffolds also provide important biochemical and biomechanical cues to direct and regulate cell fate and function through interaction with cells at the surface level. Therefore, understanding these interactions is of vital importance for the design and engineering of smart and effective scaffolds for the different TERM applications.

In our bodies, interaction of cells with their surrounding natural scaffolding, i.e. the extracellular matrix (ECM), occurs via cell structures like the cell membrane

© The Author(s), under exclusive license to Springer Nature Switzerland AG 2026

E. García-Gareta, J. M. García-Aznar, *Contemporary Tissue Engineering and Regenerative Medicine*, Series in BioEngineering,
https://doi.org/10.1007/978-3-032-03441-0_7

and the cytoskeleton [2]. Thus, it is worth revising the knowledge on these major cellular structures. The ECM is extremely complex in terms of composition and architecture, with each tissue and organ possessing an ECM with a distinct make-up and topology, which are curated to provide the optimal environment for the cells to perform their tissue functions [3]. Importantly, the ECM is not a static, inert scaffold, but it is dynamic in terms of remodelling of its components and material properties [4]. Therefore, the current knowledge on the ECM should be discussed.

Ultimately, in TERM constructs, cells interact with the scaffold of choice, with adhesion being the first event of cell-scaffold interaction. Cell adhesion to biomaterials depends on surface properties such as chemistry, wettability, topography or stiffness [5–7]. Following adhesion, cellular events such as spreading, proliferation, migration or phenotypic differentiation occur [6, 8]. Nevertheless, TERM scaffolds show three-dimensionality, and therefore, cells respond to 3D scaffold architecture, including porosity, pore size and interconnectivity.

The aim of this chapter is to discuss the cell—scaffold interactions that take place in TERM, starting with reviewing knowledge on the cellular structures that participate in these events, namely the cell membrane and the cytoskeleton. Similarly, the current knowledge on the complex ECM will be reviewed, highlighting the dynamic aspect of this natural scaffold. Finally, cell adhesion to biomaterials and cell responses to architectural traits of scaffolds will be discussed. The chapter ends with conclusions and questions to reinforce the different concepts and aspects covered in this chapter.

7.2 The Cell Membrane

The cell membrane, which acts as a barrier between the cell's interior and exterior, consists of lipids, proteins and carbohydrates (Fig. 7.1). The popular fluid mosaic model of the cell membrane describes it as a dynamic and flexible structure. The fundamental phospholipid bilayer forms a stable fencing between two aqueous

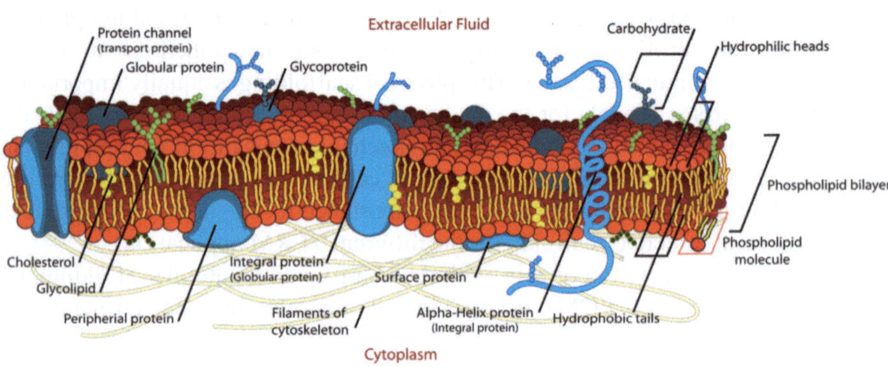

Fig. 7.1 The cell membrane according to the fluid mosaic model (free image)

compartments, i.e. the inside and the outside of the cell. The various proteins embedded within the phospholipid bilayer carry out functions such as selective molecule transport, cell-cell recognition, or cell-ECM interaction [2].

Phospholipids are the most abundant membrane lipids and have a polar head group and two hydrophobic hydrocarbon tails that are usually fatty acids [9]. Importantly, phospholipids spontaneously form bilayers in aqueous environments, with the hydrophilic heads in contact with the water at each surface of the bilayer, and the hydrophobic tails in the interior [9]. Furthermore, phospholipid bilayers have self-healing properties: since a small tear creates a free edge with water, which is not energetically favourable, the phospholipids spontaneously rearrange to remove the free edge. The distinctive phospholipid structure is responsible for the fundamental function of cell membranes as barriers between two separate aqueous compartments [9]. Moreover, the membrane is not permeable to water-soluble molecules, which includes most biological molecules as well as ions. Interestingly, phospholipid bilayers are viscous fluids, which accounts for their fluidity. The cell membranes of animal cells also contain cholesterol and glycolipids, both essential to their integrity and function [2].

As just seen, lipids are the fundamental structural blocks of the cell membrane, and approximately constitute 50% by weight. Proteins, the other 50% by weight of cell membranes, are important components that carry out specific membrane functions. According to the fluid mosaic model, proteins are inserted into the lipid bilayer and can be classified into integral or transmembrane, and peripheral [2]. Whilst the peripheral membrane proteins bind to either of the outer sides of the bilayer, integral proteins are transmembrane and usually expose parts of their structure on both sides of the membrane. Proteins fulfil various membrane functions, including transport, catalysis, cell-cell adhesion, signal reception, and structural. Transport proteins move biological proteins and ions across the membrane, whilst membrane enzymes act as catalysts in important chemical reactions such as redox or hydrolysis. Cell adhesion proteins are key for cell recognition and cell-cell adhesion. Membrane protein receptors convey signals between the cells internal and external environments. Finally, membrane proteins can also have a structural function as they connect the membrane with other elements such as the ECM or cytoskeleton filaments [2].

The last component of cell membranes are carbohydrates, which present associated to lipids, i.e. glycolipids, or proteins, i.e. glycoproteins [2]. The extracellular portions of membrane proteins tend to be glycosylated. Similarly, the carbohydrate portions of glycolipids are exposed on the external side of the plasma membrane. Therefore, carbohydrates coat the surface of the cell, which is known as the glycocalyx [10]. This glycocalyx protects the cell surface and participates in a variety of cell-cell interactions, as the oligosaccharides exposed on the cell surface act as markers to identify the different cell types of multicellular organisms [10].

7.3 The Cytoskeleton

As mentioned in previous chapters, the cytoskeleton is a dynamic network of interlinking protein filaments that is present in the cytoplasm of all cells, have important functions such as maintaining cell shape, cell locomotion, and organization of the organelles as well as their movement [2]. The filaments that comprise the cytoskeleton are actin filaments or microfilaments, intermediate filaments, and microtubules [2].

Actin filaments (F-actin) are linear polymers of the globular actin protein (G-actin) [11]. They are also called microfilaments and their structure is two strands of actin wound in a spiral of approximately 7 nm diameter, and thus are the thinnest filaments in the cytoskeleton [11]. Actin filaments form bundles of closely packed parallel fibres, or networks where the filaments are loosely orthogonally crosslinked thereby forming 3D meshworks that behave like semisolid gels (Fig. 7.2). Various proteins that crosslink actin filaments in specific patterns govern whether bundles or meshworks are formed. As a curiosity, actin is the most abundant protein in the majority of eukaryotic cells [11]. Actin filaments have important roles in determining cell shape and cell adhesion to a substrate. Furthermore, they are also key in cell movement or migration, as well as mediating specific activities within the cell, like cell cleavage during mitosis [13, 14].

Intermediate filaments, which are about 8–12 nm wide, are in-between the size of the other two filaments that comprise the cytoskeleton [15]. Whilst microfilaments are ultimately composed of actin and microtubules of tubulin, more than 50 different proteins are found in intermediate filaments. Some of these proteins are keratin, vimentin, desmin, lamin or neurofilament proteins [15]. In contrast to microfilaments or microtubules, intermediate filaments are very stable and the true skeleton of the cell. Their main function is to anchor and position the nucleus. Furthermore, they contribute towards the cell's elastic properties and ability to withstand tension [15].

The largest of the cytoskeleton fibres are microtubules with a diameter of approximately 23 nm [16]. Microtubules are hollow tubes composed of alpha and beta tubulin and are constantly assembling and disassembling [16]. They can form structures like flagella or cilia, both involved in cell movement. Most of the microtubules come from the centrosome, a cell organelle that is a microtubule organizing centre and is located near the middle of the cell, with microtubules radiating outward from it [16]. Microtubules play a vital role in shifting the daughter chromosomes to the newly forming cells during mitosis [16].

Apart from the discussed filaments, other proteins are often considered part of the cytoskeleton, such as septins, which form filaments and attachment sites for some proteins, or spectrin, which is found along the intracellular side of the cell membrane and participates in maintaining cell structure [17, 18].

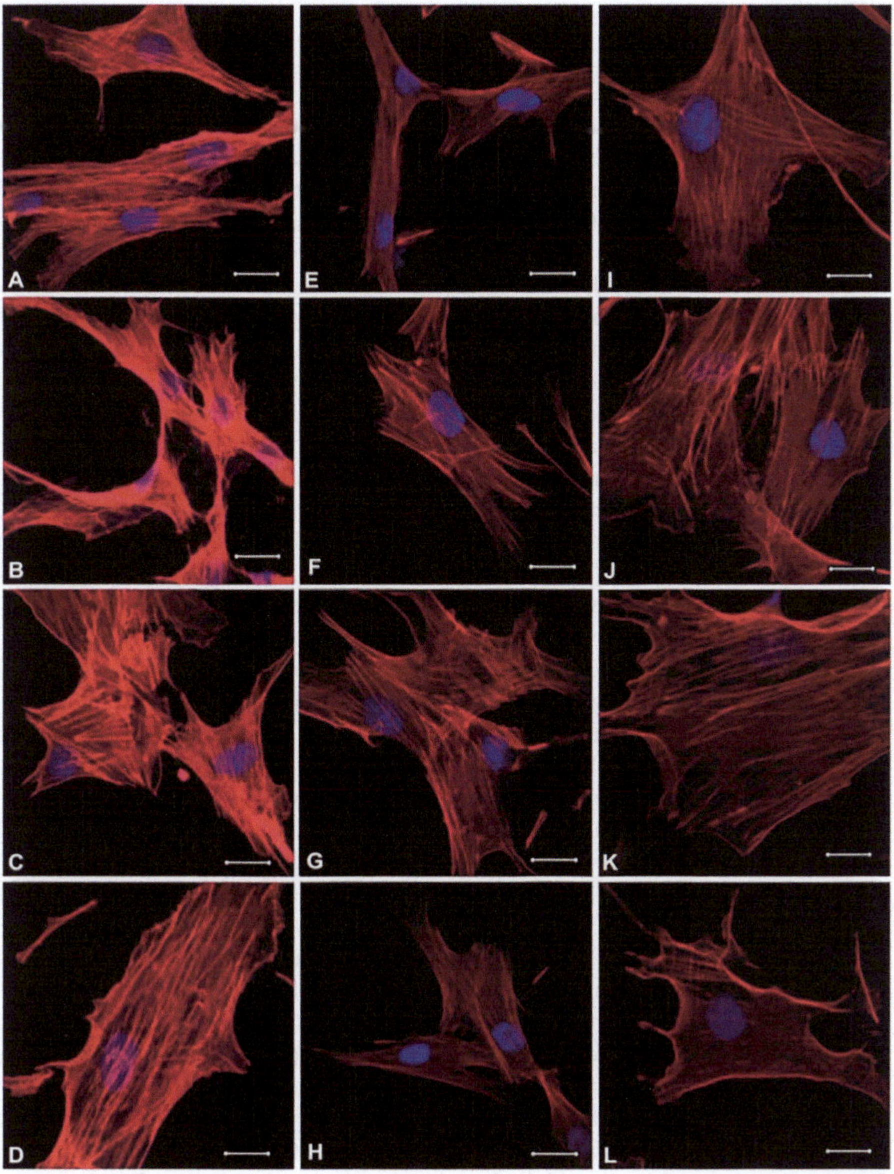

Fig. 7.2 Adipose-derived mesenchymal stem cells cultured in the presence of silica and silica-boron nanoparticles for 24 h, where the nanoparticles employed directly interact with the cytoskeleton. Cells fluorescently labelled with blue DAPI (double stranded DNA, blue) and TRITC-phalloidin (F-actin, red). (**a–d**) Cells from the 'Control 24 h' group, where F-actin fibres are longitudinally packed ending with some thickening at the periphery of the cells, whilst nuclei are located in the central part of the cells. (**e–h**) Cells from the 'Silica 24 h' group, where some isolated transversally arranged actin filaments can be seen besides longitudinally packed fibres. (**i–l**) Cells from the 'Silica-Boron 24 h' group, where transversally arranged filaments are seen to a much greater extent and numerous actin filaments show clavate growing. Scale bar = 20 μm. Image from Ogneva et al. [12]. Open access article distributed under the terms and conditions of the Creative Commons Attribution (CC BY) license

7.4 The Extracellular Matrix

The ECM is the non-cellular integrant that is present in the tissues and organs of our body [3]. Two main forms of the ECM exist, namely the interstitial ECM and the basement membrane [19, 20]. The interstitial ECM presents as a porous and interconnecting 3D network surrounding stromal cells. On the other hand, the basement membrane is a stable, sheet-like dense structure that lines the basal surface of epithelial or endothelial cells, as an example. The basement membrane is also responsible for the compartmentalization of tissues.

Initially thought to be an inert scaffold that merely provided cell support, the vital role of the ECM was glimpsed at when it was discovered that cells are anchorage-dependent and as such, they need to adhere to the ECM to survive and proliferate [3]. Over the decades, it has been shown that it provides crucial chemical, physical and mechanical cues needed for tissue morphogenesis, differentiation and homoeostasis [19]. These cues are detected by cell surface receptors, which triggers intracellular signalling cascades that result in the expression of genes that regulate apoptosis, cell proliferation, migration or differentiation, amongst other cell functions [19].

Resident tissue cells are responsible for secreting the numerous ECM components. These are modified in response to oxygen and nutrient availability or mechanical stimuli. In a nutshell, the ECM comprises of water, proteins and polysaccharides. The collagen superfamily is the major protein component of the ECM. Other proteins found in it are laminins, elastin, or fibronectin [19]. The interstitial ECM main proteins are collagen types I, III, V, etc., fibronectin and elastin, whilst the basement membrane displays collagen type IV, laminins and network-bridging proteins such as nidogen and heparan sulphate proteoglycans [21]. Important polysaccharides found in the ECM are glycosaminoglycans (GAGs) such as hyaluronan, chondroitin sulfate, or heparan sulfate [21]. However, each tissue and organ possess a unique ECM [3].

The ECM is not a static scaffold. Indeed, it is continuously reformed in terms of structure, composition, abundance, and organization of individual components via different remodelling mechanisms [4]. These changes in the ECM affect the cell microenvironment thereby changing cell behaviour. The remodelling of the ECM is highly complex and involves various mechanisms, such as: (1) ECM deposition, which tunes the biochemical and mechanical properties of the ECM by modifying the individual components as well as their abundance; (2) chemical modification at the post-translational level, which alters biochemical and structural ECM aspects; (3) proteolytic degradation, which releases bioactive ECM fragments and ECM-bound factors; and (4) force-mediated physical remodelling, which changes the alignment of ECM fibres thereby creating new paths for cell migration. Tissue homoeostasis hinges on the tight regulation of ECM remodelling, since altering the delicate remodelling balance has important effects on complex cellular signalling networks [4, 20].

7.5 Cell-Cell and Cell-Extracellular Matrix Interactions

Cells interact with other cells and the ECM through large macromolecular assembles mediated by families of proteins such as cadherins, integrins, claudins, or connexins. The main cell-cell interactions are through tight junctions, adherens junctions, desmosomes, and gap junctions. The main cell-ECM interactions occur through focal adhesions and hemidesmosomes. Cell-ECM interactions are mediated by integrins, which are also key to understand cell-scaffold interactions in TERM.

7.5.1 Tight Junctions

Tight junctions are intercellular adhesion complexes in epithelia and endothelia that regulate the paracellular permeability of water, ions, and macromolecules [22]. A tight junction creates a seal that prevents any leakage of the content through the cell membranes involved. The major transmembrane proteins occludin, claudins, and junction adhesion molecule (JAM) proteins mediate tight junctions, which join the cytoskeletons of adjacent cells [22].

Tight junctions are vital to provide epithelial and endothelial cells with barrier function, where they contribute to maintain the polarity of cells and the osmotic balance. Tight junctions are also important in maintaining the blood-brain barrier. Finally, they also function as a means to hold cells together [22].

7.5.2 Adherens Junctions

Adherens junctions are cell-cell adhesion complexes with important functions, such as regulation of the actin cytoskeleton, or transcriptional and intracellular signalling regulation [23]. Adherens junctions are essential for tissue homeostasis and morphogenesis [24].

Cadherins are the main family of transmembrane proteins involved in the formation of adherens junctions [23, 24]. The cytoplasmic domain of cadherins binds common cytoplasmic molecules called catenins, which in turn link with various molecules, including cytoskeletal proteins and their regulators [23]. Adherens junctions allow cell movement during tissue development whilst maintaining tissue architecture [24].

7.5.3 Desmosomes

Desmosomes are one of the stronger cell-to-cell contacts and are found in cells of tissues that experience intense mechanical forces, such as keratinocytes in the epidermis or cardiomyocytes in cardiac muscle [25]. Therefore, desmosomes ensure the integrity of tissues that are under mechanical stress.

In terms of composition, desmosome-intermediate filament complexes, which are networks of cadherins, linker proteins, and intermediate filaments, form the desmosomes, which are highly ordered and specialized structures [26]. Desmosomes mechanically integrate cells within tissues by means of cell–cell adhesion and cytoskeletal linkages.

7.5.4 Gap Junctions

Gap junctions are a special type of cell-to-cell contact that consist of intercellular channels arranged in clusters. Direct diffusion of small metabolites, molecules and ions between adjacent cells is allowed through these intercellular channels, thereby allowing biochemical and electrical coupling between cells [27]. Gap junctions are formed by the integral membrane proteins connexins arranged in hexameric assemblies [27]. Another family of proteins called pannexins have also been identified in gap junctions of vertebrates [27].

Cellular communication through gap junctions is essential in cell synchronization, differentiation, growth, and metabolic coordination of avascular tissues such as the epidermis and lens. Since mutations in genes encoding connexins cause a variety of genetic disorders, a critical role in tissue homeostasis for this type of cell-to-cell contact is suggested [27].

7.5.5 Focal Adhesions

Focal adhesions (FA) are responsible for forming mechanical links between the ECM and bundles of actin fibres of the cytoskeleton [13, 14] (Fig. 7.3). FA are large macromolecular assemblies or multi-protein structures where integrins play a vital role (Fig. 7.3a). Cell adhesion to the ECM through FA allows cells to crawl during migration [13]. Thus, FA have a key role in cell motility and spreading. In stationary cells, they play a vital role in cell anchorage and maintaining cell shape [13].

The highly orchestrated process of forming a FA begins with activation and clustering of integrins, which then bind to proteins of the ECM, thereby recruiting various intracellular proteins such as talin, paxillin, VASP (vasodilator-stimulated phosphoprotein) or zyxin, that in turn eventually bind to the actin fibres [14]. Figure 7.4 offers a detailed visual scheme of the FA formation process. It has been

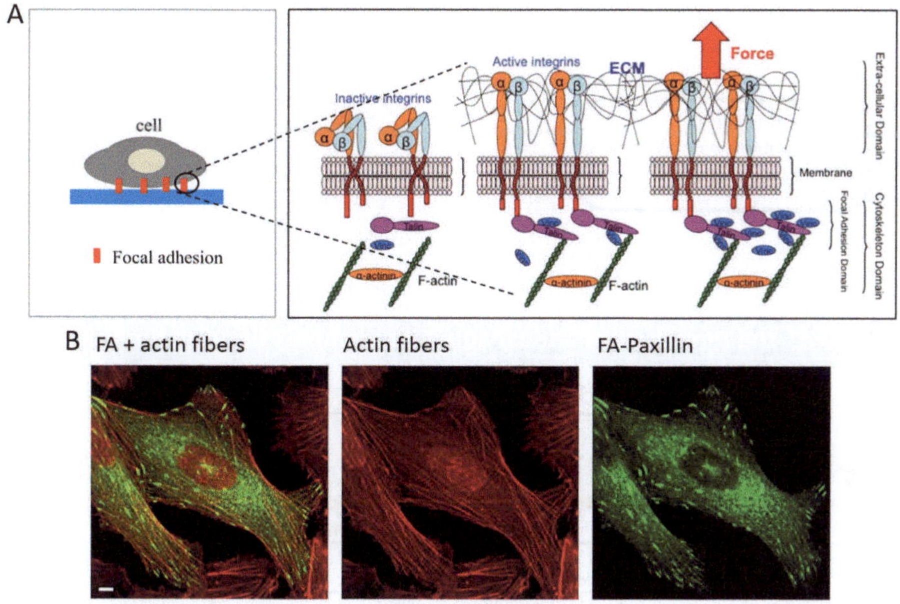

Fig. 7.3 (**a**) Scheme of FA showing integrin binding to the ECM, intracellular proteins such as talin or vinculin (Vinc) involved in linking to F-actin, which is stabilized by α-actinin. (**b**) Immunocytochemistry of actin fibers (red fluorescence) and paxillin (green fluorescence) of U2OS cells (epithelial cell line) cultured in a 2D substrate. Scale = 5 μm. Images adapted from Khalili et al. [13] and Legerstee et al. [14]. Open access articles distributed under the terms and conditions of the Creative Commons Attribution (CC BY) license

shown that the cytosolic proteins of FA possess a three-layered nanostructure: (1) an integrin signalling layer (ISL) which is found within ~10–20 nm to the cell membrane, (2) an actin-regulatory layer (ARL) located ~50–60 nm from the cell membrane, and (3) a force transduction layer (FTL) in between the ISL and ARL [14, 28]. It is important to keep in mind that FA continuously form and disassemble throughout the cell, and have an average lifetime of about 1 h [14].

As mentioned, FA are of paramount importance in cell adhesion and migration, which are vital in embryonic development, the immune system, or metastasis [13].

7.5.6 Hemidesmosomes

Hemidesmosomes are specialized multiprotein complexes that facilitate the attachment of epithelial cells to the underlying basement membrane in tissues like skin, cornea, the gastrointestinal tract, or the amnion [29]. Hemidesmosomes are very small stud-like structures that are similar to desmosomes when observed under an electron microscope [29]. However, whilst desmosomes bind adjacent cells,

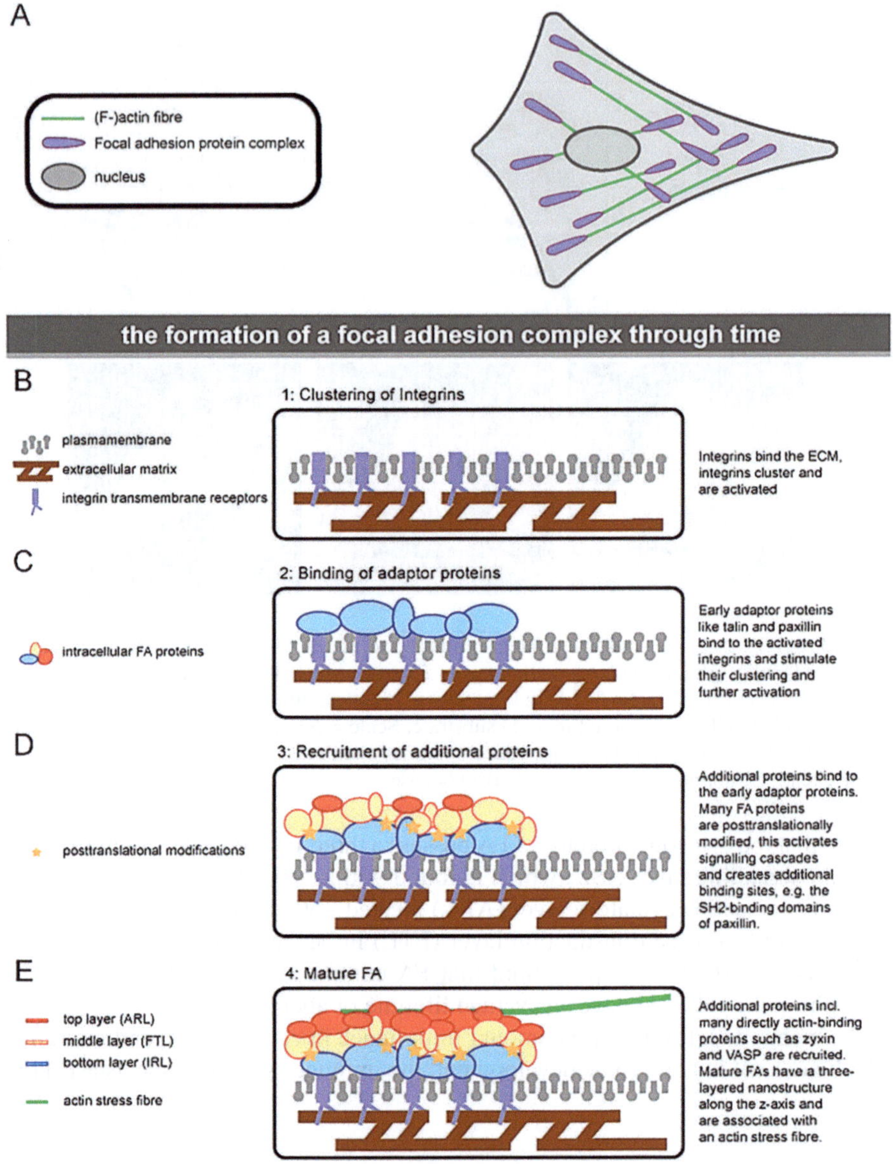

Fig. 7.4 Scheme of the formation of FA. (**a**) Schematic representation of FA in a cell. (**b–e**) Overview of the main steps involved in the formation of FA. Image from Legerstee et al. [14]. Open access article distributed under the terms and conditions of the Creative Commons Attribution (CC BY) license

hemidesmosomes link the basement membrane and the internal keratin intermediate filaments, which are stress-resilient [30]. Integrins are key in binding to the ECM, especially integrin α6β4 that binds to the basement membrane protein laminin-332 [30]. Hemidesmosomes provide epithelial cells with cues needed for their polarization and spatial organization, both important for tissue architecture.

7.5.7 Integrins

The ubiquitous family of cell surface receptors called integrins recognise short amino acid motifs in numerous ECM proteins, growth factors, cytokines, blood proteins, and cell surface proteins [31, 32]. Cell attachment mediated by integrins regulates important cellular events such as survival, migration, growth, proliferation, differentiation, signalling, or apoptosis. They also participate in all processes critical to inflammation, immunity, infection, and angiogenesis [33]. More specifically, integrins are heterodimeric transmembrane proteins with α and β domains that are arranged in various dimeric pairings [33]. Integrins transduce mechanochemical cues and bi-directional signalling across the plasma membrane by dynamically linking the intracellular actin cytoskeleton and the ECM [34]. In FA, integrins connect to the F-actin filaments through a multiprotein complex which includes talin, filamin, vinculin, paxillin, or tensin [14].

As mentioned, integrins recognize short amino acid sequences such as the thoroughly characterized RGD sequence, which is present in fibronectin, vitronectin, fibrinogen, and many others. In collagen and some laminins, these RGD sequences are exposed upon denaturation or proteolytic cleavage, thereby allowing cells to bind these RGD ligands by means of RGD-binding receptors [31]. There are other amino acid sequences used by integrins other than RGD. For instance, the collagen-binding integrins recognize the triple-helical GFOGER sequence in the major collagens [31].

The biology of integrins is extremely complex and not within the remit of this book. However, they are the principal family of receptors used by animal cells to bind to the ECM (Table 7.1), and as such play a critical role in cell-scaffold interaction in TERM.

7.6 Cell Adhesion to Biomaterials

In the TERM process, the initial response of cells to a scaffold is cell adhesion, which is a complex phenomenon influenced by multiple factors [36]. Three phases can be distinguished in the cell adhesion process, namely water-surface interaction, protein adsorption to the scaffold surface, and cell-biomaterial interaction.

When a scaffold is either cultured with cells for in vitro models or implantation, or implanted in the body, the initial response is water molecule adsorption to the

Table 7.1 Ligand binding specificities of human integrins for ECM components and ECM-related proteins. *COMP* cartilage oligomeric matrix protein, *VEGF* vascular endothelial growth factor, *LAP-TGF-β* transforming growth factor beta latency-associated peptide, *MMP* matrix metalloproteinase, *FGF* fibroblast growth factor. Adapted from [35]

Integrin	Ligands of ECM and ECM-related proteins
α1β1	Laminin, collagen
α2β1	Laminin, collagen, thrombospondin, tenascin
α3β1	Laminin, thrombospondin
α4β1	Thrombospondin, fibronectin, osteopontin
α5β1	Fibronectin, osteopontin, fibrillin, thrombospondin, COMP
α6β1	Laminin, thrombospondin
α7β1	Laminin
α8β1	Tenascin, fibronectin, osteopontin, vitronectin, nephronectin
α9β1	Tenascin, osteopontin, VEGF-C, VEGF-D
α10β1	Laminin, collagen
α11β1	Collagen
αVβ1	LAP-TGF-β, fibronectin, osteopontin
αLβ2	–
αMβ2	–
αXβ2	Collagen
αDβ2	Fibronectin, vitronectin
αIIbβ3	Thrombospondin, fibronectin, vitronectin
αVβ3	Vitronectin, thrombospondin, fibrillin, tenascin, fibronectin, osteopontin, COMP, MMP, FGF-2, LAP-TGF-β
α6β4	Laminin
αVβ5	Osteopontin, vitronectin, LAP-TGF-β
αVβ6	LAP-TGF-β, fibronectin, osteopontin
α4β7	Fibronectin, osteopontin
αEβ7	–
αVβ8	LAP-TGF-β

scaffold surface [37]. Cell culture media is an aqueous solution, as are physiological fluids. Water molecule adsorption occurs within the first few nanoseconds [37]. It is important to take into account that water behaves differently with different biomaterials: it can either spread on the surface (hydrophilic material) or form a droplet (hydrophobic material). Therefore, surface wettability is a factor that would influence cell adhesion. To what extend water-surface interactions affect the subsequent protein adsorption and cell-biomaterial interactions are currently under investigation [38, 39]. Interestingly, it has been suggested that intermediate water bound to a biomaterial may be an indicator of the bioinertness of the material [38], suggesting the importance of this first stage of cell adhesion to scaffolds.

Following water molecules adsorption, proteins adsorb to the scaffold surface [36]. In vivo, proteins are observed on a biomaterial surface under a second after implantation. In seconds to minutes, a monolayer of protein adsorbs to most surfaces, with the small proteins being the first to adsorb due to their rapid transport to the surface. Over time, these proteins are replaced by larger ones that have a greater affinity towards the biomaterial surface. This stage is critical and therefore, we will elaborate on it in the following section.

Finally, cell-biomaterial interaction occurs, starting with cell attachment to the surface, which is influenced by the adsorbed protein layer as well as surface properties [36]. As the protein adsorption occurs well before cells arrive at the surface, the cells primarily see the adsorbed protein layer rather than the biomaterial surface itself. Following cell attachment, other cellular events ensue, such as cell spreading, migration, proliferation and differentiation [36], which are influenced by surface properties like nano and microscale roughness or wettability [5–7].

In terms of cell adhesion, if the biomaterial presents any elements of the ECM, for example in the case of natural protein-based polymers, integrins recognise the target amino acid motifs in these proteins and FA form [40–42]. When the scaffold surface is composed of other materials, such as metals or ceramics for example, following water and protein adsorption, integrins recognise target sequences in the adsorbed proteins and subsequently FA form [43]. Of course, in the scaffolds composed of natural proteins, adsorbed proteins also present amino acids motifs recognised by integrins. In terms of where these adsorbed proteins come from, in vitro, cell culture media is typically supplemented with serum that contains proteins like fibronectin or vitronectin [44], which contain the popular RGD sequence already discussed [45]. Likewise, in vivo, serum proteins containing integrin-recognition sequences adsorb onto the implant's surface [36].

7.6.1 Protein Adsorption

Both the proteins and the surface properties influence protein adsorption, and are summarized in Tables 7.2 and 7.3.

The majority of proteins are large amphipathic molecules, i.e. they possess both hydrophobic and hydrophilic elements, which makes them intrinsically surface-active [47]. Intermolecular forces, e.g. Coulombic, van der Waals, or Lewis acid-base, as well as hydrophobic interactions, conformational entropy or restricted mobilities are behind the adsorption of proteins to biomaterials surfaces [47].

Table 7.2 Properties of proteins that affect protein adsorption and their effect. Adapted from [46]

Property	Effect
Size	Larger proteins can have more sites of contact with the scaffold surface
Charge	Proteins near their isoelectric point generally adsorb more readily
Structure stability	Proteins with less intra-molecular cross-linking can unfold to a greater extent and form more contact points with the scaffold surface
Unfolding rate	Proteins that rapidly unfold can form contacts with the scaffold surface more rapidly

Table 7.3 Properties of scaffold surface that affect protein adsorption and their effect. Adapted from [46]

Property	Effect
Topography	Greater surface area increases interaction with proteins
Composition	Chemical composition of the scaffold surface will determine the types of intermolecular forces governing interaction with proteins
Wettability	Hydrophobic surfaces tend to bind more protein
Heterogeneity	A non-uniform surface results in areas that can interact differently with proteins
Potential	Surface potential influences the distribution of ions in solution and interaction with proteins

Furthermore, intramolecular forces leading to an alteration of protein conformation also affect the adsorption process [47].

The surface of biomaterial scaffolds present properties that greatly affect protein adsorption [46, 48]. Topography is one of such properties and determines the surface area that interacts with proteins. A larger surface area increases the interaction sites for protein adsorption [46]. Thus, by tailoring the roughness, porosity, pore size, or particle size at the surface level, the surface area can be increased in order to maximize protein adsorption. Similarly, the chemical composition of the surface influences protein adsorption through either electrostatic or hydrophobic interactions [46]. For instance, the electrostatic force plays a key role in protein adsorption. Charged ions or groups present on the scaffold surface can bond the charged functional groups of proteins, which include amino (-NH_2), carbonyl (-CO), carboxyl (-COOH) or aromatic groups [46]. For example, in calcium phosphate ceramics, the charged ion Ca^{2+} and group PO_4^{3-} are considered to be the protein binding sites [46]. Hydrophobic interactions are also very important [48]. Generally, a hydrophobic surface adsorbs proteins more strongly and to a greater extent than a hydrophilic surface [48]. Proteins unfold and spread their hydrophobic core over the scaffold surface to reduce the net hydrophobic surface area exposed to the usually polar (aqueous) solvent [48].

7.6.2 *Surface Chemistry*

The relationship between biomaterial surface chemistry and cell adhesion plays a crucial role in determining the success of various TERM strategies. Surface chemistry influences the interactions between the material and the surrounding biological environment, which includes cells [37, 49]. The chemical composition, charge, and functional groups on the biomaterial surface can directly affect cell adhesion and subsequent cell events like proliferation, migration or phenotypic differentiation [37, 49, 50].

The presence of specific functional groups on the biomaterial surface, such as amino (-NH_2), hydroxyl (-OH), carboxyl (-COOH), and thiol (-SH) groups, can affect surface chemistry [50, 51]. These groups can participate in chemical interactions with biological molecules such as water molecules or proteins, and cells. Functional groups have an effect on the wettability of the material's surface, thereby affecting its interaction with water, proteins, and cells [51]. The surface charge of biomaterials, influenced by the ionization state of functional groups, can influence interactions with charged components in the biological environment. For instance, positively charged surfaces may attract negatively charged proteins. In the case of polymer-based scaffolds, the type of polymer used and its molecular structure influence surface properties. For example, the polymer's backbone, side chains, and pendant groups can affect its biocompatibility [51].

When designing scaffolds, researchers aim to tailor surface chemistry to promote desired cellular responses [51]. For instance, the introduction of bioactive molecules onto the biomaterial surface can enhance cell adhesion by providing specific binding sites for cell receptors like integrins [42]. Thus, a common strategy is to produce biocompatible coatings composed of bioactive molecules, such as peptides, proteins, or functional groups, on the biomaterial surface to enhance specific cellular responses [51]. For example, the immobilization of cell-adhesive peptides can promote cell adhesion [41]. Additionally, controlling the hydrophilicity or hydrophobicity of the surface can influence the wettability and protein adsorption, thereby affecting cell adhesion and spreading [52]. Understanding and manipulating the intricate interplay between biomaterial surface chemistry and cell adhesion are essential for developing scaffolds that not only integrate seamlessly with biological tissues but also elicit the desired cellular responses for optimal performance and biocompatibility.

7.6.3 Surface Wettability

The literature shows that wettability of a scaffold surface is a key factor to influence cell-scaffold interaction and the subsequent cell behaviors [53, 54]. Wettability can be defined as the ability of a liquid to maintain contact with a solid surface, and is the result of intermolecular interactions. Wettability of a surface is heavily influenced by its functional groups or roughness, among others [53]. As already seen, wettability can control protein adsorption and subsequently affect cell activities [53]. Typically, most animal cells prefer a surface of moderate hydrophilicity for adhesion and growth, whereas surfaces of superhydrophilicity (contact angle below 5°) and superhydrophobicity (contact angle above 150°) are not favorable for cell attachment and growth [54].

Surface wettability can be modified and tailored by various methods. For instance, Al-Azzam and colleagues used oxygen plasma treatment and graphene oxide coating to create hydrophilic regions on a hydrophobic surface [52]. Both techniques were effective at patterning the surface's wettability, which resulted in micropatterning of cell adhesion and subsequent growth [52]. Another example is the work by Frost and co-workers, where surface wettability was modified by coating with ECM components without significantly altering surface roughness [8].

7.6.4 Surface Topography

Material surface topography or roughness has a direct influence on cellular morphology, proliferation, phenotype expression, and implant integration with the host tissue. The response of cells to roughness is different depending on the cell type. In a study by Lee et al. the behaviour of MG63 osteoblast-like cells seeded and cultured on polycarbonate membrane surfaces with different micropore sizes (0.2–8.0 μm in diameter) was investigated [55]. Cell adhesion and proliferation were progressively inhibited with increasing micropore size. Interestingly, in isolated cells, increasing micropore size produced improved protein synthesis and alkaline phosphatase (ALP) activity. This early study showed that surface topography plays an important role in MG63 osteoblast-like cells behaviour [55].

A more recent work by Hou and colleagues systematically studied the mechanotransduction and fate determination of human mesenchymal stem cells (MSCs) cultured on surface roughness gradients produced on glass substrates by tilted dip-coating in catecholic polyglycerol (Fig. 7.5) [56].

The authors reported that interfacial roughness mediated cell adhesion with cell spreading area decreasing with increasing roughness. They also observed that actin fibres bundles decreased at high roughnesses (Fig. 7.6). Furthermore, the authors

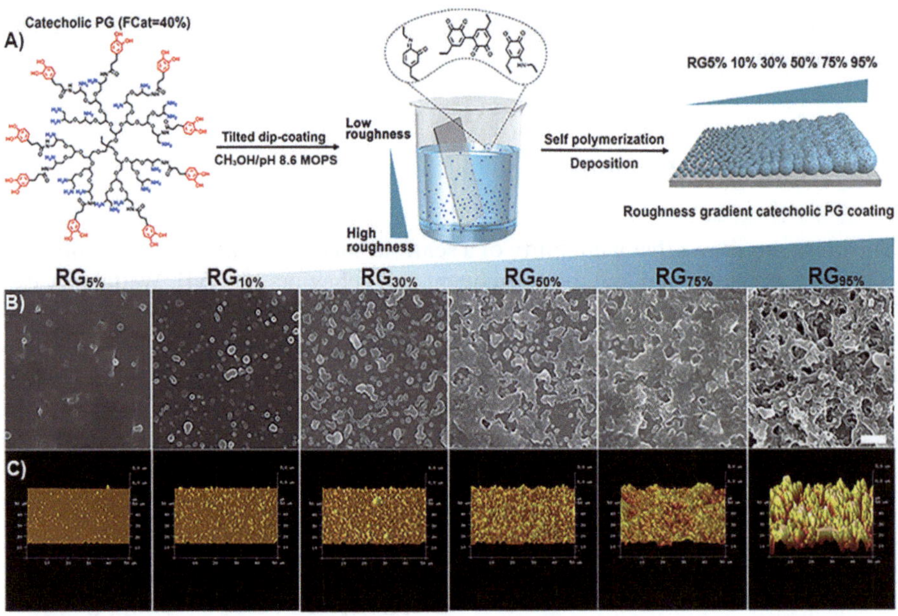

Fig. 7.5 (**a**) Fabrication of roughness gradients on glass substrates. (**b**) Scanning electron microscopy (SEM) and (**c**) atomic force microscopy (AFM) of the surface. Scale bar = 5 μm for SEM images and 10 μm for AFM images demonstrating increasing surface roughness. Image adapted from Hou et al. [56]. Open access article distributed under the terms and conditions of the Creative Commons Attribution (CC BY) license

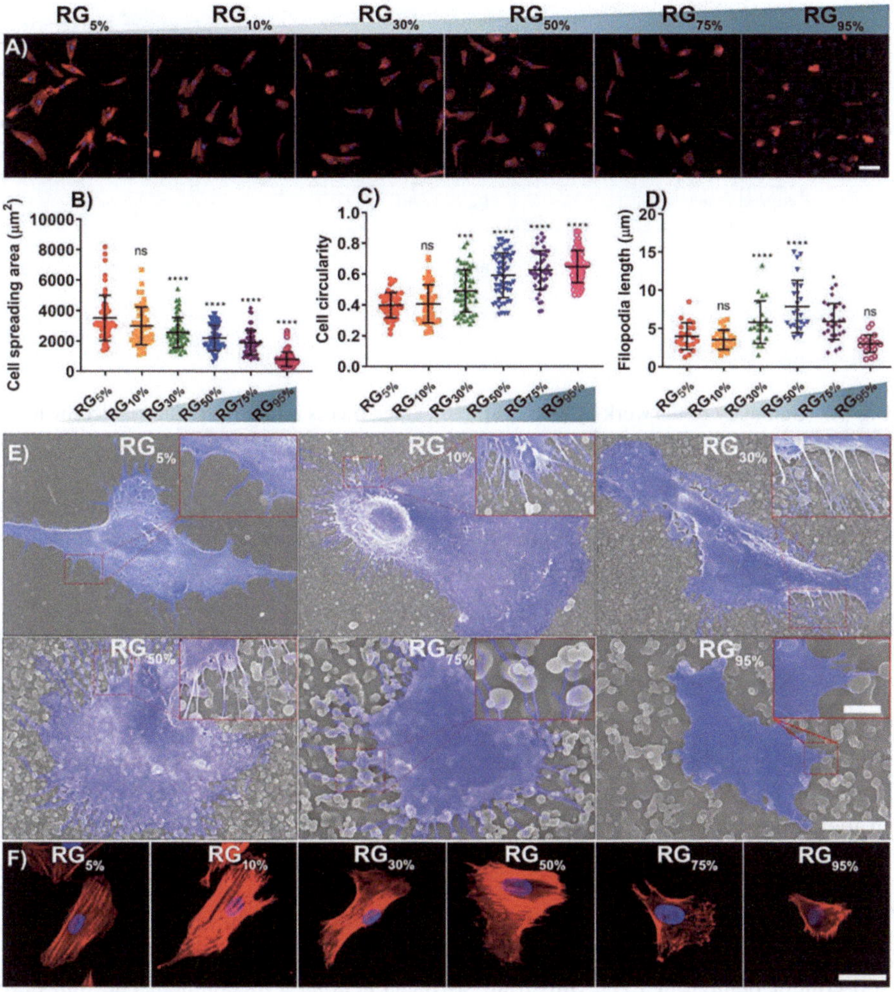

Fig. 7.6 (**a**) Representative images of actin filaments after 24 h of culture. (**b**) Cell spreading area, (**c**) circularity and (**d**) filopodia length of MSCs. (**e**) Representative SEM images of adhered cells on roughness gradient surface (scale bar = 15 μm, inset bar = 5 μm). (**f**) Representative images of the upper-side of MSCs with actin filaments stained (scale bar = 50 μm). Image adapted from Hou et al. [56]. Open access article distributed under the terms and conditions of the Creative Commons Attribution (CC BY) license

found that MSCs sensed roughness through FA as well as actomyosin cytoskeleton. It was also reported that cell nuclear mechanics and transcriptional activity was regulated by the interfacial roughness. Finally, fate determination of MSCs was mediated by the surface roughness.

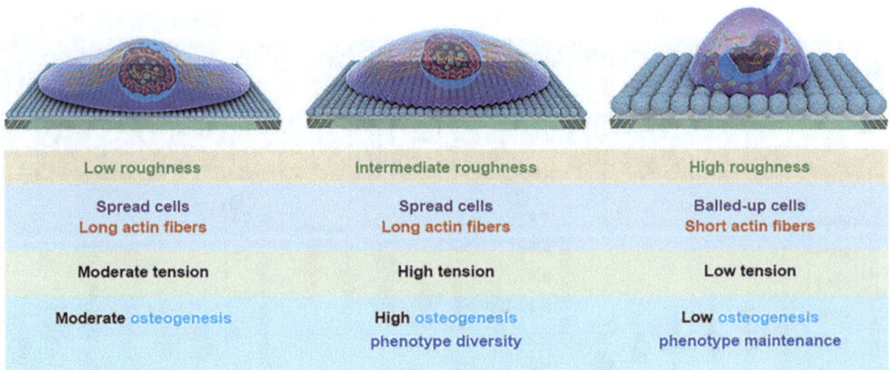

Fig. 7.7 Summary of the work by Hou et al. [56]. Open access article distributed under the terms and conditions of the Creative Commons Attribution (CC BY) license

The important results from this study are summarised in Fig. 7.7 and clearly show the influence of surface roughness on cell behaviour [56]. Therefore, this parameter needs to be optimised for effective TERM strategies.

7.6.5 *Surface Stiffness*

The relationship between biomaterial surface stiffness and cell adhesion is a critical aspect in the field of TERM. Cells are highly sensitive to the mechanical properties of their microenvironment. In the human body, ECM stiffness (Young's Modulus) widely varies between different tissues and organs, with the brain showing E = 1–3 kPa, muscle E = 23–42 kPa, blood vessel E = 1.16–860 MPa, tendon E = 136–820 MPa, or bone E = 15–40 GPa [57]. Therefore, it is not surprising that the stiffness of biomaterial surfaces can significantly influence cell adhesion and subsequent behaviour [57]. Substrate stiffness acts as a mechanical cue that cells can sense and respond to through various signalling pathways [57]. Generally, cells tend to exhibit increased adhesion, spreading, and proliferation on stiffer surfaces [58, 59]. This phenomenon is often attributed to the modulation of focal adhesions, integrins, and cytoskeletal elements within cells in response to the mechanical properties of the substrate. An example is the work by Voigt et al. where substrates with stiffness ranging from $E = 0.39 \pm 0.03$ kPa to $E = 51.4 \pm 4.5$ kPa were fabricated and HeLa cells cultured on them [59]. It was observed that stiff substrates facilitated cell adhesion and cells displayed a higher cell spreading area on stiffer substrates [59].

This understanding is leveraged to design biomaterials with tunable stiffness to mimic the mechanical properties of specific tissues in the human body [57]. For instance, materials with stiffness resembling that of bone tissue may promote enhanced adhesion and differentiation of osteogenic cells [57]. The intricate interplay between biomaterial surface stiffness and cell adhesion underscores the importance of considering mechanical cues in the design of scaffolds for TERM.

7.7 Cell Responses to Scaffold Architecture

TERM scaffolds are three-dimensional and have critical architectural traits such as porosity, a pore size range, pore interconnectivity, or permeability. On a macroscopic level, scaffolds provide boundaries for tissue regrowth, whilst on a microscopic level they provide a local framework for cell growth and tissue formation and remodelling, allowing cell attachment, infiltration and proliferation within a controllable microenvironment. A typical problem faced when using scaffolds is the rapid cell attachment and proliferation on the outer edge of the scaffold, which restricts cell penetration to the scaffold centre, resulting in non-homogeneous tissue growth. Porosity plays a critical role here, since cell growth at the centre of the scaffold is limited due to insufficient diffusion of oxygen and nutrients when the pores are too small, whilst larger pores compromise the mechanical stability of the scaffold and its ability to provide physical support for the cells and the newly grown tissue [60].

In terms of the optimum pore size range, this changes with the type of tissue. It has been reported that the ideal pore size range for skin regeneration is 20–125 μm [61–63], whilst the ideal pore sizes reported for bone regeneration tend to be larger, i.e. 200–350 μm was shown to be optimal for osteoblast proliferation [64]. An interesting study by Wo and colleagues explored the relationship between overall scaffold porosity and pore size range [65]. In this study, pore size distribution and porosity of Ti-6Al-4 V scaffolds were controlled by 3D printing fabrication. The effects of pore size distribution at a constant porosity, or porosity distribution at a constant pore size on adhesion, proliferation, and differentiation of the mouse embryonic osteoblast precursor cells (MC3T3-E1) were investigated separately. The average pore size and porosity of the fabricated scaffolds were: (1) 301 ± 9 μm and 58.8 ± 1.8%, (2) 300 ± 9 μm and 43.4 ± 1.3%, (3) 501 ± 11 μm and 58.3 ± 1.2%, (4) 499 ± 12 μm and 42.7 ± 1.1%, and (5) 804 ± 10 μm and 58.9 ± 1.3%, respectively. SEM images confirmed active attachment of cells that oriented with the direction of metal rod. MC3T3-E1 cells grown on the scaffold with an average porosity of 804 μm and porosity of 58.9% displayed significantly higher proliferation compared with each group after 3 days incubation ($p < 0.05$). The gene expression pattern of ALP (alkaline phosphatase), OCN (osteocalcin), Col1 (collagen type I), and Runx2 (Runt-related transcription factor 2) showed that these were downregulated when pore size increased or porosity decreased ($p < 0.05$) [65].

Pore interconnectivity is another architectural feature of TERM scaffolds that is often neglected or ignored. Interconnecting pathways through porous scaffolds are essential for nutrient and oxygen supply, removal of metabolites, cell infiltration, and tissue ingrowth. The use of the term "interconnectivity" does not usually describe the transport attributes of these pathways, and does not include the potential to support tissue synthesis [66]. A study by Ashworth and co-workers provided a critical analysis of the methods reported for the elucidation of scaffold transport pathways. The authors tested collagen scaffolds and found that depending on the method, the proportion of accessible pore space dramatically changed. Therefore, the authors investigated a method for transport pathway parameterization based on percolation theory, which was used to investigate the invasion behaviour of primary fibroblasts in relation to pore wall alignment and pore space accessibility. The authors found that both properties affected fibroblast invasion efficiency. The also reported that cells attached and migrated along the pore walls, adopting an elongated morphology [66].

In TERM, scaffold permeability refers to the ability of fluids, nutrients, and oxygen to pass through the pores of the scaffold [67]. A scaffold with high permeability allows for easier transport of these substances, which is crucial for cell survival, proliferation, and tissue regeneration [67]. Factors like porosity, pore size, and pore interconnectivity influence permeability, and a balance between these factors is needed for optimal outcomes [67]. For instance, Chao and colleagues showed that by optimizing the initial design of the microstructure, including permeability, the performance of porous scaffolds can be controlled [68].

As just seen, the architectural features of 3D scaffolds affect cell behaviour, from adhesion and attachment, to infiltration and gene expression, and therefore, they need to be carefully optimised.

7.8 Conclusions

Cells and scaffolds interact at the surface level through cellular structures like the cell membrane and the cytoskeleton, with the integrin family of cell membrane receptors playing a central role in linking the cytoskeleton to the scaffold surface. Cell adhesion is the initial response of cells to a scaffold, and is followed by cellular events such as spreading, migration, proliferation, or differentiation. Commonly, cells bind to a layer of adsorbed proteins on the scaffold surface. Protein adsorption is a dynamic process influenced by both protein properties and scaffold surface characteristics such as topography, composition, wettability or potential. Therefore, the scaffold surface characteristics are of key importance for cell adhesion. Nevertheless, TERM scaffolds are three-dimensional and present important architectural properties that affect cell behaviour, such as porosity, pore size range, pore interconnectivity, or permeability.

This chapter highlights the complexity of cell-biomaterial interactions, and how the scaffold properties are of vital importance in maximising and tailoring these interactions.

Questions

- Name the main components of the cell membrane.
- Name the three types of filaments that form the cytoskeleton and the main differences between them.
- Discuss four mechanisms involved in the remodelling of the ECM.
- What is the main role of focal adhesions?
- What are integrins and why are they important in cell-scaffold interactions?
- Name the three phases that can be distinguished in the cell adhesion process.
- Which properties affect protein adhesion onto the scaffold surface? Briefly discuss them.
- What is surface wettability and how does it affect cell adhesion?
- Why and how does surface stiffness affect cell adhesion?
- Name and briefly discuss three architectural features that affect cell behaviour.

References

1. Langer, R., & Vacanti, J. P. (1993). Tissue engineering. *Science, 260*, 920–926. https://doi.org/10.1126/science.8493529
2. Alberts, B., Johnson, A., Lewis, J., Raff, M., Roberts, K., & Walter, P. (2002). *Molecular biology of the cell* (4th ed.). Garland Science.
3. García-Gareta, E. (2019). *Collagen, from tissue culture to biomaterials, tissue engineering, and beyond* (1st ed.). Cambridge Scholars Publishing.
4. Bonnans, C., Chou, J., & Werb, Z. (2014). Remodelling the extracellular matrix in development and disease. *Nature Reviews Molecular Cell Biology, 15*, 786–801. https://doi.org/10.1038/nrm3904
5. Lloyd, A. (2003). Engineering surfaces to enhance cell adhesion: Surface engineering. *Materials Today, 6*, 19. https://doi.org/10.1016/S1369-7021(03)01120-9
6. Majhy, B., Priyadarshini, P., & Sen, A. K. (2021). Effect of surface energy and roughness on cell adhesion and growth—facile surface modification for enhanced cell culture. *RSC Advances, 11*, 15467–15476. https://doi.org/10.1039/D1RA02402G
7. Recek, N., Resnik, M., Motaln, H., Lah-Turnšek, T., Augustine, R., Kalarikkal, N., Thomas, S., & Mozetič, M. (2016). Cell adhesion on Polycaprolactone modified by plasma treatment. *International Journal of Polymeric Science, 2016*, 7354396. https://doi.org/10.1155/2016/7354396
8. Frost, O. G., Owji, N., Thorogate, R., Kyriakidis, C., Sawadkar, P., Mordan, N., Knowles, J. C., Lali, F., & Garcia-Gareta, E. (2021). Cell morphology as a design parameter in the bioengineering of cell-biomaterial surface interactions. *Biomaterials Science, 9*, 8032–8050. https://doi.org/10.1039/d1bm01149a
9. Dai, Y., Tang, H., & Pang, S. (2021). The crucial roles of phospholipids in aging and lifespan regulation. *Frontiers in Physiology, 12*, 775648. https://doi.org/10.3389/fphys.2021.775648
10. Jin, J., Fang, F., Gao, W., Chen, H., Wen, J., Wen, X., & Chen, J. (2021). The structure and function of the Glycocalyx and its connection with blood-brain barrier. *Frontiers in Cellular Neuroscience, 15*, 739699. https://doi.org/10.3389/fncel.2021.739699
11. Dominguez, R., & Holmes, K. C. (2011). Actin structure and function. *Annual Review of Biophysics, 40*, 169–186. https://doi.org/10.1146/annurev-biophys-042910-155359

12. Ogneva, I. V., Buravkov, S. V., Shubenkov, A. N., & Buravkova, L. B. (2014). Mechanical characteristics of mesenchymal stem cells under impact of silica-based nanoparticles. *Nanoscale Research Letters, 9*, 1–10. https://doi.org/10.1186/1556-276X-9-284
13. Khalili, A. A., & Ahmad, M. R. (2015). A review of cell adhesion studies for biomedical and biological applications. *International Journal of Molecular Sciences, 16*, 18149–18184. https://doi.org/10.3390/ijms160818149
14. Legerstee, K., & Houtsmuller, A. B. (2021). A layered view on focal adhesions. *Biology, 10*, 1189. https://doi.org/10.3390/biology10111189
15. Herrmann, H., & Aebi, U. (2016). Intermediate filaments: Structure and assembly. *Cold Spring Harbor Perspectives in Biology, 8*, a018242. https://doi.org/10.1101/cshperspect.a018242
16. Goodson, H. V., & Jonasson, E. M. (2018). Microtubules and microtubule-associated proteins. *Cold Spring Harbor Perspectives in Biology, 10*, a022608. https://doi.org/10.1101/cshperspect.a022608
17. Mostowy, S., & Cossart, P. (2012). Septins: The fourth component of the cytoskeleton. *Nature Reviews. Molecular Cell Biology, 13*, 183–194. https://doi.org/10.1038/nrm3284
18. Liem, R. K. H. (2016). Cytoskeletal integrators: The spectrin superfamily. *Cold Spring Harbor Perspectives in Biology, 8*. https://doi.org/10.1101/cshperspect.a018259
19. Kular, J. K., Basu, S., & Sharma, R. I. (2014). The extracellular matrix: Structure, composition, age-related differences, tools for analysis and applications for tissue engineering. *Journal of Tissue Engineering, 5*. https://doi.org/10.1177/2041731414557112
20. García-Gareta, E., Pérez, M. Á., & García-Aznar, J. M. (2022). Decellularization of tumours: A new frontier in tissue engineering. *Journal of Tissue Engineering, 13*. https://doi.org/10.1177/20417314221091682
21. García-Gareta, E. (2023). Protein-based materials for wound healing. In F. R. A. Maia, J. M. Oliveira, & R. L. Reis (Eds.), *Handbook of the extracellular matrix: Biologically-derived materials* (pp. 1–28). Springer International Publishing. https://doi.org/10.1007/978-3-030-92090-6_20-1
22. Zihni, C., Mills, C., Matter, K., & Balda, M. S. (2016). Tight junctions: From simple barriers to multifunctional molecular gates. *Nature Reviews. Molecular Cell Biology, 17*, 564–580. https://doi.org/10.1038/nrm.2016.80
23. Meng, W., & Takeichi, M. (2009). Adherens junction: Molecular architecture and regulation. *Cold Spring Harbor Perspectives in Biology, 1*. https://doi.org/10.1101/cshperspect.a002899
24. Harris, T. J. C., & Tepass, U. (2010). Adherens junctions: From molecules to morphogenesis. *Nature Reviews. Molecular Cell Biology, 11*, 502–514. https://doi.org/10.1038/nrm2927
25. Müller, L., Hatzfeld, M., & Keil, R. (2021). Desmosomes as signaling hubs in the regulation of cell behavior. *Frontiers in Cell and Development Biology, 9*. https://doi.org/10.3389/fcell.2021.745670
26. Harmon, R. M., & Green, K. J. (2013). Structural and functional diversity of desmosomes. *Cell Communication & Adhesion, 20*, 171–187. https://doi.org/10.3109/15419061.2013.855204
27. Meşe, G., Richard, G., & White, T. W. (2007). Gap junctions: Basic structure and function. *The Journal of Investigative Dermatology, 127*, 2516–2524. https://doi.org/10.1038/sj.jid.5700770
28. Kanchanawong, P., Shtengel, G., Pasapera, A. M., Ramko, E. B., Davidson, M. W., Hess, H. F., & Waterman, C. M. (2010). Nanoscale architecture of integrin-based cell adhesions. *Nature, 468*, 580–584. https://doi.org/10.1038/nature09621
29. Borradori, L., & Sonnenberg, A. (1999). Structure and function of hemidesmosomes: More than simple adhesion complexes. *The Journal of Investigative Dermatology, 112*, 411–418. https://doi.org/10.1046/j.1523-1747.1999.00546.x
30. Walko, G., Castañón, M. J., & Wiche, G. (2015). Molecular architecture and function of the hemidesmosome. *Cell and Tissue Research, 360*, 363–378. https://doi.org/10.1007/s00441-014-2061-z
31. Barczyk, M., Carracedo, S., & Gullberg, D. (2010). Integrins. *Cell and Tissue Research, 339*, 269–280. https://doi.org/10.1007/s00441-009-0834-6

32. Gailit, J., Clarke, C., Newman, D., Tonnesen, M. G., Mosesson, M. W., & Clark, R. A. F. (1997). Human fibroblasts bind directly to fibrinogen at RGD sites through integrin $\alpha v\beta 3$. *Experimental Cell Research, 232*, 118–126. https://doi.org/10.1006/excr.1997.3512
33. Mezu-Ndubuisi, O. J., & Maheshwari, A. (2021). The role of integrins in inflammation and angiogenesis. *Pediatric Research, 89*, 1619–1626. https://doi.org/10.1038/s41390-020-01177-9
34. Kechagia, J. Z., Ivaska, J., & Roca-Cusachs, P. (2019). Integrins as biomechanical sensors of the microenvironment. *Nature Reviews. Molecular Cell Biology, 20*, 457–473. https://doi.org/10.1038/s41580-019-0134-2
35. Takada, Y., Ye, X., & Simon, S. (2007). The integrins. *Genome Biology, 8*, 215. https://doi.org/10.1186/gb-2007-8-5-215
36. Przekora, A. (2019). The summary of the most important cell-biomaterial interactions that need to be considered during in vitro biocompatibility testing of bone sca ff olds for tissue engineering applications. *Materials Science & Engineering C, 97*, 1036–1051.
37. Roach, P., Eglin, D., Rohde, K., & Perry, C. C. (2007). Modern biomaterials: a review—Bulk properties and implications of surface modifications. *Journal of Materials Science. Materials in Medicine, 18*, 1263–1277. https://doi.org/10.1007/s10856-006-0064-3
38. Tanaka, M., Morita, S., & Hayashi, T. (2021). Role of interfacial water in determining the interactions of proteins and cells with hydrated materials. *Colloids Surfaces B Biointerfaces, 198*, 111449. https://doi.org/10.1016/j.colsurfb.2020.111449
39. Dargaville, B. L., & Hutmacher, D. W. (2022). Water as the often neglected medium at the interface between materials and biology. *Nature Communications, 13*. https://doi.org/10.1038/s41467-022-31889-x
40. Davidenko, N., Schuster, C. F., Bax, D. V., Farndale, R. W., Hamaia, S., Best, S. M., & Cameron, R. E. (2016). Evaluation of cell binding to collagen and gelatin: a study of the effect of 2D and 3D architecture and surface chemistry. *Journal of Materials Science. Materials in Medicine, 27*. https://doi.org/10.1007/s10856-016-5763-9
41. García-Gareta, E., Levin, A., & Hook, L. (2019). Engineering the migration and attachment behaviour of primary dermal fibroblasts. *Biotechnology and Bioengineering*. https://doi.org/10.1002/bit.26927
42. Tzeranis, D. S., Soller, E. C., Buydash, M. C., So, P. T. C., & Yannas, I. V. (2016). In situ quantification of surface chemistry in porous collagen biomaterials. *Annals of Biomedical Engineering, 44*, 803–815. https://doi.org/10.1007/s10439-015-1445-x
43. Howlett, C. R., Evans, M. D. M., Walsh, W. R., Johnson, G., & Steele, J. G. (1994). Mechanism of initial attachment of cells derived from human bone to commonly used prosthetic materials during cell culture. *Biomaterials, 15*, 213–222. https://doi.org/10.1016/0142-9612(94)90070-1
44. Hayman, E. G., Pierschbacher, M. D., Suzuki, S., & Ruoslahti, E. (1985). Vitronectin—A major cell attachment-promoting protein in fetal bovine serum. *Experimental Cell Research, 160*, 245–258. https://doi.org/10.1016/0014-4827(85)90173-9
45. Greiling, D., Clark, R. A., Atkinson, S. J., d'Ortho, M. P., Yamada, K. M., Zardi, L., & Murphy, G. (1997). Fibronectin provides a conduit for fibroblast transmigration from collagenous stroma into fibrin clot provisional matrix. *Journal of Cell Science., 110*(7), 861–870. https://doi.org/10.1002/14651858.CD004837.pub3
46. Wang, K., Zhou, C., Hong, Y., & Zhang, X. (2012). A review of protein adsorption on bioceramics. *Interface Focus, 2*, 259–277. https://doi.org/10.1098/rsfs.2012.0012
47. Hlady, V., & Buijs, J. (1996). Protein adsorption on solid surfaces. *Current Opinion in Biotechnology, 7*, 72–77. https://doi.org/10.1016/S0958-1669(96)80098-X
48. Wertz, C. F., & Santore, M. M. (2001). Effect of surface hydrophobicity on adsorption and relaxation kinetics of albumin and fibrinogen: Single-species and competitive behavior. *Langmuir, 17*, 3006–3016. https://doi.org/10.1021/la0017781
49. Anselme, K., Ploux, L., & Ponche, A. (2010). Cell/material interfaces: Influence of surface chemistry and surface topography on cell adhesion. *Journal of Adhesion Science and Technology, 24*, 831–852. https://doi.org/10.1163/016942409X12598231568186

50. Keselowsky, B. G., Collard, D. M., & García, A. J. (2004). Surface chemistry modulates focal adhesion composition and signaling through changes in integrin binding. *Biomaterials, 25*, 5947–5954. https://doi.org/10.1016/j.biomaterials.2004.01.062
51. Chen, J. W., Lim, K., Bandini, S. B., Harris, G. M., Spechler, J. A., Arnold, C. B., Fardel, R., Schwarzbauer, J. E., & Schwartz, J. (2019). Controlling the surface chemistry of a hydrogel for spatially defined cell adhesion. *ACS Applied Materials & Interfaces, 11*, 15411–15416. https://doi.org/10.1021/acsami.9b04023
52. Al-Azzam, N., & Alazzam, A. (2022). Micropatterning of cells via adjusting surface wettability using plasma treatment and graphene oxide deposition. *PLoS One, 17*, e0269914. https://doi.org/10.1371/journal.pone.0269914
53. Arima, Y., & Iwata, H. (2007). Effect of wettability and surface functional groups on protein adsorption and cell adhesion using well-defined mixed self-assembled monolayers. *Biomaterials, 28*, 3074–3082. https://doi.org/10.1016/j.biomaterials.2007.03.013
54. Dowling, D. P., Miller, I. S., Ardhaoui, M., & Gallagher, W. M. (2010). Effect of surface wettability and topography on the adhesion of osteosarcoma cells on plasma-modified polystyrene. *Journal of Biomaterials Applications, 26*, 327–347. https://doi.org/10.1177/0885328210372148
55. Lee, S. J., Choi, J. S., Park, K. S., Khang, G., Lee, Y. M., & Lee, H. B. (2004). Response of MG63 osteoblast-like cells onto polycarbonate membrane surfaces with different micropore sizes. *Biomaterials, 25*, 4699–4707. https://doi.org/10.1016/j.biomaterials.2003.11.034
56. Hou, Y., Xie, W., Yu, L., Camacho, L. C., Nie, C., Zhang, M., Haag, R., & Wei, Q. (2020). Surface roughness gradients reveal topography-specific mechanosensitive responses in human mesenchymal stem cells. *Small, 16*. https://doi.org/10.1002/smll.201905422
57. Yi, B., Xu, Q., & Liu, W. (2022). An overview of substrate stiffness guided cellular response and its applications in tissue regeneration. *Bioactive Materials, 15*, 82–102. https://doi.org/10.1016/j.bioactmat.2021.12.005
58. El-Mohri, H., Wu, Y., Mohanty, S., & Ghosh, G. (2017). Impact of matrix stiffness on fibroblast function. *Materials Science and Engineering: C, 74*, 146–151. https://doi.org/10.1016/j.msec.2017.02.001
59. Voigt, J. L., Timmer, J., Pennarola, F., Christian, J., Meng, N., Blumberg, J. W., Schwarz, U. S., Grimm, D., & Cavalcanti-Adam, E. A. (2023). Substrate stiffness and particle properties influence cellular uptake of nanoparticles and viruses from the ventral side. *Advanced Functional Materials, 34*, 2304674. https://doi.org/10.1002/adfm.202304674
60. Karageorgiou, V., & Kaplan, D. (2005). Porosity of 3D biomaterial scaffolds and osteogenesis. *Biomaterials, 26*, 5474–5491. https://doi.org/10.1016/j.biomaterials.2005.02.002
61. Ikada, Y. (2006). *Tissue engineering: Fundamentals and applications* (1st ed.). Elsevier.
62. Sharma, V., Patel, N., Kohli, N., Ravindran, N., Hook, L., Mason, C., & García-Gareta, E. (2016). Viscoelastic, physical, and bio-degradable properties of dermal scaffolds and related cell behaviour. *Biomedical Materials, 11*. https://doi.org/10.1088/1748-6041/11/5/055001
63. O'Brien, F. J., Harley, B. A., Yannas, I. V., & Gibson, L. J. (2005). The effect of pore size on cell adhesion in collagen-GAG scaffolds. *Biomaterials, 26*, 433–441. https://doi.org/10.1016/j.biomaterials.2004.02.052
64. Abbasi, N., Hamlet, S., Love, R. M., & Nguyen, N.-T. (2020). Porous scaffolds for bone regeneration. *Journal of Science: Advanced Materials and Devices, 5*, 1–9. https://doi.org/10.1016/j.jsamd.2020.01.007
65. Wo, J., Huang, S., Wu, D., Zhu, J., Li, Z., & Yuan, F. (2020). The integration of pore size and porosity distribution on Ti-6A1-4V scaffolds by 3D printing in the modulation of osteo-differentation. *Journal of Applied Biomaterials & Functional Materials, 18*, 2280800020934652. https://doi.org/10.1177/2280800020934652

66. Ashworth, J. C., Mehr, M., Buxton, P. G., Best, S. M., & Cameron, R. E. (2016). Parameterizing the transport pathways for cell invasion in complex scaffold architectures. *Tissue Engineering Part C Methods, 22*, 409–417. https://doi.org/10.1089/ten.tec.2015.0483
67. Dias, M. R., Fernandes, P. R., Guedes, J. M., & Hollister, S. J. (2012). Permeability analysis of scaffolds for bone tissue engineering. *Journal of Biomechanics, 45*, 938–944. https://doi.org/10.1016/j.jbiomech.2012.01.019
68. Chao, L., Jiao, C., Liang, H., Xie, D., Shen, L., & Liu, Z. (2021). Analysis of mechanical properties and permeability of trabecular-like porous scaffold by additive manufacturing. *Frontiers in Bioengineering and Biotechnology, 9*, 779854. https://doi.org/10.3389/fbioe.2021.779854

Chapter 8
Mass Transfer, Vascularization, and Innervation of Scaffolds and Constructs

Abstract Mass transfer is a limiting factor when maintaining large tissue constructs in vitro. Once implanted, because of diffusion constraints, only cells located within 100–200 μm of the nearest capillary will receive the necessary nutrients. Therefore, tissue engineering and regenerative medicine (TERM) products intended for vascularized tissues should be vascularized prior to implantation or incorporate in their composition materials and factors that promote rapid vascularization upon implantation. Similarly, sensing of stimuli is of paramount importance for successful tissue regeneration. This chapter highlights the significance of both vascularization and innervation of TERM constructs and models. The different strategies used to achieve vascularized and innervated TERM materials are discussed. Our body also contains avascular tissues like cornea or cartilage, which are briefly discussed too. An overview on the anatomy and physiology of blood vessels, vascularization, nerves, and innervation is also presented to understand the complexity of incorporating nerves and blood vessels into term strategies. This chapter ends with conclusions and questions to strengthen the acquired knowledge.

8.1 Introduction

The majority of our body tissues need the nutrients and oxygen that are dutifully delivered by our blood vessel network [1–3]. Likewise, waste products like carbon dioxide produced by cell metabolism are removed by blood vessels and delivered to the organs that deal with their elimination from our body. The diffusion limit of oxygen is 100–200 μm. Therefore, tissue growth beyond 100–200 μm requires formation of new blood vessels [2, 3]. This has important implications for tissue engineering and regenerative medicine (TERM) strategies. In fact, one of the biggest challenges in TERM is effective mass transfer, which is an important limiting factor in large constructs and models built and maintained in vitro [4]. The challenge remains when large tissue constructs are implanted in vivo, since due to diffusion

© The Author(s), under exclusive license to Springer Nature Switzerland AG 2026

E. García-Gareta, J. M. García-Aznar, *Contemporary Tissue Engineering and Regenerative Medicine*, Series in BioEngineering,
https://doi.org/10.1007/978-3-032-03441-0_8

constraints, only the cells located within the mentioned distance to the nearest capillary will survive. Furthermore, after implantation, host blood vessels ingrow into the construct at a rate of several tenths of micrometres per day [5]. This means that several weeks are necessary for complete vascularization of scaffolds and constructs of just millimetres in size. Therefore, TERM scaffolds and constructs intended for vascularized tissues should be vascularized prior to implantation, meaning that an effective capillary network is formed within them [4]. However, as already mentioned, mass transfer in vitro remains a critical challenge [4].

Similarly, complex and advanced tissue models require vascularization. This is the case of organoids, for example, where there is a real challenge to vascularize them in order to keep them in culture for long periods of time [6]. Establishment of a vasculature is also necessary in models looking a pathological phenomena like cancer cell extravasation or blood-brain barrier dysfunction in Alzheimer's disease [7, 8].

Different strategies and approaches are being investigated for vascularization of scaffolds, constructs, and tissue models. For instance, four main approaches are used in scaffolds and constructs: (1) scaffold design, (2) inclusion and delivery of angiogenic factors, (3) in vitro prevascularization, and (4) in vivo prevascularization [1, 4]. The first two strategies rely on blood vessel ingrowth after implantation, whilst in the other two, the implanted material is connected to the host vasculature via microsurgery. Nevertheless, in the in vivo prevascularization strategy, the scaffold or construct needs to be pre-implanted to allow invasion from the host blood vessels before implantation in the target site [1]. Some of the mentioned approaches are also used in the vascularization of tissue models.

As with the continuous supply of nutrients, the sensing of stimuli is of paramount importance for successful tissue regeneration, remodelling, and functioning [9, 10]. For example, skeletal muscle comprises basic contractile units of muscle fibres, where each muscle fibre, known as myofiber, is innervated by a single motor neuron. These individual myofibers contract upon neuron activation, which eventually transforms into muscle movement [11]. Similarly, the central nervous system regulates bone development and skeletal remodelling. It has been shown that nerve reinnervation plays a major role in fracture healing, as well as in transmitting sensations and proprioception [9, 11]. These two examples illustrate the importance of innervation of tissue constructs and different strategies have been researched for this purpose.

The aim of this chapter is to highlight the importance of both vascularization and innervation of TERM constructs and models, as well as discuss the different strategies currently used. Avascular tissues will be briefly discussed too. The chapter also offers an overview on the anatomy and physiology of blood vessels, vascularization, nerves, and innervation. The chapter will end with conclusions on this subject and questions to strengthen the acquired knowledge.

8.2 Blood Vessels and Vascularization

Our circulatory system is composed of blood vessels, which have the vital function of transporting blood, which contains nutrients and waste products, to and from almost every tissue and organ of the human body [1–3]. Approximately, 8% of an adult human body mass corresponds to blood vessels, which stretch more than 100,000 km in total length [2]. Considering that the Earth's circumference is 40,075 km, we can get an idea of the size and relevance of the blood vessel network, where three distinct structures can be distinguished: macrovessels, i.e. arteries and veins, which branch out into microvessels, which are arterioles and venules that finally branch into capillaries of 2–12 μm diameter [1] (Fig. 8.1). About 80% of the total length is capillaries, which facilitate the actual distribution of nutrients to the tissues in the body [1].

In terms of blood vessel components, the lumen is lined with endothelial cells (ECs) that attach to an underlying basement membrane primarily composed of laminin, collagen IV, nidogen, collagen XV, collagen XVIII, fibronectin, and perlecan [13]. Situated below the basement membrane lies the intima, which acts as a barrier separating ECs from the internal elastic laminae (IEL), constructed from multiple layers of contractile vascular smooth muscle cells (SMCs) interspersed with elastic fibres and collagen-rich ECM [13]. In healthy vessels, the intima is typically minimal, transitioning directly to the media at the IEL. The media consists of concentric lamellar units comprising elastic fibres and SMCs divided by interlaminar matrix collagens, microfibrils, proteoglycans, glycoproteins, and ground substance.

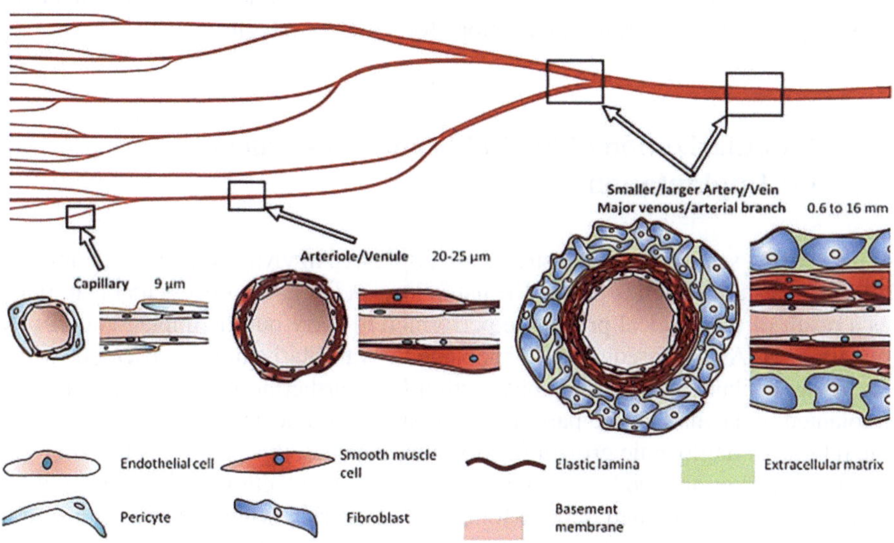

Fig. 8.1 Scheme of blood vessels and their relative sizes and composition. Image from Schöneberg et al. [12], open access article distributed under the terms and conditions of the Creative Commons Attribution (CC BY) license

However, the composition of media components may vary. For instance, arteries tend to have higher levels of collagens and elastin compared to veins. Beyond the SMCs layer of larger vessels lies an adventitial layer extending beyond the external elastic laminae, encompassing fibroblasts and other cells and a diverse range of ECM proteins such as fibrillar collagens I and III, fibronectin, and chondroitin sulfate and dermatan sulfate proteoglycans [13]. Under pathological circumstances, vascular matrix proteins experience proteolytic breakdown, producing bioactive fragments that impact the remodelling of the vascular wall matrix [13].

Blood vessel formation occurs in three stages [1]. Initially, in a process called vasculogenesis, de novo vessels form by differentiation of ECs precursors into ECs that proliferate to give rise to a primitive capillary plexus, that is later remodelled into a more mature and complex network in a process called angiogenesis [1, 2]. During angiogenesis, ECs degrade their surrounding matrix by using matrix metalloproteinases (MMPs) and migrate into the newly created gaps where new capillaries form [1, 2]. Finally, in a process called arteriogenesis, the pre-existing small vessels are structurally enlarged as well as remodelled into larger blood vessels [1]. Vasculogenesis was thought to occur only during embryonic development, however, it has been shown that post-natal vasculogenesis also occurs for example from bone-marrow derived endothelial progenitor cells in response to certain signals or tissue ischemia [14].

It is important to highlight that vessel maturation must accompany vessel formation [1]. In this process cells such as pericytes or SMCs supress ECs growth and stabilize vessels. Without this vessel maturation process, the resulting vascular network would be disorganized and leaky, among other issues. On the other hand, if vessel maturation kicks in too early, the resulting vascular network will be undeveloped [1]. This has important implications for TERM strategies.

8.3 Vascularization of Scaffolds and Constructs for Implantation

When TERM scaffolds and constructs are implanted in vivo, a typical inflammatory response ensues, which is part of the natural wound healing process that is initiated as a result of the surgical procedure performed to carry out the implantation of the material [1]. As a consequence, a spontaneous vascularization is observed, which occurs as explained in the previous section [1]. Furthermore, when constructs are implanted, the cells that are part of it generate a hypoxic environment that induces the release of angiogenic growth factors. Nevertheless, this process is too slow and further strategies are needed. As mentioned in the introduction of this chapter, these strategies can be classified into the following categories, which will be discussed in detail: (1) scaffold design, (2) inclusion and delivery of angiogenic factors, (3) in vitro prevascularization, and (4) in vivo prevascularization [1, 4].

8.3.1 Scaffold Design

As seen in previous chapters, scaffold design and tailoring in terms of architectural, mechanical, and compositional characteristics are essential for optimising and modulating cell-scaffold interaction. Similarly, these are key to maximise vascularization [15–18].

A study by Walthers and colleagues illustrated the importance of scaffold architecture for angiogenesis [16]. Specifically, they looked at scaffold macroporosity. The authors introduced 250 μm pores into multi-layered electrospun polycaprolactone scaffolds with macroporosity levels ranging from 15% to 25%, aiming to promote angiogenesis. The survival of green fluorescent protein (GFP)-expressing SMCs was assessed after 2 weeks of implantation. While host cellular infiltration remained consistent across scaffolds with varying macroporosities, the development of blood vessels increased with higher macroporosity levels. Scaffolds featuring 25% macropores exhibited the highest number of GFP-expressing SMCs, aligning with the most significant degree of angiogenesis extending up to 1 mm from the outermost layer. This indicated that the 25% macroporous scaffolds surpassed a critical threshold of macropore connectivity, thereby accelerating angiogenesis and enhancing the survival of implanted cells in a tissue-engineered smooth muscle construct [16].

The work by Mason and co-workers shows the importance of scaffold stiffness for angiogenesis [17]. Collagen solutions subjected to glycation before polymerization yielded collagen gels with a threefold enhancement in compressive modulus, while the collagen architecture was largely unchanged. Using these scaffolds, endothelial cell spreading, as well as the number and length of angiogenic sprouts and overall spheroid outgrowth, increased with matrix stiffness. Notably, differences in sprout length persisted even upon inhibition of the receptor for advanced glycation end products [17].

The inclusion of bioactive materials is a popular approach to increase vascularization of TERM scaffolds and constructs [2]. Natural polymers and ECM are the bioactive materials of choice to add to scaffolds, and examples are fibrin, collagen or decellularized native tissues [2, 15, 18–24]. Furthermore, bioactive glasses have also been investigated as they have the ability to induce angiogenesis [25–27]. For example, Liu and colleagues incorporated phosphosilicate calcium bioactive glass in alginate and gelatin bioinks for 3D bioprinting of full thickness skin substitutes, which showed upregulation of angiogenic genes in vitro and enhanced vascularization in wounds in vivo [28].

Fibrin is a critical protein in blood clotting, essential for wound healing and preventing excessive bleeding [29, 30]. Fibrinogen, a soluble protein in blood plasma, is converted into fibrin by the enzyme thrombin during blood clotting. Fibrin molecules then aggregate to form a mesh-like network, providing a scaffold for platelets and other blood components to adhere to, ultimately forming a stable clot [29, 30]. Thus, fibrin is the first bio-matrix that the human body builds in response to injury. Importantly for TERM strategies, fibrin is a pro-angiogenic natural polymer and as such, it has been shown to increase vascularization of TERM scaffolds. For instance, the work by Kohli and colleagues showed that composite scaffolds of fibrin, alginate and calcium-phosphate (CaP) had intrinsic pro-angiogenic capabilities [19]. The composite scaffolds were tested in a non-sentient in vivo chick chorioallantoic membrane (CAM) model showing that they were pro-angiogenic and biocompatible (Fig. 8.2). It was concluded that not only the scaffolds composition but also their high porosity contributed towards their angiogenic properties [19], as it was shown

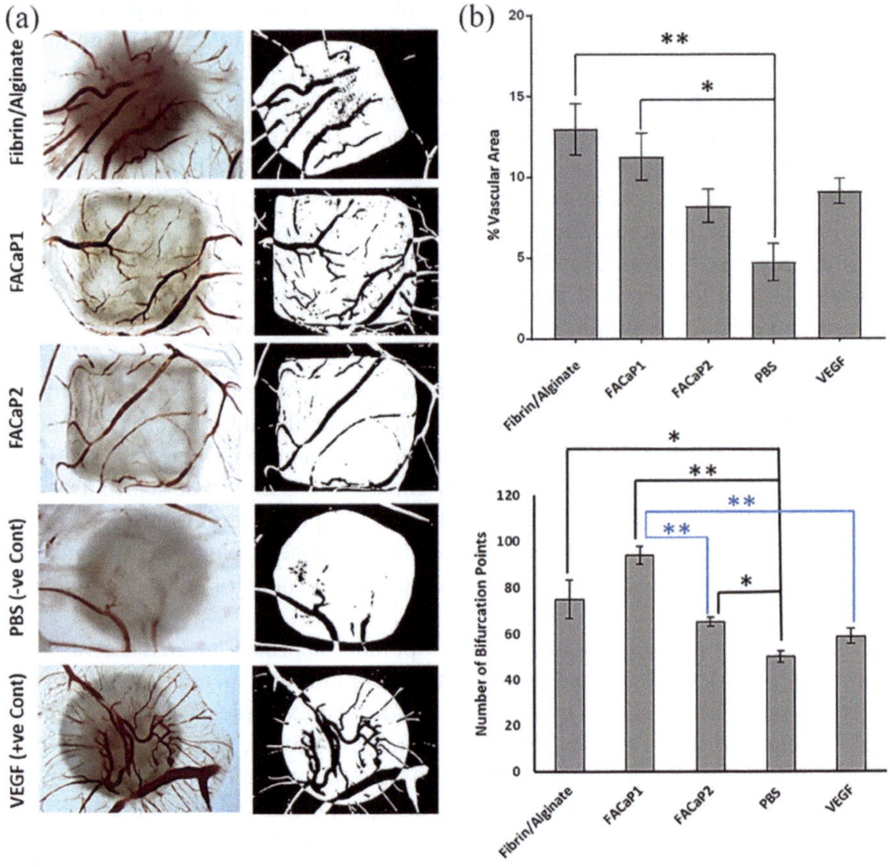

Fig. 8.2 (**a**) Representative macroscopic (left) and binary images (right) of the fibrin-based scaffolds (FACaP means scaffolds of fibrin, alginate and CaP) and controls tested in a CAM model. The negative control (-ve Cont) is a filter disc soaked in phosphate buffered saline (PBS), whilst the positive control (+ve Cont) is a filter disc soaked in vascular endothelial growth factor (VEGF). The blood vessels are observed in red (left panel) and black colour over white background of the scaffold (right panel). (**b**) Percentage vascular area and bifurcation points of blood vessels within the materials. Data show mean ± SEM of n = 6 for each material (except for the number of bifurcation points for FA and FACaP1 where n = 5). *$p < 0.05$. **$p < 0.01$. Image from Kohli et al. [19], open access article distributed under the terms and conditions of the Creative Commons Attribution (CC BY) license

using the same CAM model that composition and architecture of scaffolds are important for angiogenesis [15]. Nevertheless, the mechanical properties of fibrin are quite poor, and therefore, it is often combined with more mechanically robust materials like polyesters or silk fibroin [20, 31, 32]. Finally, fibrin can also be used in bioinks for bioprinting of pro-angiogenic TERM constructs [23].

As seen in previous chapters, collagen is the most abundant component of the ECM of tissues, including the ECM of blood vessels [33]. Furthermore, collagen is also the main constituent of decellularized native tissues [34–36]. It has been shown that collagen I interacts with ECs during both embryonic and pathological angiogenesis [2]. The formation of endothelial capillary-like lumens relies on the engagement of endothelial $\alpha2\beta1$ integrin with a specific binding site on collagen I, potentially triggering signalling pathways involving p38 MAPK and the deactivation of focal adhesion kinase (FAK) [2]. Given the pivotal role of collagen in the initial phases of angiogenesis, this knowledge has been utilized to stimulate angiogenesis in TERM scaffolds. An example is the work by He and co-workers, where diabetic foot ulcers were targeted as they remain a significant clinical challenge, with neovascularization being crucial in all stages of tissue repair [24]. To address this challenge, a novel multi-functional pro-angiogenic molecule, LXW7-DS-SILY, was developed by combining the cyclic peptide LXW7, which binds to integrin $\alpha v\beta3$ on ECs, with collagen-binding peptides (SILY) on a dermatan sulfate (DS) glycosaminoglycan backbone. This molecule was then used to modify collagen-based ECM scaffolds. In vitro studies confirmed that LXW7-DS-SILY significantly enhanced endothelial progenitor cell (EPC) attachment and growth on the ECM scaffolds, while in vivo experiments demonstrated increased EPC survival in ischemic conditions, leading to improved wound healing, enhanced neovascularization, and modulation of collagen fibrillogenesis [24]. A last example on decellularized tissues is the work by Shariatzadeh and colleagues, on placenta membranes [37]. Decellularized and de-epithelialized placenta membranes have found widespread use as scaffolds and grafts in TERM due to their exceptional pro-angiogenic and biomechanical properties, coupled with low immunogenicity [38]. In this study, an optimized combination of enzymatic solutions and mechanical scrapping effectively removed the amniotic epithelium and chorion trophoblastic layer, leaving the basement membranes intact on both sides without causing damage to the spongy layer in between. Both sides of the basement membranes could support ECs growth and promoted angiogenesis in vivo, concluding that this bi-sided substrate holds promise for applications in skin, cardiac, vascularized composite allografts, and microvascular tissue engineering [37].

8.3.2 Inclusion and Delivery of Angiogenic Factors

The integration of angiogenic factors into tissue-engineered constructs has been widely acknowledged to improve vascularization post-implantation [39]. These factors can target various stages of blood vessel formation, enhancing the development of a tissue-engineered graft's vasculature [39]. Growth factors like vascular endothelial growth factor (VEGF) or basic fibroblast growth factor (bFGF) promote the mobilization and recruitment of ECs, expediting angiogenesis [39–41]. An example is the work by Zigdon-Giladi and colleagues, who fabricated electrospun nanofibrous scaffolds made of polycaprolactone (PCL) and polyethylene glycol (PEG) that supported the release of recombinant human VEGF, thereby significantly enhancing the migration of ECs and the formation of blood vessels [41]. However, excessive delivery of these factors can lead to the formation of disorganized, leaky, and haemorrhagic vessels, necessitating precise dosage control, especially considering VEGF's potential to cause severe vascular leakage and hypotension [1]. Additionally, it is crucial to stabilize newly formed vessels, often achieved through the recruitment of SMCs or pericytes and the production of an ECM. Key growth factors involved in vessel stabilization include platelet-derived growth factor (PDGF), transforming growth factor β (TGF-β), and angiopoietin 1 (Ang1) [1]. For instance, do Amaral et al. observed and reported that PDGF stabilised the vascularization of collagen-glycosaminoglycan scaffolds in vitro by upregulation of angiogenic genes that are involved in vessel maturation [42]. Since both vessel formation and stabilization are vital for establishing a functional vascular network within a tissue-engineered graft, delivering complementary sets of factors that promote both processes may be an ideal strategy.

Apart from directly delivering factors that promote vessel formation or maturation, alternative approaches have been explored, which involves delivering other substances, such as sonic hedgehog homolog (SHH), hypoxia-inducible factor 1 (HIF-1), therapeutic ions, or bone morphogenetic proteins (BMP-2, -4, or -6), which stimulate nearby cells to produce angiogenic factors [1, 43]. As an example, the work by Perez et al. who used Cobalt (Co) ion as a pro-angiogenic factor together with BMP-2 in hydrogel fibre scaffolds, ultimately achieving bone regeneration through the sequential delivery of both factors [43]. Their approach aimed at regenerating bone through angiogenesis [43]. The indirect method to stimulate cells to produce angiogenic factors offers several advantages. Firstly, cellular secretion of angiogenic factors is often regulated, ensuring their concentration remains within physiological levels and can adapt to different stages of vessel formation. Secondly, the production of these factors creates microgradients, crucial for capillary morphogenesis. Lastly, stimulation with indirect factors often leads to the secretion of multiple angiogenic factors, regulating both vessel formation and maturation.

8.3.3 *In Vitro Prevascularization*

An intriguing method for enhancing vascularization is in vitro prevascularization [1]. The concept relies on the ability of ECs to form prevascular structures under specific in vitro conditions. During in vitro prevascularization, ECs are introduced into scaffolds, leading to the development of a prevascular network within the construct [2]. Upon implantation, this network can spontaneously connect with the host's growing vasculature, providing nutrients to the construct [1, 2]. Unlike the traditional method where host blood vessels must infiltrate the entire construct, here they only need to reach the outer regions until they encounter the prevascular structures. Although this accelerates vascularization from weeks to days, perfusion may not be as rapid post-implantation if the vascular network is not microsurgically connected [1]. Nonetheless, future advancements in this area may involve establishing a vascular axis within the in vitro construct, potentially enabling surgical connection to the host vasculature.

The importance of ECs in vascularization for TERM applications is well recognized, yet two significant challenges remain regarding their use as a cell source [2]. Firstly, obtaining terminally differentiated ECs, such as human primary aortic endothelial cells (HAEC) and human primary dermal microvascular endothelial cells (HDMVEC), involves limited expansion potential and necessitates invasive procedures for isolation from patients. Moreover, variability among donors, tissue specificity, and diminished functionality in in vitro cultures further restrict their applicability in TERM. Secondly, ECs alone cannot establish a sustained, self-renewing, and fully functional vascular network. Hence, addressing the need for supportive cells like pericytes, fibroblasts, and vascular SMCs is crucial. Vascular SMCs are also widely used as a source of cells for vascularization in TERM. However, they present similar issues to those discussed for ECs in terms of isolation and expansion. Consequently, a wide range of cell types has garnered interest as potential alternatives, including pluripotent stem cells such as embryonic stem cells (ESCs) and induced pluripotent stem cells (iPSCs), as well as various multipotent adult stem/progenitor cells like mesenchymal stem cells (MSCs), umbilical cord blood-derived stromal cells (UCBSCs), and endothelial progenitor cells (EPCs) [2]. These cell types offer advantages in terms of their ability to be expanded in vitro due to their high self-renewal capacity, thereby facilitating clinically relevant applications [2]. Additionally, each cell source possesses unique differentiation and angiogenesis potential, serving as viable alternatives to primary ECs and vascular SMCs.

There are multiple studies in the literature where the abovementioned cells were investigated for prevascularization of scaffolds [44–47]. The work by Mishra and colleagues serves as an example [44]. This study introduced a new composite scaffold consisting of poly(propylene fumarate) (PPF) and fibrin, aimed at fostering the growth of vascularized neobone tissue. The authors investigated the impact of prevascularization, involving in vitro pre-culture before implantation, utilizing human MSCs and human umbilical vein endothelial cells (HUVECs), on the vascularization of these scaffolds in vivo. Five conditions were examined: no pre-culture (NP), 1-week pre-culture (1P), 2-week pre-culture (2P), 3-week pre-culture (3P), and scaffolds without cells (control, C). Subcutaneous implantation of the scaffolds was carried out in an immunodeficient mouse model for 9 days. Comparing the NP and 3P groups, a significantly higher vascular density was observed in vivo. Immunohistochemical staining for human CD-31 expression demonstrated the expansion of vascular networks with prolonged pre-culture time. Furthermore, perfused lectin staining in human CD-31 positive vessels indicated the perfusion of these vascular networks with mouse blood [44]. These findings underscore the supportive role of in vitro prevascularization in promoting in vivo vascularization within scaffolds.

8.3.4 In Vivo Prevascularization

An approach to enhance vascularization of TERM scaffolds and constructs is through in vivo prevascularization, also known as tissue prefabrication, which involves two distinct phases [1, 47, 48]. Initially, a tissue-engineered implant is placed into a region containing a suitable artery or vascular axis for microsurgical transfer. This may involve encasing the tissue-engineered graft in axially vascularized tissue like muscle, or directly implanting an artery into the graft. Over several weeks, vascularization occurs at the initial implant site, leading to the formation of a microvascular network in the tissue-engineered implant, nourished by the vascular axis. Following this stage, the tissue-engineered implant, along with the microvascular network and supplying artery, is harvested and re-implanted at the defect site. Microsurgical vascular-anastomosis techniques are then employed to connect the vascular axis to the local vasculature, ensuring instant perfusion of the entire construct. The in vivo prevascularization approach offers the advantage of immediate perfusion post-implantation, but its main disadvantage is that it requires two separate surgeries [1].

An example of this approach is the work by Debski and colleagues, who used both adipose-derived stem cells (ADSCs) and surgical prefabrication methods to stimulate angiogenesis within PCL cylindrical scaffolds fabricated by fused deposition modelling [47]. The study's objective was to compare the angiogenic efficacy of ADSC-seeded scaffolds combined with scaffold prefabrication against that of non-seeded, non-prefabricated scaffolds. To induce prefabrication, functional blood vessels were integrated into the scaffold using a flow-through pedicle system. Three groups were studied: non-seeded, ADSC-seeded, and osteogenic-induced ADSC-seeded scaffolds. In each group, one rat received two scaffolds implanted in the inguinal region. On the right side, a scaffold was subcutaneously implanted around the inferior epigastric vessels (classic prefabrication group). On the left side, the inferior epigastric vessels were inserted into the prefabricated scaffold using the flow-through pedicle system (flow-through prefabrication group). Two months post-implantation results showed that the mean vessel densities were 10- and five-fold higher in the ADSC-seeded and osteogenic-induced ADSC-seeded scaffolds with flow-through prefabrication, respectively, compared to the non-seeded classic prefabricated group ($p < 0.001$). Microcomputed tomography imaging revealed the presence of functional vessels within the scaffold. The authors concluded that ADSC-seeded scaffolds with prefabrication exhibited significantly enhanced scaffold vasculogenesis, suggesting their potential utility in clinical tissue engineering applications [47].

8.4 Vascularization of Tissue Models

TERM also concerns the engineering of tissue models for the study of diseases, physiological processes, or investigation of new therapies. Complex and advanced tissue models require vascularization. These include organoids [6], or models looking a pathological phenomena like cancer cell extravasation or blood-brain barrier dysfunction in Alzheimer's disease [7, 8]. Particularly, vascularization of organoids has emerged as a prominent challenge in the field of TERM.

8.4.1 Vascularization of Organoids

Organoids are an excellent example of technological breakthrough achieved over the last two decades [49–51]. They are defined as self-organized 3D cell aggregates containing organ specific cell types that recapitulate the structure, composition, physiology and function of their in vivo counterpart [49]. Organoids mirror numerous pivotal processes in organogenesis, rendering them invaluable as models in developmental biology. Furthermore, organoids have found widespread application in exploring a spectrum of genetic disorders and diseases [50]. Additionally, the integration of organoids into disease modelling has paved the way for innovative approaches in drug screening and personalized medicine [49, 51].

Despite remarkable progress, several limitations constrain the broader application and impact of organoids [49]. One significant limitation is the absence of a perfusable vasculature [6]. In 3D culture, organoids rely solely on passive diffusion for nutrient, oxygen, and waste exchange. However, due to limitations in mass transfer, a necrotic core develops over time [6]. Introducing vasculature could significantly enhance size and lifespan, potentially enabling organoids to emulate mature organs and facilitating their use in organ replacement. Moreover, vasculature incorporation allows for proper regional patterning and intercellular communication, thereby more accurately recapitulating in vivo physiology and functionality. Additionally, vasculature endothelium offers paracrine signalling and facilitates basement membrane interactions with other cell types, potentially enhancing organoid maturation [6].

Several approaches are being investigated for achieving vascularization of organoids, including co-culture with ECs, co-differentiation with mesodermal progenitor cells, and mechanical stimulation [6]. However, perfusion of organoid models remains an important challenge, which would be solved by anastomosis of the internal organoid vasculature and the external vessels present in the ECM [6–8]. For this purpose, the engineering of perfusable capillary beds is of key importance, and can be achieved by producing vascularized hydrogels and using microfluidic platforms (Fig. 8.3) [7, 8]. Hydrogel design is an important approach, where pro-angiogenic polymers like the previously discussed fibrin are used and seeded with cells such as ECs to achieve a microvascular network within the fibrin-based hydrogel [7].

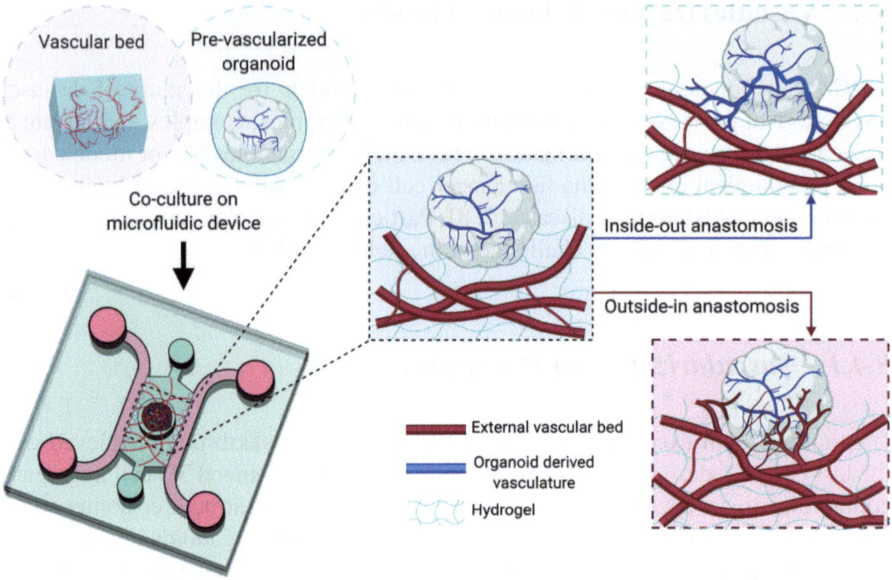

Fig. 8.3 Diagram illustrating potential methods for forming connections between a pre-vascularized organoid and an in vitro vascular bed when co-cultured on a microfluidic chip. Upon placement of the organoid, functional links between vessels within the hydrogel and the organoid's native vasculature can occur either 'inside-out' or 'outside-in'. In the former, the organoid-derived vasculature expands into the surrounding ECM and connects with the pre-existing microvascular network in the hydrogel. In the latter, angiogenic sprouts emerge from ECs within the vascular bed, penetrating the organoids and linking with the organoid-derived vessels. Image from Zhang et al. [6], open access article distributed under the terms and conditions of the Creative Commons Attribution (CC BY) license

Pre-patterning techniques have also been used in order to form networks of 3D channels with pre-determined geometry in hydrogels. These channels are then seeded with ECs that eventually develop a monolayer on the channels' walls [6]. Although important advances have been made, improvements are needed concerning tuning of the mechanical properties of the natural hydrogels used, cell sources of ECs, and eventually long-term cultures of anastomosed organoids to functional capillary beds.

8.5 Avascular Tissues in Tissue Engineering and Regenerative Medicine

Our body has tissues that do not contain blood vessels and are referred to as avascular tissues. Examples are cornea and articular cartilage [10, 52]. TERM strategies for regeneration of these tissues need to take into account their avascular nature, and consider the factors present in their microenvironment that drive tissue formation and homeostasis [10, 52]. For instance, given the avascular aspect of articular cartilage, the metabolism of chondrocytes and tissue homeostasis depend on nutrients, oxygen and regulatory factors diffusion from the subchondral bone and synovial fluid through the ECM, which is a fibrous network of mainly collagen type II, proteoglycans, and water [52]. Factors present in the chondrocyte microenvironment that are vital for tissue homeostasis and function include oxygen tension, mechanical stimulation, and cytokines and growth factors [52, 53]. The articular cartilage microenvironment is hypoxic due to the avascular nature of the tissue, and it has been shown that hypoxia promotes cartilage ECM production and accumulation [52]. Chondrocytes regularly experience compressive loads, shear, as well as tensile stress [53]. Mechanical stress is necessary for chondrogenesis in fetal development as well as for maintaining cartilage homeostasis in the adult body [53]. Cytokines such as transforming growth factor β (TGF-β), BMPs and insulin-like growth factor-1 (IGF-1) stimulate the synthesis of ECM by chondrocytes during development [52]. Strategies for regeneration of lost articular cartilage need to take into account these various microenvironmental factors to achieve the distinct functional characteristics of this highly specialised connective tissue.

8.6 The Nervous System, Nerves and Innervation

The nervous system comprises the central nervous system (CNS), which includes the brain and spinal cord, and the peripheral nervous system (PNS), encompassing the network of nerves throughout the body [54]. Alongside its role in higher-order functions, the CNS integrates sensory input to indirectly regulate the activities of tissues and organs. This coordination occurs via connections between the CNS and PNS nerves, facilitated by axonal projections from neurons [54]. Axonal connections play a crucial role in the nervous system's involvement in tissue and organ development, function, and regulatory processes, as well as in instances of regeneration and pathology [54]. The autonomic nervous system (ANS), responsible for involuntary functions, operates through its parasympathetic, sympathetic, and enteric components [55]. These systems, originating from the spinal cord, regulate responses to rest and stress in coordination with the CNS [55]. The enteric nervous system (ENS) operates within the gastrointestinal tract, influencing gut function autonomously while still being modulated by parasympathetic and sympathetic activity [56]. Conversely, the somatic nervous system (SNS) governs voluntary control over skeletal muscle and facilitates sensory input transmission to the CNS [57].

Neurons, which are electrically responsive cells, convey signals across the body [58, 59]. They utilize a combination of electrical and chemical elements for information transmission. Neurons establish connections with other neurons at synapses and with effector organs or cells at neuroeffector junctions [58]. A neuron comprises a soma, its central cell body, and at least one neurite branch [59]. Neurites conveying signals away from the soma are termed axons, while those relaying signals towards the soma are dendrites. The soma houses the nucleus and typical cellular organelles such as mitochondria. Notably, the rough endoplasmic reticulum, known as Nissl substance in neurons, is prominently featured as it synthesizes a significant amount of membrane for the neurites [59]. Most axons and lengthy dendrites are enveloped in a protective sheath called myelin, which acts as an electrical insulator, enhancing impulse conduction speed. Myelin forms as non-neuronal glial cells wrap tightly around neurites, displacing their cytoplasm to the periphery, leaving only their membrane in the inner layer. Myelin covers neurites partially, leaving periodic gaps known as nodes of Ranvier, crucial for impulse transmission. In the CNS, myelin is produced by oligodendrocytes, while in the PNS, Schwann cells generate myelin bundles for individual axon internodes [59].

Axons from multiple neurons aggregate to form a nerve [59]. Each axon within a nerve is surrounded by a connective tissue layer called the endoneurium, which secretes endoneurial fluid, thereby safeguarding axons during injury [59]. Numerous axons assemble into fascicles, each fascicle encased in the perineurium layer. Lastly, the outermost layer, the epineurium, envelops multiple fascicles and nerve blood vessels [59]. Various nerve fiber types have been categorized, with some participating in both motor and sensory pathways, while others exclusively serve sensory pathways [58].

Innervation is key in regulating and modulating the function of tissues and organs [60]. For instance, autonomic nerves govern various functions like vasoconstriction, pancreatic secretions, urination, gut motility, and saliva production by processing sensory information about environmental and physiological conditions within the organs [60]. There is growing acknowledgment of the vital role innervation plays in organ development and regeneration. For example, autonomic nerves aid in organogenesis, wound healing, and tissue regrowth by preserving phenotypes and functions within stem cell niches, and by providing growth and transcription factors crucial for maintaining migrating cells in wounds. Studies have reported significant roles for nerves during development, such as pancreatic sympathetic innervation affecting islet cytoarchitecture and functional maturation, and parasympathetic innervation regulating tubulogenesis in the salivary gland [60].

8.7 Innervation in Tissue Engineering and Regenerative Medicine

Sensing of stimuli is intrinsic to tissue function and therefore, in order to regenerate tissues innervation of tissue-engineered products is necessary [60]. However, innervation is generally overlooked when designing and fabricating TERM scaffolds and constructs.

Typically, implanted tissue-engineered materials lack pre-existing neural networks and rely on the host's natural nerve growth to integrate with the native nerve supply [60, 61]. This phenomenon has been observed in previous endeavours involving engineered urinary bladders, skin, and intestines [61]. While there have been studies on transplanting nerve grafts into target organs and some attempts to create pre-innervated engineered tissue, these efforts remain limited, with insufficient emphasis on fostering targeted and appropriate connections between the host nervous system and the tissue-engineered materials [60]. The overall lack of attention to innervation in TERM primarily stems from the immense complexity of many tissues and organs under investigation. Additionally, there is a need for greater understanding and recognition within the field regarding the vital role innervation plays in the function of native tissues and organs, as well as the potential benefits for tissue-engineered constructs.

Nevertheless, the literature shows encouraging examples of studies where innervation of scaffolds was investigated using innovative approaches in scaffold design. A recent example is the work by Westphal and colleagues, who employed electrospun polyvinylidene-trifluoroethylene (PVDF-TrFE) due to its biocompatibility, mechanical properties, and piezoelectric characteristics to encourage the elongation of Schwann cells (SC) and extension of sensory neurons (SN) [62]. Various electrospinning parameters were evaluated to determine the optimal thickness, porosity, and tensile strength of PVDF-TrFE electrospun scaffolds for engineered skin tissue applications. The PVDF-TrFE scaffolds were electrically activated by mechanical deformation using low-intensity pulsed ultrasound (LIPUS) waves, serving as a non-invasive method to stimulate the piezoelectric properties of the scaffold and deliver electric potential to cells. Utilizing this therapeutic approach, the integration of neurites in tissue-engineered skin substitutes was assessed, including neurite alignment, elongation, and penetration into PVDF-TrFE scaffolds. Results indicated that LIPUS stimulation promoted cell alignment on aligned scaffolds and significantly increased the elongation of SCs and extension of SNs, both individually and in coculture, on aligned scaffolds, while it decreased elongation and extension on unaligned scaffolds (Fig. 8.4). This trend was also observed in the analysis of cell penetration depth into the scaffolds, where LIPUS enhanced the penetration of SCs, SNs, and cocultures. Overall, this study highlights the significant potential of non-invasive electric stimulation [62].

An early example of investigation of innervation following scaffold implantation is the study by Agrawal and co-workers [61], who undertook a retrospective analysis to examine the presence of nervous tissue in two animal models: abdominal wall reconstruction in rats and esophageal reconstruction in canines, both utilizing ECM scaffolds as inductive frameworks. Specifically, the ECM scaffolds were decellularized urinary bladder matrix. Mature nerve fibers, immature nerves, and Schwann cells were observed within the remodelled ECM after 28 days in the rat model and after 91 days in the canine model. Furthermore, a microscopic and morphological investigation into primary cultured neurons seeded on ECM scaffolds revealed that the ECM substrate facilitated neuronal survival and outgrowth. Lastly, the rapid degradation of the ECM scaffold led to the generation of matricryptic peptides, which in vitro, prompted concentration-dependent migration of terminal SCs. These findings suggested that utilizing decellularized ECM scaffolds enables tissue reconstruction, particularly in cases where innervation plays a crucial functional role [61].

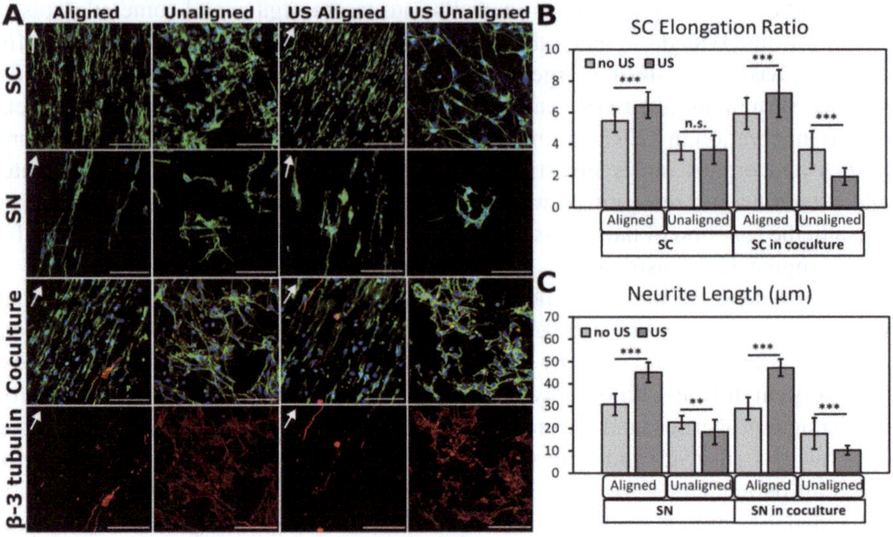

Fig. 8.4 Ultrasound stimulation of SCs and SNs on electrospun PVDF-TrFE scaffolds. (**a**) Confocal microscopy images of SCs, SNs and coculture on aligned and unaligned PVDF-TrFE scaffolds stimulated with ultrasound. Cytoskeleton stained with fluorescein phalloidin (green), nuclei stained with DAPI (blue), and SNs labelled with beta-III tubulin in cocultures. White arrows indicate direction of fiber alignment. Scale bar = 100 μm. (**b**) SC elongation and (**c**) neurite length on aligned and unaligned scaffolds with and without ultrasound treatment. Data show mean ± SD. ** $p \leq 0.005$; *** $p \leq 0.0005$. *n.s.* no significance. Image from Westphal et al. [62], open access article distributed under the terms and conditions of the Creative Commons Attribution (CC BY) license

Finally, there are examples of combined strategies for achieving both vascularization and innervation of scaffolds intended for regeneration of tissues that require both phenomena. This is the case of bone, for instance, which is a highly vascularized and innervated organ where growth of both blood vessels and nerves is necessary for its regeneration, healing and remodelling [9, 11]. To facilitate the regeneration of bone, simultaneous ingrowth of blood vessels and nerves is indispensable, yet often overlooked. To tackle this challenge, Qin et al. developed a "pre-angiogenic" bioprinted scaffold laden with HUVECs drawing from the bioactivities of silicate bioceramics and the instructive influence of vascular cells on both neurogenesis and bone repair [46]. In comparison to conventional cell-free scaffolds, these HUVEC-laden scaffolds, demonstrated the ability to sustain cell survival and growth for up to 3 weeks. Furthermore, these long-lasting scaffolds exhibited persistent angiogenic capabilities both in vitro and in vivo. The pre-angiogenic scaffolds promoted the differentiation of neural cells towards neurogenesis and mesenchymal stem cells towards osteogenesis, owing to the combined effects of released bioactive ions and the neuro-attractant properties of vascular cells. The resulting augmented bone regeneration, characterized by both vascularization and innervation, defined a novel paradigm termed "vascular-innervated" bone regeneration [46]. This underscores the potential of "vascular-innervated scaffolds" as the forefront of biomaterials innovation for the intricate regeneration of complex tissues and organs.

8.8 Conclusions

Mass transfer is an important limiting factor when maintaining large tissue constructs in vitro for implantation. Once implanted, because of diffusion limits, only cells located within 100–200 μm of the nearest capillary will receive the necessary nutrients for their survival. Therefore, ideally, TERM scaffolds and constructs targeting vascularized tissues should be vascularized prior to implantation or incorporate in their composition materials and factors that promote rapid vascularization upon implantation. Similarly, complex and advanced tissue models such as organoids require a perfusable vasculature for prolonged tissue culture and their potential use in organ replacement. Several strategies are discussed, such as (1) the importance of scaffold design in terms of both architecture and composition, (2) the inclusion of pro-angiogenic factors that upon their release in vivo they either promote vascularization of the implant or induce the local cells to produce angiogenic factors, (3) in vitro pre-vascularization by means of incorporating endothelial cells, smooth muscle cells, endothelial progenitor cells, or pluripotent or multipotent stem cells, and (4) in vivo pre-vascularization by implanting the TERM material in a region containing a suitable artery or vascular axis, followed by re-implantation of the vascularized implant in the target site using microsurgical vascular-anastomosis techniques. All the discussed approaches have their advantages and disadvantages and the challenge of producing vascularized TERM implants and complex models still remains. Finally, there are avascular tissues in our body, like cartilage or cornea, and TERM strategies for their regeneration must take into account their avascular nature and look at the microenvironmental factors involved in tissue formation and homeostasis.

Similar to the continuous supply of nutrients, sensing of stimuli is of paramount importance for successful tissue regeneration, remodelling, and functioning. The importance of innervation has been shown to be vital for effective tissue and organ regeneration. Unfortunately, TERM strategies generally overlook innervation when producing TERM scaffolds and constructs. Nevertheless, there are studies that target innervation by scaffold design, such as using piezoelectric scaffolds or naturally-derived ECM materials.

This chapter highlights the importance of including vascularization and innervation of scaffolds and constructs in TERM strategies.

Questions

- What function have blood vessels?
- Which is the minimum distance from the nearest capillary that cells need to be to receive nutrients?

- Name four strategies employed in the vascularization or scaffolds and constructs intended for implantation.
- Name two natural polymers used in the design of pro-angiogenic scaffolds.
- Name two growth factors that promote the mobilization and recruitment of endothelial cells.
- Name three cell types used in in vitro prevascularization of constructs.
- What would be the advantages of introducing a perfusable vasculature in organoids?
- Name examples of avascular tissues of our body.
- What function have neurons?
- Name an approach used in the innervation of scaffolds and constructs.

References

1. Rouwkema, J., Rivron, N. C., & van Blitterswijk, C. A. (2008). Vascularization in tissue engineering. *Trends in Biotechnology, 26*, 434–441. https://doi.org/10.1016/j.tibtech.2008.04.009
2. Yang, G., Mahadik, B., Choi, J. Y., & Fisher, J. P. (2020). Vascularization in tissue engineering: Fundamentals and state-of-art. *Progress in Biomedical Engineering, 2*, 012002. https://doi.org/10.1088/2516-1091/ab5637
3. Masson-Meyers, D. S., & Tayebi, L. (2021). Vascularization strategies in tissue engineering approaches for soft tissue repair. *Journal of Tissue Engineering and Regenerative Medicine, 15*, 747–762. https://doi.org/10.1002/term.3225
4. Lovett, M., Lee, K., Edwards, A., & Kaplan, D. L. (2009). Vascularization strategies for tissue engineering. *Tissue Engineering. Part B, Reviews, 15*, 353–370. https://doi.org/10.1089/ten.teb.2009.0085
5. Clark, E. R., & Clark, E. L. (1939). Microscopic observations on the growth of blood capillaries in the living mammal. *The American Journal of Anatomy, 64*, 251–301. https://doi.org/10.1002/aja.1000640203
6. Zhang, S., Wan, Z., & Kamm, R. D. (2021). Vascularized organoids on a chip: strategies for engineering organoids with functional vasculature. *Lab on a Chip, 21*, 473–488. https://doi.org/10.1039/D0LC01186J
7. Ko, E. C., Spitz, S., Pramotton, F. M., Barr, O. M., Xu, C., Pavlou, G., Zhang, S., Tsai, A., Maaser-Hecker, A., Jorfi, M., Choi, S. H., Tanzi, R. E., & Kamm, R. D. (2023). Accelerating the in vitro emulation of Alzheimer's disease-associated phenotypes using a novel 3D blood-brain barrier neurosphere co-culture model. *Frontiers in Bioengineering and Biotechnology, 11*. https://doi.org/10.3389/fbioe.2023.1251195
8. Kim, S., Wan, Z., Jeon, J. S., & Kamm, R. D. (2022). Microfluidic vascular models of tumor cell extravasation. *Frontiers in Oncology, 12*. https://doi.org/10.3389/fonc.2022.1052192
9. Wan, Q.-Q., Qin, W.-P., Ma, Y.-X., Shen, M.-J., Li, J., Zhang, Z.-B., Chen, J.-H., Tay, F. R., Niu, L.-N., & Jiao, K. (2021). Crosstalk between bone and nerves within bone. *Advancement of Science, 8*, 2003390. https://doi.org/10.1002/advs.202003390
10. Mohan, R. R., Kempuraj, D., D'Souza, S., & Ghosh, A. (2022). Corneal stromal repair and regeneration. *Progress in Retinal and Eye Research, 91*, 101090. https://doi.org/10.1016/J.PRETEYERES.2022.101090
11. Poh, P. S. P., Woodruff, M. A., & García-Gareta, E. (2020). Polymer-based composites for musculoskeletal regenerative medicine. In N. E. Vrana, H. Knopf-Marques, & J. Barthes (Eds.), *Biomaterials for soft tissue repair and regeneration* (pp. 33–82). Woodhead Publishing. https://doi.org/10.1016/B978-0-08-102906-0.00003-9

12. Schöneberg, J., De Lorenzi, F., Theek, B., Blaeser, A., Rommel, D., Kuehne, A. J. C., Kießling, F., & Fischer, H. (2018). Engineering biofunctional in vitro vessel models using a multilayer bioprinting technique. *Scientific Reports, 8*, 10430. https://doi.org/10.1038/s41598-018-28715-0
13. Xu, J., & Shi, G.-P. (1842). Vascular wall extracellular matrix proteins and vascular diseases. *Biochimica et Biophysica Acta, 2014*, 2106–2119. https://doi.org/10.1016/j.bbadis.2014.07.008
14. Takahashi, T., Kalka, C., Masuda, H., Chen, D., Silver, M., Kearney, M., Magner, M., Isner, J. M., & Asahara, T. (1999). Ischemia-and cytokine-induced mobilization of bone marrow-derived endothelial progenitor cells for neovascularization. *Nature Medicine, 5*(4), 434–438.
15. Kohli, N., Sawadkar, P., Ho, S., Sharma, V., Snow, M., Powell, S., Woodruff, M. A., Hook, L., & García-Gareta, E. (2020). Pre-screening the intrinsic angiogenic capacity of biomaterials in an optimised ex ovo chorioallantoic membrane model. *Journal of Tissue Engineering, 11*. https://doi.org/10.1177/2041731420901621
16. Walthers, C. M., Nazemi, A. K., Patel, S. L., Wu, B. M., & Dunn, J. C. Y. (2014). The effect of scaffold macroporosity on angiogenesis and cell survival in tissue-engineered smooth muscle. *Biomaterials, 35*, 5129–5137. https://doi.org/10.1016/j.biomaterials.2014.03.025
17. Mason, B. N., Starchenko, A., Williams, R. M., Bonassar, L. J., & Reinhart-King, C. A. (2013). Tuning three-dimensional collagen matrix stiffness independently of collagen concentration modulates endothelial cell behavior. *Acta Biomaterialia, 9*, 4635–4644. https://doi.org/10.1016/j.actbio.2012.08.007
18. Solovieva, E. V., Fedotov, A. Y., Mamonov, V. E., Komlev, V. S., & Panteleyev, A. A. (2018). Fibrinogen-modified sodium alginate as a scaffold material for skin tissue engineering. *Biomedical Materials, 13*. https://doi.org/10.1088/1748-605X/aa9089
19. Kohli, N., Sharma, V., Orera, A., Sawadkar, P., Owji, N., Frost, O. G., Bailey, R. J., Snow, M., Knowles, J. C., Blunn, G. W., & García-Gareta, E. (2021). Pro-angiogenic and osteogenic composite scaffolds of fibrin, alginate and calcium phosphate for bone tissue engineering. *Journal of Tissue Engineering, 12*. https://doi.org/10.1177/20417314211005610
20. Ren, J., Kohli, N., Sharma, V., Shakouri, T., Keskin-Erdogan, Z., Saifzadeh, S., Brierly, G. I., Knowles, J. C., Woodruff, M. A., & García-Gareta, E. (2021). Poly-ε-caprolactone/fibrin-alginate scaffold: A new pro-angiogenic composite biomaterial for the treatment of bone defects. *Polymers, 13*. https://doi.org/10.3390/polym13193399
21. Sawadkar, P., Mohanakrishnan, J., Rajasekar, P., Rahmani, B., Kohli, N., Bozec, L., & García-Gareta, E. (2020). A synergistic relationship between Polycaprolactone and natural polymers enhances the physical properties and biological activity of scaffolds. *ACS Applied Materials & Interfaces, 12*, 13587–13597. https://doi.org/10.1021/acsami.9b19715
22. Sawadkar, P., Mandakhbayar, N., Patel, K. D., Buitrago, J. O., Kim, T. H., Rajasekar, P., Lali, F., Kyriakidis, C., Rahmani, B., Mohanakrishnan, J., Dua, R., Greco, K., Lee, J. H., Kim, H. W., Knowles, J., & García-Gareta, E. (2021). Three dimensional porous scaffolds derived from collagen, elastin and fibrin proteins orchestrate adipose tissue regeneration. *Journal of Tissue Engineering, 12*. https://doi.org/10.1177/20417314211019238
23. Zhou, Y., Fan, Y., Chen, Z., Yue, Z., & Wallace, G. (2022). Catechol functionalized ink system and thrombin-free fibrin gel for fabricating cellular constructs with mechanical support and inner micro channels. *Biofabrication, 14*, 015004. https://doi.org/10.1088/1758-5090/ac2ef8
24. He, S., Walimbe, T., Chen, H., Gao, K., Kumar, P., Wei, Y., Hao, D., Liu, R., Farmer, D. L., Lam, K. S., Zhou, J., Panitch, A., & Wang, A. (2022). Bioactive extracellular matrix scaffolds engineered with proangiogenic proteoglycan mimetics and loaded with endothelial progenitor cells promote neovascularization and diabetic wound healing. *Bioactive Materials, 10*, 460–473. https://doi.org/10.1016/j.bioactmat.2021.08.017
25. Crush, J., Hussain, A., Seah, K. T. M., & Khan, W. S. (2021). Bioactive glass: Methods for assessing angiogenesis and osteogenesis. *Frontiers in Cell and Development Biology, 9*. https://doi.org/10.3389/fcell.2021.643781

26. Li, H., Wu, Z., Zhou, Y., & Chang, J. (2019). Bioglass for skin regeneration. In E. García-Gareta (Ed.), *Biomaterials for skin repair and regeneration* (pp. 225–250). Woodhead Publishing. https://doi.org/10.1016/B978-0-08-102546-8.00008-X
27. Kohli, N., & Garcia-Gareta, E. (2020). Bioactive nanoceramics. In *Nanomaterials for theranostics and tissue engineering* (pp. 233–257). Elsevier. https://doi.org/10.1016/B978-0-12-817838-6.00009-7
28. Liu, Y., Liu, X., Guo, H., Wang, X., Li, A., Qiu, D., & Gu, Q. (2024). 3D bioprinting bioglass to construct vascularized full-thickness skin substitutes for wound healing. *Materials Today Bio, 24*, 100899. https://doi.org/10.1016/j.mtbio.2023.100899
29. Ahmed, T. A. E., Dare, E. V., & Hincke, M. (2008). Fibrin: A versatile scaffold for tissue engineering applications. *Tissue Engineering. Part B, Reviews, 14*, 110306231744007. https://doi.org/10.1089/ten.teb.2007.0435
30. Brown, A. C., & Barker, T. H. (2014). Fibrin-based biomaterials: Modulation of macroscopic properties through rational design at the molecular level. *Acta Biomaterialia, 10*, 1502–1514. https://doi.org/10.1016/j.actbio.2013.09.008
31. Van Lieshout, M., Peters, G., Rutten, M., Baaijens, F., & Knitted, A. (2006). Fibrin-covered Polycaprolactone scaffold for tissue engineering of the aortic valve. *Tissue Engineering, 12*, 481–487. https://doi.org/10.1089/ten.2006.12.481
32. Ramakrishnan, R., Chouhan, D., Sreelatha, V., Arumugam, S., Mandal, B. B., & Krishnan, L. K. (2022). Silk fibroin-based bioengineered scaffold for enabling Hemostasis and skin regeneration of critical-size full-thickness heat-induced burn wounds. *ACS Biomaterials Science & Engineering*. https://doi.org/10.1021/acsbiomaterials.2c00328
33. García-Gareta, E. (2019). *Collagen, from tissue culture to biomaterials, tissue engineering, and beyond* (1st ed.). Cambridge Scholars Publishing.
34. Chaves, C., Alshomer, F., Alhujayri, A. K., & Kalaskar, D. M. (2019). Decellularized dermal tissue substitutes. In E. García-Gareta (Ed.), *Biomaterials for skin repair and regeneration* (pp. 103–124). Woodhead Publishing. https://doi.org/10.1016/B978-0-08-102546-8.00004-2
35. García-Gareta, E., Abduldaiem, Y., Sawadkar, P., Kyriakidis, C., Lali, F., & Greco, K. V. (2020). Decellularised scaffolds: Just a framework? Current knowledge and future directions. *Journal of Tissue Engineering, 11*. https://doi.org/10.1177/2041731420942903
36. Gilpin, A., & Yang, Y. (2017). Decellularization strategies for regenerative medicine: From processing techniques to applications. *BioMed Research International, 2017*, 9831534. https://doi.org/10.1155/2017/9831534
37. Shariatzadeh, S., Shafiee, S., Zafari, A., Tayebi, T., Yazdanpanah, G., Majd, A., Haj-Mirzaian, A., Bahrami, S., & Niknejad, H. (2021). Developing a pro-angiogenic placenta derived amniochorionic scaffold with two exposed basement membranes as substrates for cultivating endothelial cells. *Scientific Reports, 11*, 22508. https://doi.org/10.1038/s41598-021-01922-y
38. Kyriakidis, C., Lali, F., Greco, K. V., & García-Gareta, E. (2021). Chronic leg ulcers: Are tissue engineering and biomaterials science the solution? *Bioengineering, 8*, 62. https://doi.org/10.3390/bioengineering8050062
39. Omorphos, N. P., Gao, C., Tan, S. S., & Sangha, M. S. (2021). Understanding angiogenesis and the role of angiogenic growth factors in the vascularisation of engineered tissues. *Molecular Biology Reports, 48*, 941–950. https://doi.org/10.1007/s11033-020-06108-9
40. Jin, H., Quesada, C., Aliabouzar, M., Kripfgans, O. D., Franceschi, R. T., Liu, J., Putnam, A. J., & Fabiilli, M. L. (2021). Release of basic fibroblast growth factor from acoustically-responsive scaffolds promotes therapeutic angiogenesis in the hind limb ischemia model. *Journal of Controlled Release, 338*, 773–783. https://doi.org/10.1016/J.JCONREL.2021.09.013
41. Zigdon-Giladi, H., Khutaba, A., Elimelech, R., Machtei, E. E., & Srouji, S. (2017). VEGF release from a polymeric nanofiber scaffold for improved angiogenesis. *Journal of Biomedical Materials Research Part A, 105*, 2712–2721. https://doi.org/10.1002/jbm.a.36127
42. Amaral, R. J. F. C., Cavanagh, B., O'Brien, F. J., & Kearney, C. J. (2019). Platelet-derived growth factor stabilises vascularisation in collagen–glycosaminoglycan scaffolds in vitro.

Journal of Tissue Engineering and Regenerative Medicine, 13, 261–273. https://doi.org/10.1002/term.2789
43. Perez, R. A., Kim, J. H., Buitrago, J. O., Wall, I. B., & Kim, H. W. (2015). Novel therapeutic core–shell hydrogel scaffolds with sequential delivery of cobalt and bone morphogenetic protein-2 for synergistic bone regeneration. *Acta Biomaterialia, 23*, 295–308. https://doi.org/10.1016/J.ACTBIO.2015.06.002
44. Mishra, R., Roux, B. M., Posukonis, M., Bodamer, E., Brey, E. M., Fisher, J. P., & Dean, D. (2016). Effect of prevascularization on in vivo vascularization of poly(propylene fumarate)/fibrin scaffolds. *Biomaterials, 77*, 255–266. https://doi.org/10.1016/J.BIOMATERIALS.2015.10.026
45. Eke, G., Mangir, N., Hasirci, N., MacNeil, S., & Hasirci, V. (2017). Development of a UV crosslinked biodegradable hydrogel containing adipose derived stem cells to promote vascularization for skin wounds and tissue engineering. *Biomaterials, 129*, 188–198. https://doi.org/10.1016/j.biomaterials.2017.03.021
46. Qin, C., Zhang, H., Chen, L., Zhang, M., Ma, J., Zhuang, H., Huan, Z., Xiao, Y., & Wu, C. (2023). Cell-laden scaffolds for vascular-innervated bone regeneration. *Advanced Healthcare Materials, 12*, 2201923. https://doi.org/10.1002/adhm.202201923
47. Dębski, T., Kurzyk, A., Ostrowska, B., Wysocki, J., Jaroszewicz, J., Święszkowski, W., & Pojda, Z. (2020). Scaffold vascularization method using an adipose-derived stem cell (ASC)-seeded scaffold prefabricated with a flow-through pedicle. *Stem Cell Research & Therapy, 11*, 34. https://doi.org/10.1186/s13287-019-1535-z
48. Okuda, T., Uysal, A. C., Tobita, M., Hyakusoku, H., & Mizuno, H. (2010). Prefabrication of tissue engineered bone grafts: An experimental study. *Annals of Plastic Surgery, 64*. https://journals.lww.com/annalsplasticsurgery/fulltext/2010/01000/prefabrication_of_tissue_engineered_bone_grafts_.25.aspx
49. Park, S. E., Georgescu, A., & Huh, D. (2019). Organoids-on-a-chip. *Science, 364*, 960–965. https://www.science.org
50. Mollica, P. A., Booth-Creech, E. N., Reid, J. A., Zamponi, M., Sullivan, S. M., Palmer, X.-L., Sachs, P. C., & Bruno, R. D. (2019). 3D bioprinted mammary organoids and tumoroids in human mammary derived ECM hydrogels. *Acta Biomaterialia, 95*, 201–213. https://doi.org/10.1016/j.actbio.2019.06.017
51. Chaudhary, S., & Chakraborty, E. (2022). Hydrogel based tissue engineering and its future applications in personalized disease modeling and regenerative therapy. *Beni-Suef University Journal of Basic and Applied Sciences, 11*, 3. https://doi.org/10.1186/s43088-021-00172-1
52. Liu, Y., Shah, K. M., & Luo, J. (2021). Strategies for articular cartilage repair and regeneration. *Frontiers in Bioengineering and Biotechnology, 9*. https://doi.org/10.3389/fbioe.2021.770655
53. Zhao, Z., Li, Y., Wang, M., Zhao, S., Zhao, Z., & Fang, J. (2020). Mechanotransduction pathways in the regulation of cartilage chondrocyte homoeostasis. *Journal of Cellular and Molecular Medicine, 24*, 5408–5419. https://doi.org/10.1111/jcmm.15204
54. Mai, J., & Paxinos, G. (2011). *The_Human_Nervous_System* (3rd ed.). Academic.
55. Mccorry, L. K. (2007). Physiology of the autonomic nervous system. *American Journal of Pharmaceutical Education, 71*, 78.
56. Rao, M., & Gershon, M. D. (2016). The bowel and beyond: the enteric nervous system in neurological disorders. *Nature Reviews. Gastroenterology & Hepatology, 13*, 517–528. https://doi.org/10.1038/nrgastro.2016.107
57. Akinrodoye, M. A., & Lui, F. (2024). *Neuroanatomy, somatic nervous system*. StatPearls Publishing.
58. Ludwig, P. E., Reddy, V., & Varacallo, M. (2024). *Neuroanatomy, neurons*. Elsevier.
59. Ashley, K., & Lui, F. (2024). *Physiology, nerve*. Wiley-Blackwell.

60. Das, S., Gordián-Vélez, W. J., Ledebur, H. C., Mourkioti, F., Rompolas, P., Chen, H. I., Serruya, M. D., & Cullen, D. K. (2020). Innervation: The missing link for biofabricated tissues and organs. *npj Regenerative Medicine, 5*, 11. https://doi.org/10.1038/s41536-020-0096-1
61. Agrawal, V., Brown, B. N., Beattie, A. J., Gilbert, T. W., & Badylak, S. F. (2009). Evidence of innervation following extracellular matrix scaffold-mediated remodelling of muscular tissues. *Journal of Tissue Engineering and Regenerative Medicine, 3*, 590–600. https://doi.org/10.1002/term.200
62. Westphal, J. A., Bryan, A. E., Krutko, M., Esfandiari, L., Schutte, S. C., & Harris, G. M. (2024). Innervation of an ultrasound-mediated PVDF-TrFE scaffold for skin-tissue engineering. *Biomimetics, 9*. https://doi.org/10.3390/biomimetics9010002

Chapter 9
Bioreactors for Tissue Engineering and Regenerative Medicine

Abstract In tissue engineering cells are combined with biomaterial scaffolds and stimuli to build implantable constructs and tissue models. These stimuli include biochemical and biophysical cues that are essential for cell function. Biophysical stimuli can be delivered by bioreactor systems, particularly mechanical cues. Furthermore, bioreactors are essential to achieve effective mass transfer of oxygen and nutrients, especially in large tissue constructs. Hence, bioreactors become powerful tools for tissue engineering applications like engineering of constructs for transplantation, recellularization of decellularized tissues and organs, or cell expansion for cell therapies. There are important design considerations to bear in mind when building a bioreactor, and several types, each with its advantages and disadvantages, are available. The present chapter offers an overview of bioreactors used in TERM applications.

9.1 Introduction

Tissue engineering combines cells with biomaterial scaffolds and stimuli to build implantable constructs and tissue models [1]. Stimuli are essential to provide cells with biochemical and biophysical cues for their function [2]. In vivo, cells inhabit a specific microenvironment where they continuously receive these stimuli in the form of growth factors, ions, cytokines, matrikines, exosomes, pH signals, oxygen tension, shear stress, interstitial fluid flow, compression forces, or electrical signals, to name a few [3, 4]. Therefore, to produce biomimetic constructs and models, these stimuli must be recreated and provided to our target cells. Whilst providing cells with biochemical cues is, a priori, simpler, as they can be directly administered into the cell culture media or immobilized in the scaffold, the addition of biophysical cues to culture systems is technically more complex. In this context, bioreactors become valuable tools for tissue engineering and regenerative medicine (TERM) [2, 5, 6]. Furthermore, as mentioned in the previous chapter, for large constructs, mass transfer to the cells located near or at the core becomes critical to avoid a necrotic

© The Author(s), under exclusive license to Springer Nature Switzerland AG 2026

E. García-Gareta, J. M. García-Aznar, *Contemporary Tissue Engineering and Regenerative Medicine*, Series in BioEngineering,
https://doi.org/10.1007/978-3-032-03441-0_9

and hypoxic core [7, 8]. In the absence of a perfusable vasculature, a bioreactor becomes essential to achieve effective mass transfer in large constructs.

Bioreactors for TERM can be defined as devices that create a controlled and dynamic environment for cell and tissue culture, thereby replicating the physiological conditions of specific tissues or organs [2]. These tools enable the creation of a setting that mimics the natural microenvironment of cells and tissues, promoting their growth and maturation [2]. Importantly, these devices allow continuous supply of oxygen and nutrients and steady removal of the waste products naturally produced by cell metabolism [9]. Furthermore, these devices provide cells in constructs with biophysical stimuli like shear stress, interstitial fluid flow, centrifugal forces, compression forces, or electrical signals among others [9]. Within the TERM field, bioreactors are used to engineer tissues for transplantation, to recellularize decellularized tissue matrices, to study the effects of drugs and other treatments on tissues in a more realistic dynamic in vitro setting, to create models of diseases, or to expand and differentiate cells for use in cell therapies [10–17].

There are important design considerations to bear in mind when building a bioreactor, like maintenance of sterility, use of non-cytotoxic materials, uniform distribution of nutrients, or control of culture conditions [2, 9]. Ideally, the bioreactor system should be highly efficient without compromising user-friendliness or operational costs.

The aim of this chapter is to provide an overview of bioreactors in the context of TERM applications. Bioreactor functions and design considerations are first presented, followed by a comprehensive description of the main bioreactor types used in TERM. The chapter ends with conclusions on this subject and questions to strengthen the acquired knowledge on bioreactors for TERM.

9.2 Bioreactor Functions

Bioreactors used in TERM have several important functions that include advanced 3D tissue culture, cell expansion, application of controlled stimuli, and mass transport.

Bioreactors in TERM are primarily, and at a basic level, cell culture devices that provide advanced 3D culture conditions over static systems [10, 11, 18]. The advantages of using these dynamic systems are clear, particularly in terms of mass transfer, wate products removal, and application of biomechanical forces [9]. In static culture, concentration gradient-dependent movement of solutes within the construct relies on diffusion [9]. However, in dynamic culture systems like those provided by bioreactors mass transfer is improved by convection, facilitated by fluid flow [9]. Importantly, bioreactors enable the cultivation of 3D tissue constructs, which are more representative of in vivo tissues than 2D cultures [10, 17]. Furthermore, they enhance mass transport of nutrients and oxygen within the tissue constructs, addressing a major challenge in thick, clinically relevant tissue engineering applications [19]. Moreover, bioreactors facilitate the expansion and proliferation of cells, providing a sufficient cell population for tissue engineering applications, particularly clinical ones where cells are needed in large quantities [19]. Finally, bioreactors allow for the controlled application of biochemical and physical stimuli, such as nutrients, oxygen, mechanical forces, electrical stimulation and growth factors, to influence cell behaviour and tissue development [9].

9.3 Bioreactor Design Considerations

The design of a bioreactor for TERM applications requires careful consideration of several fundamental principles to ensure an environment that supports cell survival, growth, differentiation, and the formation of functional tissue (Fig. 9.1) [2, 9].

First and foremost is the need to establish and maintain a sterile environment throughout the culture period. This requires the use of biocompatible materials that can be easily sterilized for all components that come into direct contact with the cells and culture medium. Another critical design consideration is ensuring efficient mass transport of nutrients, oxygen, and signalling molecules, while simultaneously facilitating the removal of metabolites [9]. The design must promote uniform distribution of the culture medium throughout the scaffold's pore network to overcome the limitations of diffusion, especially in larger and denser constructs [20]. Importantly, the design should allow for precise control over the generated mechanical forces within physiologically relevant ranges [11, 20]. Furthermore, the design should facilitate homogeneous cell distribution, as this is crucial for the subsequent

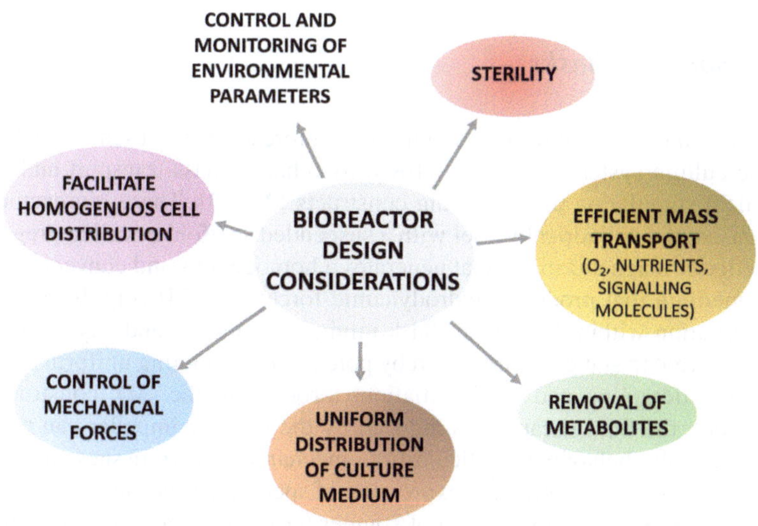

Fig. 9.1 Visual scheme of the different aspects to consider when designing bioreactors for TERM applications

development of a structurally and functionally uniform tissue construct [19]. Finally, maintaining precise control and monitoring of environmental parameters such as temperature, oxygen levels, or pH is critical for sustaining cell viability and optimal metabolic activity [21].

Important challenges in the design of bioreactors include reproducibility, as the designed bioreactor system should allow for the reproducible and controlled generation of tissue constructs. The design should also allow for scalability to produce large quantities of tissue constructs for clinical applications. Bioreactor design and operation can be complex, requiring expertise in both engineering and biology. Furthermore, the more complex is and more parts a bioreactor contains, the more expensive it becomes, thereby limiting their accessibility for some researchers. The challenge is to design highly efficient systems that integrate all the necessary parts for its correct functioning and monitoring, whilst making it user-friendly and cost-efficient.

9.4 Bioreactor Types

Several types of bioreactors are available for TERM applications, including spinner flasks, rotating wall vessels, perfusion bioreactors, hydrostatic pressure bioreactors, compression bioreactors, and electrical stimulation bioreactors (Table 9.1). Each type presents advantages and disadvantages, which are summarized in Table 9.1. The following sub-sections elaborate on the various aspects of the different bioreactor types.

9.4.1 Spinner Flasks

Benefiting from a simple design, spinner flask bioreactors represent a fundamental dynamic culture system utilized in TERM to enhance mass transport and provide mechanical stimulation to developing constructs [22, 23]. These bioreactors typically consist of a cylindrical vessel with a suspended scaffold or cell aggregate, and an impeller or magnetic stirrer that generates a homogenous and convective flow of culture medium that produces hydrodynamic forces [23, 24] (Fig. 9.2). The controlled agitation within the spinner flask minimizes nutrient and oxygen gradients that are inherent in static cultures, thereby potentially promoting uniform cell distribution, proliferation, and differentiation throughout the 3D construct [35]. Furthermore, the fluid shear stress induced by the rotating impeller can positively influence cellular behaviour, particularly in mechanosensitive tissues such as bone and cartilage, by upregulating the expression of specific genes and promoting ECM deposition [23, 34]. The application of spinner flask bioreactors in tissue engineering spans a wide range of tissues, including bone, cartilage, and vascular constructs [5, 6, 22, 23, 36]. By providing a more physiologically relevant environment

Table 9.1 Bioreactor types. Summary of main features, mechanical forces, advantages, and disadvantages. Table compiled from [5, 6, 21–33]

Bioreactor type	Main features	Physical stimuli	Advantages	Disadvantages
Spinner flasks	– Cylindrical container with stirring element and side arms for media exchange – Seeded scaffolds are often suspended – Convective flow of culture medium	– Shear stress – Other hydrodynamic forces	– Simple design and operation – Cost-effective – Improved mass transfer compared to static culture – Apply mechanical cues absent in static cultures – Widely commercially available	– Only permit ECM production at the construct's surface – Mixing can cause turbulent shear that may damage cells – Potential for nutrient and mechanical gradients leading to heterogeneous cell distribution – Relatively open system design: risk of contamination
Rotating wall vessels	– Two concentric cylinders with rotating outer wall – Low-shear, simulated microgravity environment – Create a centrifugal force that keeps constructs suspended	– Low shear stress due to laminar flow – Centrifugal forces – Other hydrodynamic forces	– Unique culture milieu – Recreate conditions of microgravity – Gentle culture conditions that are particularly advantageous for delicate cell types and the formation of intricate tissue architectures	– Specialised design and operation: more expensive – Precise control of hydrodynamic forces – Potential for scaffold collision with vessel walls – Visualising and accessing the cultures for real-time monitoring is difficult
Perfusion bioreactors	– Continuous circulation of culture medium through a 3D construct	– Fluid shear stress – Interstitial fluid flow	– Enhanced nutrient and oxygen transport, and waste removal – Precise control and monitoring of culture parameters – Provide mechanical stimulation – Improved cell uniformity	– Complex design and operation: more expensive – Potential for generating non-physiological levels of shear stress that can damage or detach cells – Difficulty in achieving uniform perfusion and mechanical forces – Complexity of optimizing flow parameters

(continued)

Table 9.1 (continued)

Bioreactor type	Main features	Physical stimuli	Advantages	Disadvantages
Hydrostatic pressure bioreactors	– Apply hydrostatic pressure exerted by a fluid at rest to constructs – Can be cyclic or static	– Hydrostatic pressure	– Improved tissue quality and enhanced cell function – Preconditioning of constructs before implantation by making them more robust and functional – Provide a platform for understanding mechanobiology	– Complex design and operation: more expensive – Maintenance of precise and consistent pressure levels necessitates sophisticated control systems and careful monitoring – Scaling up for large-scale tissue production presents challenges related to pressure uniformity and efficient nutrient delivery
Compression bioreactors	– Apply controlled mechanical loads in the form of static or dynamic compression	– Compression forces	– Mimic physiological loading in mechanically demanding tissues like cartilage and bone – Can be used to study the basic science of mechanotransduction	– Complex design and operation: more expensive – Precise control of compression parameters: magnitude, frequency, and duration – Potential for uneven compression if not designed properly
Electrical stimulation bioreactors	– Apply electrical signals mimicking the electrical activity found in the body – Designed with electrodes	– Electrical stimulation	– Enhanced cell viability, increased tissue formation, and improved tissue functionality – Recreate the electrical environment of native tissues	– Standardization of stimulation protocols – Biocompatibility and stability of electrode materials – Integration with advanced monitoring systems

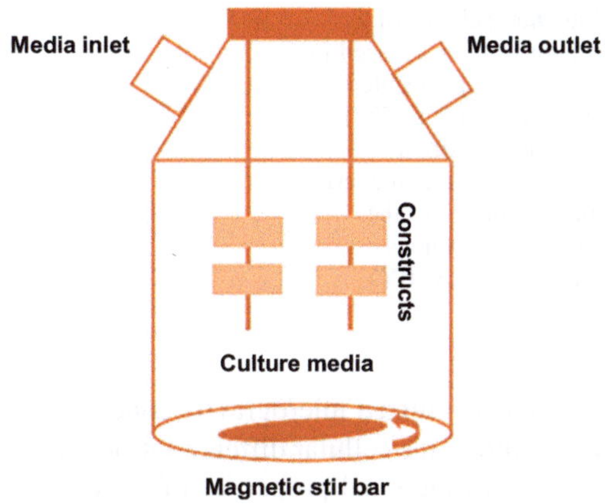

Fig. 9.2 Scheme of a spinner flask used in TERM. Image adapted from Lipreri et al. [34], open access article distributed under the terms and conditions of the Creative Commons Attribution (CC BY) license

compared to static culture, spinner flasks facilitate the development of larger and more complex tissue constructs with improved structural and functional properties.

Whilst offering significant advantages in terms of nutrient delivery and mechanical cues, these simple bioreactors have some disadvantages. Achieving uniform cell seeding within complex and highly porous scaffolds can be challenging, often leading to heterogeneous cell distribution and subsequent inconsistencies in tissue development [5, 24, 36]. Furthermore, controlling and precisely quantifying the hydrodynamic forces, particularly shear stress, experienced by the cells inside the scaffolds can be difficult, potentially leading to cellular damage or non-uniform mechanical stimulation across the construct [24, 37]. Scale-up for clinical translation can also introduce complexities in maintaining consistent culture conditions and ensuring scalability of the manufacturing process. Finally, the relatively open system design, while facilitating media exchange, can increase the risk of contamination if stringent aseptic techniques are not meticulously followed.

9.4.2 Rotating Wall Vessels

Rotating wall vessel (RWV) bioreactors provide a unique approach to tissue engineering by employing a rotating cylindrical vessel to create a low-shear, 3D culture environment [6, 24] (Fig. 9.3). The rotation of the vessel generates a centrifugal force that keeps constructs suspended, preventing them from settling or attaching to the vessel walls, thereby minimizing sedimentation and promoting uniform distribution of nutrients and oxygen [24, 39]. In RWVs, shear forces are generated by the rotation of the vessel walls, creating a gentle, controlled flow that minimizes shear stress compared to spinner flasks. This dynamic suspension culture created inside RWVs mimics aspects of the in vivo environment, particularly the low fluid shear forces experienced by cells in certain tissues, and facilitates the formation of 3D tissue-like structures with enhanced cell-cell interactions and ECM development [6,

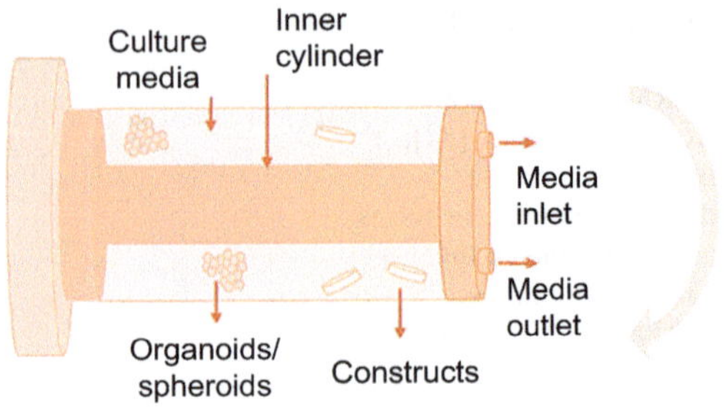

Fig. 9.3 Scheme of a rotator wall vessel used in TERM. Image adapted from Burova et al. [38], open access article distributed under the terms and conditions of the Creative Commons Attribution (CC BY) license

39]. The simulated microgravity conditions within RWVs have been shown to positively influence cellular differentiation and organization, leading to improved structural and functional properties of the engineered tissues. The application of RWV bioreactors is of particular importance in cartilage tissue engineering, as well as the generation of complex three-dimensional models like spheroids and organoids [5, 6, 15, 24, 39]. The gentle culture conditions are particularly advantageous for delicate cell types and the formation of intricate tissue architectures [15]. Furthermore, RWVs can accommodate various scaffold materials and configurations, offering flexibility in construct design.

Curiously, these bioreactors are valued tools in the research of bone regeneration in microgravity [40]. Tissue regeneration in extreme space conditions, including bone loss i.e. osteopenia, which is common among astronauts, significantly limits space exploration. The complexity of simulating microgravity hinders the research into improved therapeutic strategies for long-term space missions [40]. In this context, RWVs are valuable tools for investigating strategies to achieve bone regeneration under microgravity conditions [40].

RWVs provide a more biomimetic environment compared to static culture, but challenges remain in precisely controlling hydrodynamic forces [6, 24]. Furthermore, the specialized design and operation of RWVs can make them more technically demanding and potentially more expensive compared to simpler bioreactor systems [24]. Finally, visualizing and accessing the cultures for real-time monitoring and intervention can be less straightforward than in static or some other dynamic culture systems. Nevertheless, RWV bioreactors represent a valuable tool for advancing the development of functional and clinically relevant tissue-engineered constructs by providing a unique culture milieu.

9.4.3 Perfusion Bioreactors

Perfusion bioreactors are characterized by their ability to continuously circulate culture medium through a 3D construct, thereby ensuring a constant supply of nutrients and oxygen and removal of metabolic waste products [5, 24, 25]. The significance of perfusion bioreactors in TERM lies in their capacity to create an environment that closely mimics the natural physiological conditions experienced by tissues in the body [5, 24, 25]. This is crucial for enhancing cell growth, function, and the overall maturation of the engineered tissue. Furthermore, the continuous flow of culture medium through the porous construct generates mechanical forces, primarily in the form of fluid shear stress exerted on the cells and the scaffold matrix, as well as interstitial fluid flow within the scaffold's pore network [41]. These mechanical cues play a vital role in regulating various cellular processes, including proliferation, differentiation, and ECM production, which are essential for the development of functional biological substitutes. The ability of perfusion bioreactors to address both the mass transport limitations of static cultures and to provide crucial mechanical stimulation has positioned them at the forefront of advanced tissue engineering strategies.

A functional perfusion bioreactor system typically comprises several key components (Fig. 9.4). The culture chamber is the central element, providing a sealed and sterile space to house the cell-seeded scaffold [17, 41, 43]. The chamber design often incorporates features to securely position the scaffold and ensure that the culture medium flows through its porous structure rather than around it [42]. The material used for the chamber should be biocompatible and ideally transparent to allow for visual monitoring of the culture process [18]. A pump, often a peristaltic pump, is used to circulate the medium at a controlled flow rate without direct contact, thus maintaining sterility [18, 21, 41]. Biocompatible tubing connects the various components of the system, directing the flow of medium from the media reservoir, which holds the culture medium and allows for its replenishment or recirculation

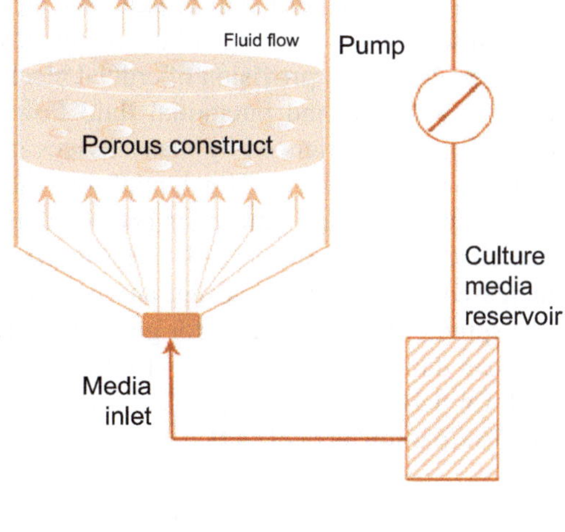

Fig. 9.4 Scheme of a perfusion bioreactor used in TERM. Image adapted from Sladkova and De Peppo [42], open access article distributed under the terms and conditions of the Creative Commons Attribution (CC BY) license

[21]. Flow control mechanisms, such as valves and programmable pumps, enable the regulation of flow rate, pressure, and flow patterns (e.g., steady, pulsatile, oscillatory) [21]. Finally, environmental control features are crucial for maintaining optimal conditions within the bioreactor. These typically include systems for temperature control, gas exchange mechanisms, and pH monitoring [21]. Some advanced systems may also incorporate humidity control or monitoring of glucose and metabolites levels [21, 44]. Finally, several culture chambers can be used in parallel [45].

Perfusion bioreactors are widely used in bone tissue engineering to enhance osteogenic differentiation of mesenchymal stem cells (MSCs) and osteoblasts, promote mineralization and the formation of bone-like ECM, and improve the mechanical properties of engineered bone constructs [5, 17, 18, 24, 46]. Studies have demonstrated their effectiveness in creating larger, more uniform, and mechanically robust bone constructs compared to static cultures [5, 17, 18]. Perfusion bioreactors are also extensively used in cartilage tissue engineering to promote chondrocyte proliferation and maintain their phenotype, enhance the synthesis and deposition of cartilage-specific ECM components, and improve the mechanical properties of engineered cartilage constructs [6]. The interstitial fluid flow in these bioreactors mimics the natural environment of cartilage, aiding in nutrient transport and providing mechanical stimulation [6]. In cardiovascular tissue engineering, perfusion bioreactors are crucial for the development of tissue-engineered blood vessels with appropriate mechanical strength and endothelialization, engineering functional heart valve substitutes, and creating myocardial patches for cardiac repair [47, 48]. Perfusion bioreactors are also being explored for neural tissue engineering to support the growth and differentiation of neural cells in 3D scaffolds and to create in vitro models for studying neural development and disease [49]. The controlled environment and enhanced nutrient delivery offered by perfusion systems can be beneficial for the complex requirements of neural tissues. The applications of perfusion bioreactors extend to other areas such as the development of skin substitutes [50], the culture of organoids for disease modelling and drug testing [14, 51], and their use in cancer research to study tumour biology and evaluate anti-cancer therapies [51].

Advantages of perfusion bioreactors include: (1) enhanced mass transport of nutrients, oxygen, and signalling molecules, (2) efficient removal of metabolic waste products, (3) provision of crucial mechanical stimulation to the cells through fluid shear stress and interstitial fluid flow, mimicking the dynamic mechanical environment experienced by many tissues in vivo, (4) improved cell seeding efficiency and uniformity, (5) potential for real-time monitoring and control, and (6) acceleration of tissue maturation and development, which is essential for their eventual clinical application [5, 18, 41]. Disadvantages of perfusion bioreactors are an increased complexity in design and operation compared with simpler bioreactors like spinner flasks, potential for generating non-physiological levels of shear stress that can damage or detach cells from the scaffold, maintaining sterility, achieving uniform perfusion and mechanical forces, higher cost, and complexity of optimizing flow parameters [21, 44].

9.4.4 Hydrostatic Pressure Bioreactors

Hydrostatic pressure bioreactors apply controlled pressure exerted by a fluid at rest to cell cultures or engineered tissues to mimic physiological conditions, potentially promoting cell growth, differentiation, and tissue formation, particularly in applications like cartilage and bone regeneration [26, 27]. These bioreactors can be used to study the effects of pressure on cell growth, differentiation, and tissue formation.

In terms of design, these bioreactors typically include chambers where cells or tissue constructs can be cultured under controlled pressure, allowing for precise control of pressure magnitude, frequency, and duration [52, 53]. With these bioreactors, it is possible to apply cyclic or continuous mechanical stimulation to cells and tissues [53]. The goal is to create an in vitro environment that closely resembles the mechanical environment of tissues in vivo.

For articular cartilage regeneration research, hydrostatic pressure can stimulate chondrocyte activity and promote cartilage tissue formation [27, 52]. An interesting example is the work by Zhao and colleagues, where hydrostatic pressure (HP) was applied with the use of a bioreactor to regulate in vitro cartilage regeneration within a hybrid photocrosslinkable hydrogel [52]. Transcriptomic analysis revealed that HP modulated in vitro cartilage regeneration by inhibiting cell senescence and apoptosis, promoting ECM synthesis, suppressing ECM catabolism, and remodelling the ECM structure. In vivo assessment showed that HP-treated regenerated cartilage further developed post-implantation, forming mature cartilage resembling native tissue, indicating substantial clinical potential [52].

For research into bone regeneration, HP can enhance osteogenesis (bone formation) and angiogenesis (blood vessel formation) [26]. Another interesting application is intervertebral disc repair where HP can be used to mimic the physiological environment of intervertebral discs, which can help in the development of therapies for disc degeneration [53].

The advantages of using HP bioreactors include: (1) improved tissue quality, as HP can lead to the formation of tissues with better structural and functional properties, (2) enhanced cell function, since pressure can stimulate cell growth, differentiation, and matrix production, (3) preconditioning of constructs before implantation by making them more robust and functional, and (4) providing a platform for understanding mechanobiology by studying the role of HP in tissue development and disease [26, 52, 54]. On the other hand, HP bioreactors also present certain disadvantages [20, 27]. For instance, the specialized equipment required can be complex and expensive. Maintaining precise and consistent pressure levels necessitates sophisticated control systems and careful monitoring, adding to the operational complexity. Furthermore, the application of high pressure may not be universally suitable for all cell types or tissue constructs, potentially causing cellular damage. Finally, scaling up these systems for large-scale tissue production can present engineering challenges related to pressure uniformity and efficient nutrient delivery within larger bioreactor volumes.

9.4.5 Compression Bioreactors

Compression bioreactors represent a crucial technology in tissue engineering, particularly for mechanically demanding tissues like cartilage and bone [5, 25, 28]. These systems are designed to apply controlled mechanical loads, such as static or dynamic compression, onto constructs or tissue explants cultured in vitro. By mimicking the physiological mechanical environment experienced by these tissues in vivo, compression bioreactors aim to enhance cellular processes including proliferation, differentiation, and ECM production [5, 25, 28]. The application of defined compressive forces can stimulate mechanotransduction pathways, leading to improved tissue development, maturation, and ultimately, the generation of functional tissue constructs for regenerative medicine applications [29].

A typical compression bioreactor consists of a chamber where the constructs are placed, a mechanical stimulation unit (e.g., a motor and a system for linear motion) that applies the compressive load, and a control unit that regulates the parameters of the mechanical loading (e.g., frequency, strain, duration) (Fig. 9.5) [6, 28, 55].

Main applications of compression bioreactors are in the engineering of tissues that experience compression forces on a regular basis, like bone and cartilage [5, 6, 25]. Importantly, these bioreactors can also be used to study the basic science of mechanotransduction [29], by which cells and tissues translate biomechanical cues, e.g. ECM stiffness or interstitial fluid flow, into cellular responses. For instance, Seo and colleagues presented the development of an interconnectable dynamic compression bioreactor for combinatorial screening of cellular behaviour within 3D hydrogels under controlled mechanical stimulation [29]. The bioreactor enabled the simultaneous application of cyclic compressive strains (0–42%) to multiple cell-laden hydrogel samples, and was utilized to investigate the influence of gelatin methacryloyl (GelMA) hydrogel concentration (5% and 10%) and dynamic compressive strain on human mesenchymal stem cell (hMSC) spreading and osteogenic differentiation. Results demonstrated that increasing hydrogel concentration inhibited cell spreading, while dynamic compression promoted it. Furthermore,

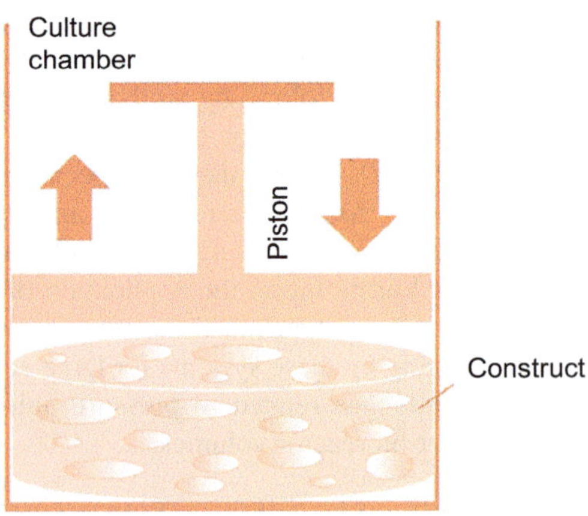

Fig. 9.5 Scheme of a compression bioreactor used in TERM. Image adapted from Sladkova and De Peppo [42], open access article distributed under the terms and conditions of the Creative Commons Attribution (CC BY) license

osteogenic differentiation was enhanced by dynamic compressive strain, with hMSCs in 5% GelMA exhibiting greater sensitivity to mechanical cues, particularly at 42% strain. The authors concluded that the developed compression bioreactor offered an efficient tool for exploring cell-matrix interactions in 3D culture [29].

Nevertheless, compression bioreactors present certain limitations. Achieving uniform and precisely controlled stress distribution across complex 3D constructs can be challenging, potentially leading to heterogeneous cellular responses within a single construct [5]. Furthermore, the integration of real-time, non-invasive monitoring of cellular behaviour and matrix deposition within these closed systems can be technically demanding, often requiring specialized sensors and imaging modalities [30]. The complexity of the mechanical loading profiles and the potential for mechanical damage to delicate constructs under high compression also necessitate careful design and optimization [5].

9.4.6 Electrical Stimulation Bioreactors

Electrical stimulation bioreactors play a crucial role in TERM by providing a controlled environment where electrical signals are applied to constructs, mimicking the electrical activity found in the body and influencing cellular behaviour like proliferation, migration, and matrix production [31, 56, 57]. This can lead to improved tissue development, healing, and regeneration. Electrical stimulation in bioreactors can help recreate the electrical environment of native tissues, which is essential for proper cell function and tissue development, particularly in those tissues that function with electrical signals, like cardiac tissue, skeletal muscle, nerves, bone, or cartilage [33, 56–59]. Electrical signals can promote cell adhesion, migration, proliferation, differentiation, and ECM deposition, all crucial for TERM applications.

Electrical bioreactors are designed with electrodes to deliver specific electrical signals, and they often incorporate other features like perfusion systems to maintain nutrient flow and waste removal [57, 60]. Similarly, some combine electrical stimulation with mechanical stimulation to create a more biomimetic environment [58].

Electrical bioreactors are used in various tissue engineering applications, including cardiac tissue engineering, bone tissue engineering, and nerve tissue engineering. For instance, bioreactors with electrodes mimicking heart's electrical activity can induce beating and improve the functionality of engineered cardiac tissues [33, 57, 60].

As already seen with other bioreactor types, electrical stimulation bioreactors can lead to enhanced cell viability, increased tissue formation, and improved tissue functionality compared to static culture conditions. While electrical bioreactors hold significant promise, several challenges remain [31, 32]. One of such challenges is standardization of stimulation protocols, since currently, there is a lack of consensus on optimal electrical stimulation parameters across different tissue types and applications [32]. Biocompatibility and stability of electrode materials is another challenge, where the safety and long-term performance of materials used to deliver electrical stimulation within the bioreactor should be ensured. Finally, integration with advanced monitoring systems involving the development of methods for real-time, non-invasive assessment of tissue development and function within the bioreactor is needed [31].

9.5 Conclusions and Future Directions

Bioreactors are important tools in TERM applications as they provide a controlled and dynamic 3D culture environment that provides effective mass transfer and biomechanical stimulation to cells and constructs. However, important challenges lie ahead. Future directions in this area of TERM include the development of novel bioreactor designs that are more efficient and versatile, whilst more user-friendly than current ones, automation, and integration of bioreactors with other technologies such as 3D printing and microfluidics, to create more sophisticated tissue engineering systems [6, 20]. For in vitro tissue models, the goal is to incorporate these dynamic systems in their development, so more biomimetic and representative models are built. For clinical translation, the ultimate goal is to produce clinically relevant tissue-engineered constructs that necessitate the use of a bioreactor system for their successful development.

Questions

- Name three functions of bioreactors for TERM applications.
- Name three design considerations to take into account when designing bioreactors for TERM.
- Name three challenges encountered in the design of TERM bioreactors.
- Name six types of bioreactors used in TERM applications.
- What are the main features of spinner flasks?
- What the advantages of rotating wall vessels?
- What is the main feature of perfusion bioreactors?
- Describe the key components of a functional perfusion bioreactor system.
- What are the main features of hydrostatic pressure bioreactors?
- Describe a typical compression bioreactor.

References

1. Langer, R., & Vacanti, J. P. (1993). Tissue engineering. *Science, 260*(5110), 920. http://science.sciencemag.org/content/260/5110/920.abstract
2. Castro, N., Ribeiro, S., Fernandes, M. M., Ribeiro, C., Cardoso, V., Correia, V., et al. (2020). Physically active bioreactors for tissue engineering applications. *Advanced Biosystems, 4*(10), 2000125.

3. Scodellaro, C., Pina, R. R., Ferreira, F. C., Sanjuan-Alberte, P., & Fernandes, T. G. (2024). Unlocking the potential of stem cell microenvironments in vitro. *Bioengineering, 11*(3), 289. https://www.mdpi.com/2306-5354/11/3/289
4. Redondo, P. A., Pavlou, M., Loizidou, M., & Cheema, U. (2017). Elements of the niche for adult stem cell expansion. *Journal of Tissue Engineering, 8*, 2041731417725464. https://doi.org/10.1177/2041731417725464
5. Watson, E., & Mikos, A. G. (2023). Advances in in vitro and in vivo bioreactor-based bone generation for craniofacial tissue engineering. *Journal of Tissue Engineering, 31*, 4. https://doi.org/10.34133/bmef.0004
6. Fu, L., Li, P., Li, H., Gao, C., Yang, Z., Zhao, T., et al. (2021). The application of bioreactors for cartilage tissue engineering: Advances, limitations, and future perspectives. *Stem Cells International, 2021*, 1–13. https://www.hindawi.com/journals/sci/2021/6621806/
7. Yang, G., Mahadik, B., Choi, J. Y., & Fisher, J. P. (2020). Vascularization in tissue engineering: Fundamentals and state-of-art. *Progress in Biomedical Engineering, 2*(1), 012002. https://doi.org/10.1088/2516-1091/ab5637
8. Lovett, M., Lee, K., Edwards, A., & Kaplan, D. L. (2009). Vascularization strategies for tissue engineering. *Tissue Engineering Reviews (Part B), 15*(3), 353–370. https://doi.org/10.1089/ten.teb.2009.0085
9. Selden C, Fuller B. Role of bioreactor technology in tissue engineering for clinical use and therapeutic target design. . 2018; 5(2): 32
10. Radi, S., EzEldeen, M., Végvári, Á., Coates, D., Jacobs, R., Bostanci, N., et al. (2025). The proteome of osteoblasts in a 3D culture perfusion bioreactor model compared with static conditions. *Scientific Reports, 15*(1), 12120.
11. Nikolits, I., Chariyev-Prinz, F., Egger, D., Liebner, F., Mytzka, N., & Kasper, C. (2024). Characterization of MSC growth, differentiation, and EV production in CNF hydrogels under static and dynamic cultures in hypoxic and normoxic conditions. *Bioengineering, 11*(10), 1050.
12. Suzuki, T., Watanabe, T., Tomiyama, F., Ito, T., & Okada, Y. (2025). Transplantation of bioengineered lung using decellularized mouse lungs and primary human endothelial cells. *Journal of Visualized Experiments, 217*, 609632.
13. Choi, M., Al Fahad, M. A., Shanto, P. C., Park, S., & Lee, B.-T. (2025). Surface modification of decellularized kidney scaffold with chemokine and AKI-CKD cytokine juice to increase the recellularization efficiency of bio-engineered kidney. *Biomaterials, 316*, 123007.
14. Zamorano, M., Aguilar-Gallardo, C., Lugo, A., Jimenez, L., Farias, J. G., & Mantalaris, A. (2025). Engineering an integrated bioprocess to produce human dental pulp stem cell-alginate-based bone organoids. *International Journal of Molecular Sciences, 26*(9), 4348.
15. Takahashi, J., Sugihara, H. Y., Kato, S., Kawasaki, S., Nagata, S., Okamoto, R., et al. (2024). Controlled aggregative assembly to form self-organizing macroscopic human intestine from induced pluripotent stem cells. *Cell Reports Methods, 4*(12), 100930.
16. Helms, F., Käding, D., Aper, T., Ruhparwar, A., & Wilhelmi, M. (2024). An arteriovenous bioreactor perfusion system for physiological in vitro culture of complex vascularized tissue constructs. *Bioengineering, 11*(11), 1147.
17. García-Gareta, E., Hua, J., & Blunn, G. W. (2015). Osseointegration of acellular and cellularized osteoconductive scaffolds: Is tissue engineering using mesenchymal stem cells necessary for implant fixation? *Journal of Biomedical Materials Research Part A, 103*(3), 1067–1076.
18. García-Gareta, E., Hua, J., Rayan, F., & Blunn, G. W. (2014). Stem cell engineered bone with calcium-phosphate coated porous titanium scaffold or silicon hydroxyapatite granules for revision total joint arthroplasty. *Journal of Materials Science. Materials in Medicine, 25*(6), 1553.
19. Sarkar, N., Bhumiratana, S., Geris, L., Papantoniou, I., & Grayson, W. L. (2023). Bioreactors for engineering patient-specific tissue grafts. *Nature Reviews Bioengineering, 1*(5), 361–377.
20. Niu, X., Xu, Z., Di, M., Huang, D., & Li, X. (2023). Bioreactor strategies for tissue-engineered osteochondral constructs: Advantages, present situations and future trends. *Composites. Part B, Engineering, 259*, 110736.

21. Yamada, S., Yassin, M. A., Schwarz, T., Mustafa, K., & Hansmann, J. (2022). Optimization and validation of a custom-designed perfusion bioreactor for bone tissue engineering: Flow assessment and optimal culture environmental conditions. *Frontiers in Bioengineering and Biotechnology, 25*, 10.
22. Teixeira, G. Q., Barrias, C. C., Lourenço, A. H., & Gonçalves, R. M. (2014). A multicompartment holder for spinner flasks improves expansion and osteogenic differentiation of mesenchymal stem cells in three-dimensional scaffolds. *Tissue Engineering. Part C, Methods, 20*(12), 984–993.
23. Melke, J., Zhao, F., Rietbergen, B., Ito, K., & Hofmann, S. (2018). Localisation of mineralised tissue in a complex spinner flask environment correlates with predicted wall shear stress level localisation. *European Cells and Materials, 1*(36), 57–68.
24. Gaspar, D. A., Gomide, V., & Monteiro, F. J. (2012). The role of perfusion bioreactors in bone tissue engineering. *Biomatter, 2*(4), 167–175. https://doi.org/10.4161/biom.22170
25. Drapal, V., Gamble, J. M., Robinson, J. L., Tamerler, C., Arnold, P. M., & Friis, E. A. (2022). Integration of clinical perspective into biomimetic bioreactor design for orthopedics. *Journal of Biomedical Materials Research Part B: Applied Biomaterials, 110*(2), 321–337.
26. Shen, H.-X., Liu, J.-Z., Yan, X.-Q., Yang, H.-N., Hu, S.-Q., Yan, X.-L., et al. (2022). Hydrostatic pressure stimulates the osteogenesis and angiogenesis of MSCs/HUVECs co-culture on porous PLGA scaffolds. *Colloids Surfaces B Biointerfaces, 213*, 112419. https://www.sciencedirect.com/science/article/pii/S0927776522001023
27. Elder, B. D., & Athanasiou, K. A. (2009). Hydrostatic pressure in articular cartilage tissue engineering: From chondrocytes to tissue regeneration. *Tissue Engineering. Part B, Reviews, 15*(1), 43–53.
28. Dursun, G., Umer, M., Markert, B., & Stoffel, M. (2021). Designing of an advanced compression bioreactor with an implementation of a low-cost controlling system connected to a mobile application. *PRO, 9*(6), 915.
29. Seo, J., Shin, J.-Y., Leijten, J., Jeon, O., Bal Öztürk, A., Rouwkema, J., et al. (2018). Interconnectable dynamic compression bioreactors for combinatorial screening of cell Mechanobiology in three dimensions. *ACS Applied Materials & Interfaces, 10*(16), 13293–13303.
30. Kluge, J. A., Leisk, G. G., Cardwell, R. D., Fernandes, A. P., House, M., Ward, A., et al. (2011). Bioreactor system using noninvasive imaging and mechanical stretch for biomaterial screening. *Annals of Biomedical Engineering, 39*(5), 1390–1402.
31. Kim, H.-S., Baby, T., Lee, J.-H., Shin, U. S., & Kim, H.-W. (2024). Biomaterials-enabled electrical stimulation for tissue healing and regeneration. *Med-X, 2*(1), 7.
32. Zimmermann, J., Farooqi, A. R., & van Rienen, U. (2024). Electrical stimulation for cartilage tissue engineering - A critical review from an engineer's perspective. *Heliyon, 10*(19), e38112.
33. Visone, R., Talò, G., Lopa, S., Rasponi, M., & Moretti, M. (2018). Enhancing all-in-one bioreactors by combining interstitial perfusion, electrical stimulation, on-line monitoring and testing within a single chamber for cardiac constructs. *Scientific Reports, 8*(1), 16944.
34. Lipreri, M. V., Baldini, N., Graziani, G., & Avnet, S. (2022). Perfused platforms to mimic bone microenvironment at the macro/Milli/microscale: Pros and cons. *Frontiers in Cell and Development Biology, 3*, 9.
35. Zhu, Y., Liu, T., Song, K., Fan, X., Ma, X., & Cui, Z. (2009). Ex vivo expansion of adipose tissue-derived stem cells in spinner flasks. *Biotechnology Journal, 4*(8), 1198–1209.
36. Godbey, W. T., Stacey Hindy, B. S., Sherman, M. E., & Atala, A. (2004). A novel use of centrifugal force for cell seeding into porous scaffolds. *Biomaterials, 25*(14), 2799–2805.
37. Jeske, R., Lewis, S., Tsai, A.-C., Sanders, K., Liu, C., Yuan, X., et al. (2021). Agitation in a microcarrier-based spinner flask bioreactor modulates homeostasis of human mesenchymal stem cells. *Biochemical Engineering Journal, 168*, 107947.
38. Iva, B., Ivan, W., & Shipley, R. J. (2019). Mathematical and computational models for bone tissue engineering in bioreactor systems. *Journal of Tissue Engineering, 1*(10), 2041731419827922. https://doi.org/10.1177/2041731419827922

39. Zhu, Y., Song, K., Jiang, S., Chen, J., Tang, L., Li, S., et al. (2017). Numerical simulation of mass transfer and three-dimensional fabrication of tissue-engineered cartilages based on chitosan/gelatin hybrid hydrogel scaffold in a rotating bioreactor. *Applied Biochemistry and Biotechnology, 181*(1), 250–266.
40. Mochi, F., Scatena, E., Rodriguez, D., Ginebra, M.-P., & Del Gaudio, C. (2022). Scaffold-based bone tissue engineering in microgravity: Potential, concerns and implications. *npj Microgravity, 8*(1), 45.
41. Dahlin, R. L., Meretoja, V. V., Ni, M., Kasper, F. K., & Mikos, A. G. (2012). Design of a high-throughput flow perfusion bioreactor system for tissue engineering. *Tissue Engineering. Part C, Methods, 18*(10), 817–820.
42. Sladkova, M., & De Peppo, G. (2014). Bioreactor systems for human bone tissue engineering. *PRO, 2*(2), 494–525.
43. Bhaskar, B., Owen, R., Bahmaee, H., Rao, P. S., & Reilly, G. C. (2018). Design and assessment of a dynamic perfusion bioreactor for large bone tissue engineering scaffolds. *Applied Biochemistry and Biotechnology, 185*(2), 555–563.
44. Simmons, A. D., Williams, C., Degoix, A., & Sikavitsas, V. I. (2017). Sensing metabolites for the monitoring of tissue engineered construct cellularity in perfusion bioreactors. *Biosens Bioelectron, 90*, 443–449. http://www.sciencedirect.com/science/article/pii/S0956566316309794
45. Egger, D., Fischer, M., Clementi, A., Ribitsch, V., Hansmann, J., & Kasper, C. (2017). Development and characterization of a parallelizable perfusion bioreactor for 3D cell culture. *Bioengineering, 4*(2), 51.
46. Schröder, M., Reseland, J. E., & Haugen, H. J. (2022). Osteoblasts in a perfusion flow bioreactor—Tissue engineered constructs of TiO2 scaffolds and cells for improved clinical performance. *Cells, 11*(13), 1995.
47. Visconti, R., Mironov, V., Kasyanov, V. A., Yost, M. J., Twal, W., Trusk, T., et al. (2006). Cardiovascular tissue engineering I perfusion bioreactors: A review. *Journal of Long-Term Effects of Medical Implants, 16*(2), 111–130.
48. Saunders, S. K., Cole, S. Y., Acuna Sierra, V., Bracamonte, J. H., Toldo, S., & Soares, J. S. (2022). Evaluation of perfusion-driven cell seeding of small diameter engineered tissue vascular grafts with a custom-designed seed-and-culture bioreactor. *PLoS One, 17*(6), e0269499.
49. Di Lisa, D., Andolfi, A., Masi, G., Uras, G., Ferrari, P. F., Martinoia, S., et al. (2024). Impact of perfusion on neuronal development in human derived neuronal networks. *APL Bioengineering, 8*(4), 046102.
50. Strüver, K., Friess, W., & Hedtrich, S. (2017). Development of a perfusion platform for dynamic cultivation of in vitro skin models. *Skin Pharmacology and Physiology, 30*(4), 180–189.
51. Avena, P., Zavaglia, L., Casaburi, I., & Pezzi, V. (2025). Perfusion bioreactor technology for organoid and Tissue Culture: A mini review. *The Oncologist, 5*(2), 17.
52. Zhao, X., Hua, Y., Wang, T., Ci, Z., Zhang, Y., Wang, X., et al. (2022). In vitro cartilage regeneration regulated by a hydrostatic pressure bioreactor based on hybrid Photocrosslinkable hydrogels. *Frontiers in Bioengineering and Biotechnology, 27*, 10.
53. Zvicer, J., & Obradovic, B. (2018). Bioreactors with hydrostatic pressures imitating physiological environments in intervertebral discs. *Journal of Tissue Engineering and Regenerative Medicine, 12*(2), 529–545.
54. Chariyev-Prinz, F., Szojka, A., Neto, N., Burdis, R., Monaghan, M. G., & Kelly, D. J. (2023). An assessment of the response of human MSCs to hydrostatic pressure in environments supportive of differential chondrogenesis. *Journal of Biomechanics, 154*, 111590.
55. Lovecchio, J., Gargiulo, P., Vargas Luna, J. L., Giordano, E., & Sigurjónsson, Ó. E. (2019). A standalone bioreactor system to deliver compressive load under perfusion flow to hBMSC-seeded 3D chitosan-graphene templates. *Scientific Reports, 9*(1), 16854.

56. Krueger, S., Riess, A., Jonitz-Heincke, A., Weizel, A., Seyfarth, A., Seitz, H., et al. (2021). Establishment of a new device for electrical stimulation of non-degenerative cartilage cells in vitro. *International Journal of Molecular Sciences, 22*(1), 394.
57. Tandon, N., Marsano, A., Cannizzaro, C., Voldman, J., & Vunjak-Novakovic, G. (2008). Design of electrical stimulation bioreactors for cardiac tissue engineering. In *2008 30th annual international conference of the IEEE engineering in medicine and biology society* (pp. 3594–3597). IEEE.
58. Van Houten, S. K., Bramson, M. T. K., & Corr, D. T. (2022). A bioreactor for controlled electrical and mechanical stimulation of developing scaffold-free constructs. *Journal of Biomechanical Engineering, 144*(9), 094501.
59. Sagita, I. D., Whulanza, Y., Dhelika, R., & Nurhadi, I. (2018). Designing electrical stimulated bioreactors for nerve tissue engineering. *AIP Conference Proceedings, 1933*, 040019.
60. Barash, Y., Dvir, T., Tandeitnik, P., Ruvinov, E., Guterman, H., & Cohen, S. (2010). Electric field stimulation integrated into perfusion bioreactor for cardiac tissue engineering. *Tissue Engineering. Part C, Methods, 16*(6), 1417–1426.

Chapter 10
Microfluidics-Based Platforms for Tissue Engineering and Regenerative Medicine

Abstract Microfluidics aims to design systems based on microchannels in which small volumes of fluids are used to achieve different applications in fuel cells, optics, life sciences, among others. Microfluidic devices are commonly used in cell biological research and especially in tissue engineering with multiple applications. In this chapter we will focus on a specific application, such as 3D cell culture to model and simulate different regenerative processes associated with disease, damage or aging. In addition, these microdevices are also an excellent system for drug testing and modeling different diseases and their corresponding therapeutic treatments. For tissue engineering, this technology has several advantages over traditional 2D and 3D cell culture methods, such as spatiotemporal controllability, fluid flow control, physiological mimicry of living tissue, and high-throughput analysis with smaller sample sizes. In addition, microdevices do not have the cost and ethical issues of animal experimentation. Therefore, this chapter will first review some of the techniques and materials used in the fabrication of microfluidic devices. Then, it will show different types of 3D cell cultures that can be developed in these chips by controlling different factors of the microenvironment. Finally, recent advances in microfluidic-based tissue engineering and some examples of successful studies in organ-on-a-chip and cancer applications are discussed.

10.1 Introduction

The development of microfluidics is closely related to many other fields of research, with important contributions from fuel cells, optics, molecular biology, and microelectronics. In this chapter, we focus on the application of microfluidics as a tool for cell culture development, given the appropriate dimensions of the volumes generated at this scale (Fig. 10.1).

Therefore, given the dimensions provided by microfluidic systems, they are especially adequate for studying cells and tissue interaction, which is crucial in tissue engineering. Certainly, different microfabrication techniques can be used to

© The Author(s), under exclusive license to Springer Nature Switzerland AG 2026

E. García-Gareta, J. M. García-Aznar, *Contemporary Tissue Engineering and Regenerative Medicine*, Series in BioEngineering,
https://doi.org/10.1007/978-3-032-03441-0_10

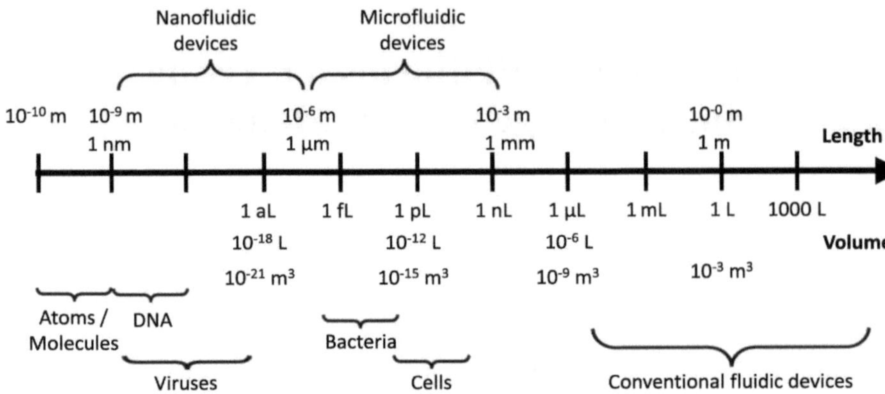

Fig. 10.1 Length and volume dimensions and possible correlations with biological applications (adopted from [1, 2])

develop different kind of microfluidic-based concepts depending on the exact application (Sect. 10.2).

One of the main goals in tissue engineering is building large pieces of tissue and whole organs, which is not an easy task, and requires a detailed understanding of the underlying cell-tissue interaction. Perhaps the greatest challenge in building a whole organ is the problem of ordering all the cells in a thick (3D) living structure [3]. Microfluidic systems offer the possibility of creating 3D cell cultures (Sect. 10.3) embedded in hydrogels that mimic our tissues (Sect. 10.4). In addition, scaffolds can be incorporated into the chips to study the interactions of our cells with the scaffold, to see if the scaffold can guide migration and stimulate cell growth and differentiation, scaffold degradation, and matrix formation (Sect. 10.4). Furthermore, the application and control of solute gradients within the microfluidic environment is critical to the study of tissue culture, facilitating the ability to mimic in vivo conditions or to study the effect of imposing concentration gradients on a cell population itself (see Sect. 10.5). Moreover, microfluidic systems provide the ability to create a dynamic culture by applying a fluid flow that provides oxygen, nutrients, and waste removal to the cell culture (see Sect. 10.6 for the applications and Sect. 10.7 for the fundamentals). Fluid flow also contributes to the generation of fluidic shear stresses on the cell surfaces and the alteration of the solute gradients by advection, enhancing the transport of different substances. More recently, organoids are being integrated into microfluidic-based chips (Sect. 10.8), as a new method to bridge the gap between in vitro 3D cultures and animal models to enable clinically relevant drug discovery and modeling of human diseases [4] (Fig. 10.2).

Among all the different designs and concepts used in microfluidics, we focus mainly on those in which the cell culture inside the chip is basically used as a bioreactor, but with much smaller dimensions than what is normally a bioreactor, as we have seen in the Chap. 9. The application of microfluidics-based cultures as a bioreactor presents many advantages with respect to a normal one: flexibility of device

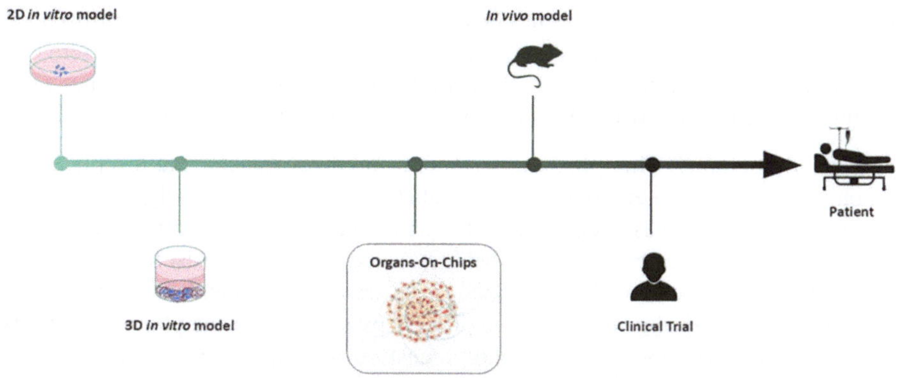

Fig. 10.2 Scheme to visualize how microfluidic-based experiments are used in comparison with other methods

design and fabrication, a low number of cells and materials is sufficient, spatio-temporal microscopy-based image quantification (4D analysis), ability to perform perfusion culture and chemical gradients, possibility to develop co-culture, and ability to recreate complex physiological conditions. But, this technique also has some disadvantages, the main limitation is the difficulty of performing analytical chemistry due to the small volumes used in the system. However, despite this limitation, microfluidic devices allow the fabrication of miniaturised microsystems that mimic biological organs at a scale not possible with macroscale equipment. This chapter finishes with some conclusions clearly identifying the possibilities provided by this technology to study cells under direct control in a fluid-based, in vivo-like, three-dimensional (3D) simulated microenvironment.

10.2 Microfabrication Techniques of Microfluidic Devices

The fabrication of microfluidic devices focused on the application of conventional silicon and glass micromachining processes, but the emergence of Micro-Electro-Mechanical Systems (MEMS) methods has opened the development of microfluidic cell culture models in an array of materials platforms [5].

Polymer materials are the most common materials used to fabricate microfluidic devices for cell culture applications due to their flexibility, optical transparency to visible/UV, easiness of molding with high fidelity, simplicity in the modification of surface properties, biocompatibility or bioactivity, durability, chemical inertness and low-toxicity [6]. The most common polymeric materials are polydimethylsiloxane (PDMS) elastomer and thermoplastics (polyethylene terephthalate—PET, polymethyl methacrylate—PMMA, polycarbonate—PC, or polyimide—PI). Certainly, there are other type of materials that can also be used to fabricate microfluidic devices, for example, rigid materials, such as, silicon and glass [6]. Currently, paper-based microfluidics is being an attractive alternative that is gaining more users [7].

There are different methods of fabrication that are widely described in many different publications [8–10]. Here some of them are shortly presented:

- Photolithography. This process, in general, involves transferring a replica of a base or master pattern onto a solid substrate which is typically an Si-wafer. Photolithography has been used as the main method for fabricating microfluidic devices. It involves exposing a substrate coated with photoresist to light in such a way that the selectively developed areas transmit the incident light, thereby etching the surface of the substrate. The steps involved in the photolithography process start with the construction of a mask [usually a "chromium (Cr)" pattern layer on a glass shield] and the casting of the silicon, glass or gallium arsenide (GaAs)-based substrate with an active photoresist substance. High-power, collimated ultraviolet (UV) light is then focused over the photoresist layer and cured, transferring the mask pattern to the photoresist (SU-8 or PMMA) on top of the substrate [8].
- Soft lithography. This technique has recently become very popular because it can be produced outside a clean room using a printing or moulding process. However, the master pattern must be produced using conventional photolithography techniques. Therefore, this process involves the following steps: making the original hard master, pouring liquid polymer into the mould, heat curing and stripping the polymer. In this way, a cast mould (replica) made of a flexible material is produced, which is then used for printing, moulding and embossing micro- and nanostructures [9]. A schematic of the soft lithography fabrication method compared to other moulding techniques is shown in Fig. 10.3. The main advantages of this method are the production of high-resolution replicas, lower cost and faster production than the old photolithography method [9].
- Injection molding. This process is based on four main steps. First, the thermoplastic is precisely melted to a liquid state inside a compressible chamber. Next, the two halves of the mould are compressed, creating a mould cavity. The thermoplastic is then injected at a controlled rate, filling the cavity. Finally, the mould is cooled and the casting is removed from the mould, as shown in Fig. 10.3.
- Hot embossing. It is another replicative technique wherein thermoplastics, or polymers that become viscous liquids at elevated temperatures, are precisely shaped using a mold, pressure, and heat. This process involves placing a thermoplastic film between two mould inserts. The mould chamber is then evacuated, compressed and heated to produce a cast. Finally, the mould is cooled and the cast is removed. The basic manufacturing process is shown in Fig. 10.3. Hot embossing has several advantages over injection moulding. Hot stamping requires the thermoplastic to flow a relatively small distance compared to injection moulding, which significantly reduces the stress generated in the material [11]. In addition, the shrinkage of the castings is reduced, making it effective in the production of delicate designs [11]. Limitations of the technology include a

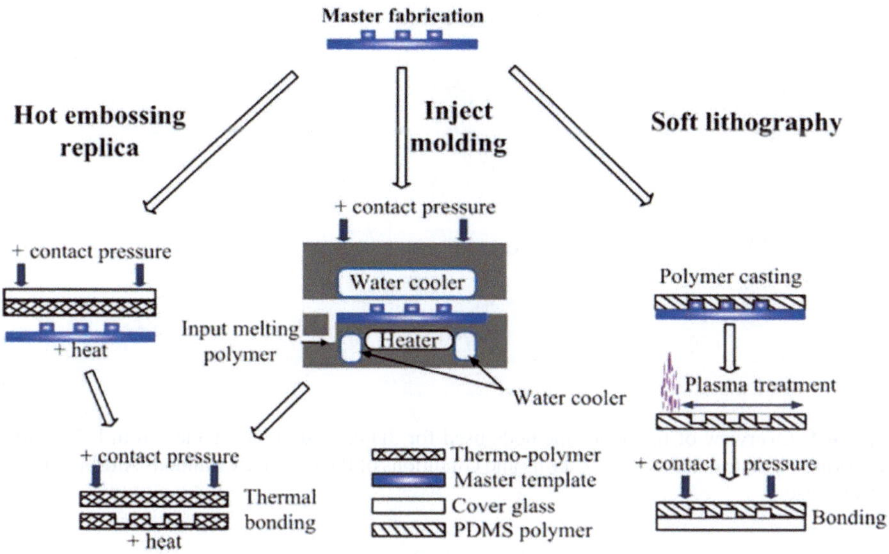

Fig. 10.3 A summary illustration of the main molding methods used in microfluidics (obtained from [10]). Open access article distributed under the terms and conditions of the Creative Commons Attribution (CC BY) license

restriction to thermoplastics and difficulty in producing complex three-dimensional structures [12].

- 3D printing. 3D printing is a layer-by-layer manufacturing technology that adds a new layer of material on top of the previous layer. The general process uses a digital 3D model that is sliced into layers that can be interpreted and created by the printer. 3D printing technologies have exciting potential for microfluidic researchers and can be easily used for prototyping [10]. There are several technologies associated with 3D printing that will be presented in the Chap. 12.

10.3 3D Cell Culture in Microfluidics

Different types of 3D cell cultures can be found in the literature that goes from the hanging drop or the transwell culture to the use of microfluidic chips [7]. Certainly, each method presents advantages and disadvantages. In general, the techniques can be split into three categories (see Fig. 10.4): hydrogel-based technology, microwell-based technology and microgel-based technology [13].

If we focus on the hydrogel-based technology, there are different strategies that can be used:

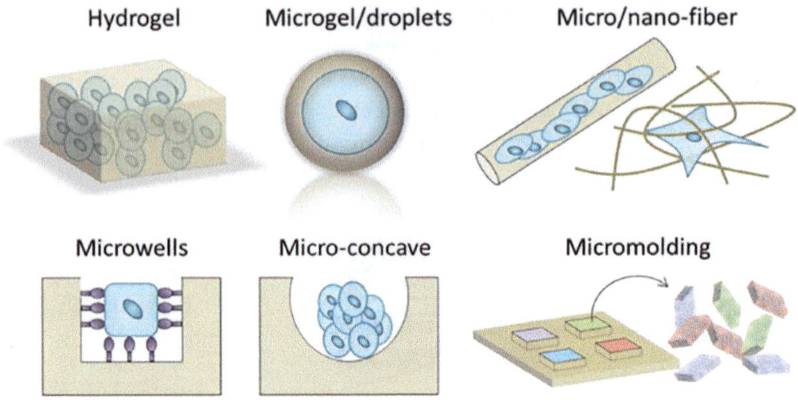

Fig. 10.4 Overview of the major methods used for 3D cell culture (obtained from [13]). Open access article distributed under the terms and conditions of the Creative Commons Attribution (CC BY) license

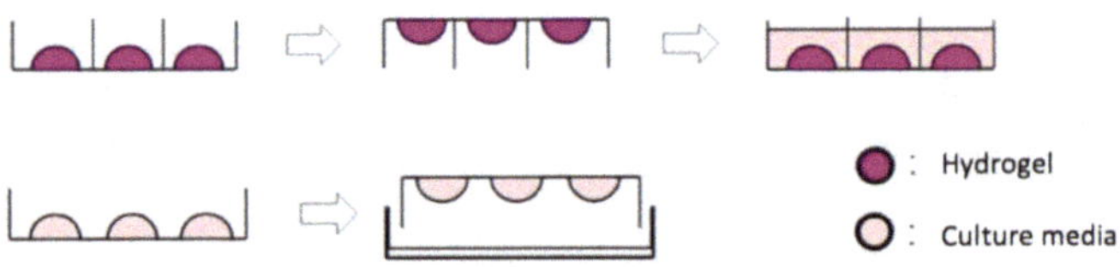

Fig. 10.5 At the top, it is shown a schematic representation of the dome cell culture, where after incubation the plate is flipped and reversed again adding culture media along the wall. At the bottom, the hanging drop method is represented with drops of cell suspension on culture media (obtained from [14]). Open access article distributed under the terms and conditions of the Creative Commons Attribution (CC BY) license

- Cells embedded in one hydrogel: cells can be embedded in hydrogels to create a 3D microenvironment that provides in vivo-like tissue stiffness and cell-matrix interactions to support correct cell phenotype formation (Fig. 10.4);
- Hanging drops or dome cultures: hanging drops are normally used for cells in suspension, but a similar strategy is commonly used with an hydrogel for creating 3D organoids or spheroids [14, 15] (Fig. 10.5);
- Cells on the top of one hydrogel: hydrogel can also be used as a substrate for cell suspension, allowing cells to grow on top of the gel (see Fig. 10.5), or for cell invasion, allowing cells to invade the interior of the gel [16];
- Cell culture inserts: the cell culture insert is used with a transwell to form a unit which is divided into two separate and interconnected chambers [17] (Fig. 10.6).

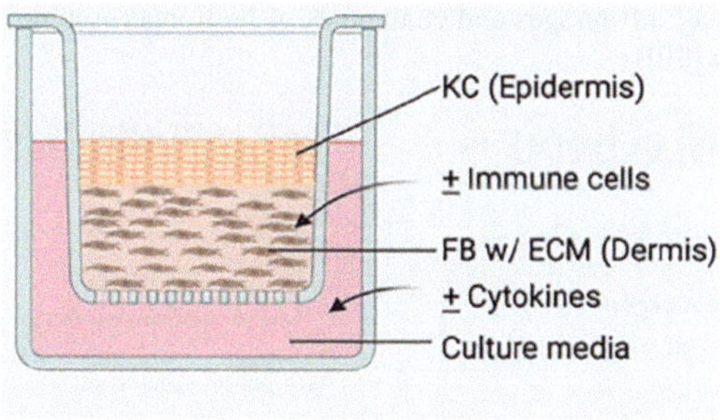

Fig. 10.6 A schematic overview of the 3D culture inserts, using a transwell to culture fibroblasts (FB) in a collagen-based hydrogel and several layers of keratinocytes (KC) on the top of the hydrogel (obtained from [17]). Open access article distributed under the terms and conditions of the Creative Commons Attribution (CC BY) license

Fig. 10.7 A schematic overview of an example of microfluidics-based approach for 3D cell cultures (obtained from [18]). Open access article distributed under the terms and conditions of the Creative Commons Attribution (CC BY) license

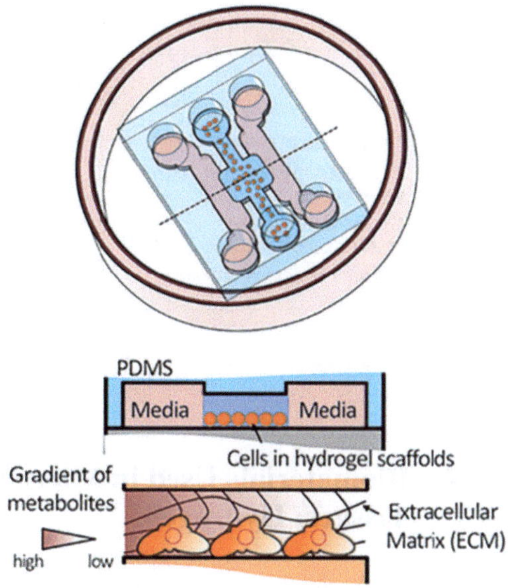

One common strategy is the use of 3D cell cultures in microfluidic devices that content a hydrogel-based chamber, simulating extracellular matrix, where cells are embedded (Fig. 10.7). Many variations of the microfluidics-based designs for 3D culture exists in the literature [19]. In this chapter, we focus mainly in those microfluidic-based 3D cell cultures that are used as microbioreactors, which is the main application in tissue engineering. In Table 10.1, a comparative analysis of using macroscopic cell cultures against microfluidic-based cultures is presented [20].

Table 10.1 Overview of advantages and challenges of both macroscopic and microfluidic-based cell cultures (based on [20])

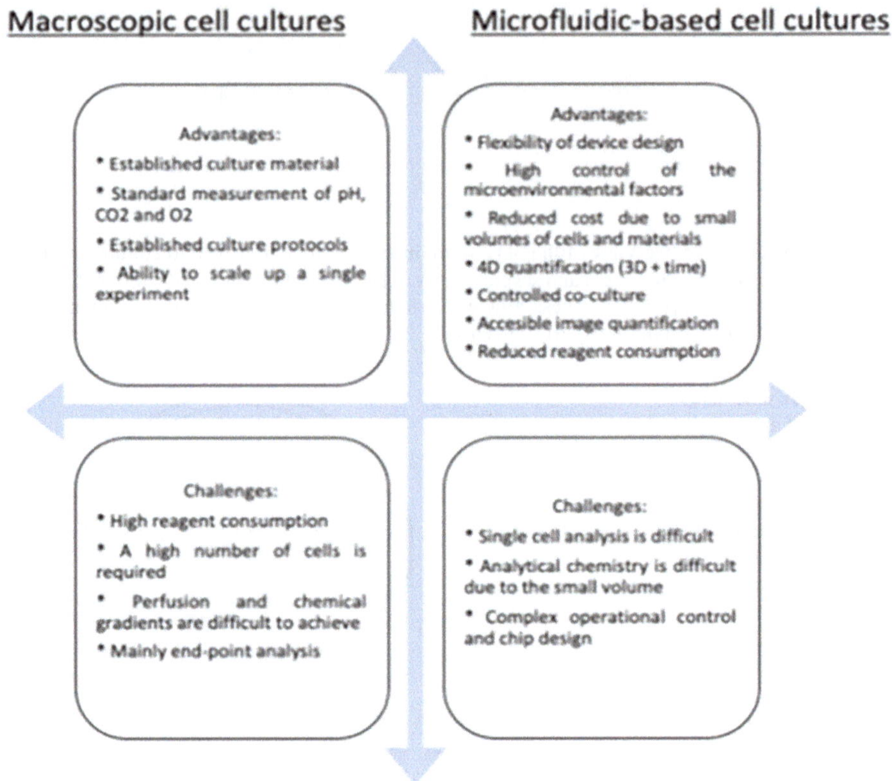

10.4 Biomaterials Used in Microfluidics to Recreate Tissues: Hydrogels

The in vitro replication of realistic tissues is desirable for their potential use in pharmacological and toxicological screening and in tissue engineering applications as a bioreactor to reduce and replace full animal models. Significant advances have been made towards this goal, developing scaffold materials to host cells in well-defined three-dimensional (3D) conditions [21]. The use of microfluidic devices with chambers filled with cell-seeded hydrogels as a scaffold is very common to be used as a microscopic bioareactor. Hydrogels are biomaterials specifically engineered to recreate the extracellular matrix (ECM), due to their swelling characteristics, the high-water content and the low elasticity [22]. Hydrogels are hydrophilic 3D cross-linked polymeric networks, consisting of interconnected microscopic pores, which can absorb biological fluid up to 99% of its volume. The amount of the absorbed fluid in the swollen hydrogel depends on the nature of the polymers and the developed polymeric network. Furthermore, the interfacial tension of hydrogels with water and biological fluids simulates the nature of the softest tissues [22]. Hydrogels as an

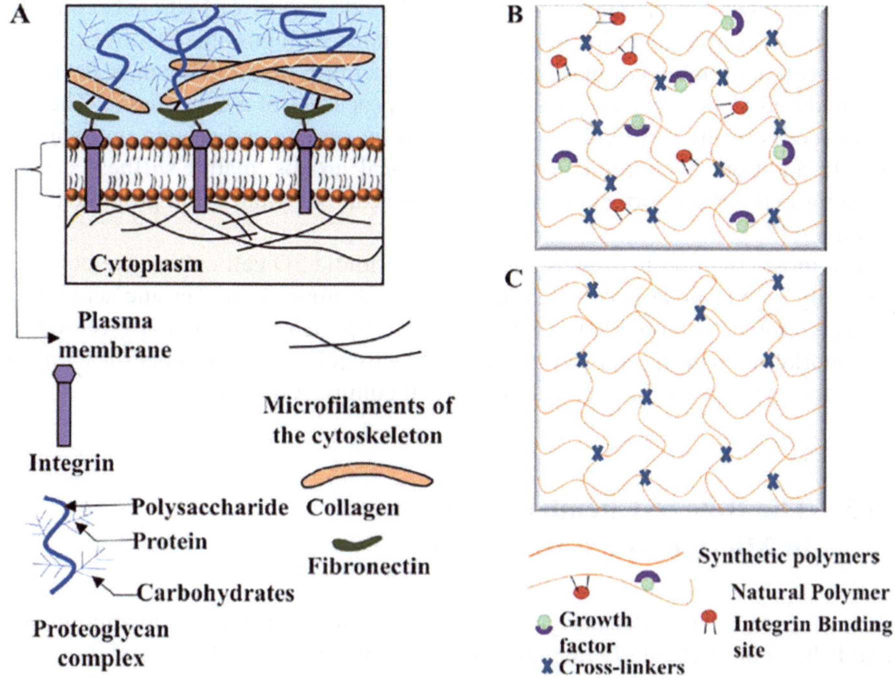

Fig. 10.8 Schematic comparison between extracellular matrix (ECM) and hydrogels. (**a**) ECM is consisted of three groups of macromolecules: structural proteins (collagens), proteoglycans, and glycoproteins (fibronectin) to form the matrix for cell attachment. Proteoglycan and structural fibrous proteins fill the cells' interstitial space to provide the mechanical and biological supports. Hydrogel scaffold from (**b**) natural polymers and (**c**) synthetic polymers. Fibrous hydrogel polymers link by interconnected microscopic pores to provide mechanical and biological cellular support (obtained from [22]). Open access article distributed under the terms and conditions of the Creative Commons Attribution (CC BY) license

artificial matrix should provide appropriate mechanical, chemical, structural and biological cues for cell adhesion, proliferation, and differentiation [9]. They are highly porous three-dimensional networks that allow the diffusion of small molecules and bioparticles, exhibit high biocompatibility, low cytotoxicity, biodegradability, controllable pore size, high permeability and aqueous nature. Therefore, hydrogels are used to develop different microfluidic component functionalities within a chip made of a more rigid material, such as 3D scaffolds, semi-permeable barriers or smart valves.

Section 5.5 of this book provides an overview of the different types of hydrogels. However, in the context of this section, to achieve a similar architecture to a mammalian ECM, hydrogels must have a hydrated protein and polysaccharide network (Fig. 10.8a). However, hydrogels derived from natural polymers have different characteristics than those derived from synthetic polymers. Like native ECM, natural polymer hydrogels contain integrin binding sites (Fig. 10.8b) to promote cellular functions. In contrast, synthetic polymer hydrogels lack integrin-binding ligands

(Fig. 10.8c), but can maintain cell viability. A wide variety of hydrogels could be prepared under mild and biocompatible conditions, with possible modification for desired cell adhesion, viscoelasticity, and degradability. Hydrogels can be designed and modified to simulate the physicochemical properties of a target tissue. The cell adhesiveness of the hydrogel can be modified by peptide and protein immobilization. In addition, by introducing new functional groups, matrix stiffness can be tuned for cellular support and pore architecture can be optimized to promote tissue formation [22].

The most common hydrogels used in microfluidic 3D cell culture are collagen (specially type I), gelatin, chitosan, fibrin, agarose, alginate, hyaluronic acid (HA), polyacrylamide, poly(ethylene glycol) (PEG) and their combination (see Akther et al. work for a wide a summary of different hydrogels with their properties, limitations, and applications in microfluidic 3D cell culture systems).

10.5 The Role of Chemical Gradients in Microfluidic-Based Chips

The existence of biomolecular gradients is very important in the regulation of cellular behaviour, particularly in guiding cell movement in many biological processes such as embryonic development, cancer spread, immune response, wound healing and osteogenesis [23]. Although there are different methods for generating biomolecular gradients in vitro, it is difficult to accurately control the diffusion of molecules and maintain it in a relatively stable state in time and space [23]. Microfluidic technology has offered a wide variety of strategies for the development of chemotaxis tests; it can generate highly dynamic, controllable, and micron-scale gradients in a small integrated chip that exposes cells to one or more customized spatiotemporal biomolecular gradients [23]. Here we focus on hydrogel-based microfluidic devices, where a chamber is filled with a hydrogel and two channels border the two extremes of the hydrogel, with a different biomolecule concentration in each channel (see for example the chip on Fig. 10.7 or Fig. 10.9).

To ensure the formation of a chemical gradient within the hydrogel chamber, where transport is dominated by the diffusion of that chemical factor, it is necessary to ensure that there is no flux flow due to a pressure gradient. For example, in the chip shown in Fig. 10.9, we must ensure that the height of the fluid channels (green) are identical to avoid convective effects.

The diffusion coefficient of a specific biomolecule through a hydrogel is dependent on the biomolecule size and the porous characteristics of the hydrogel [24]. Hence, the standard diffusion coefficient of a biomolecule of radius r within a continuous fluid, can be calculated by the Stokes-Einstein equation as

$$D_\infty = \frac{k_B T}{6\pi\eta r}$$

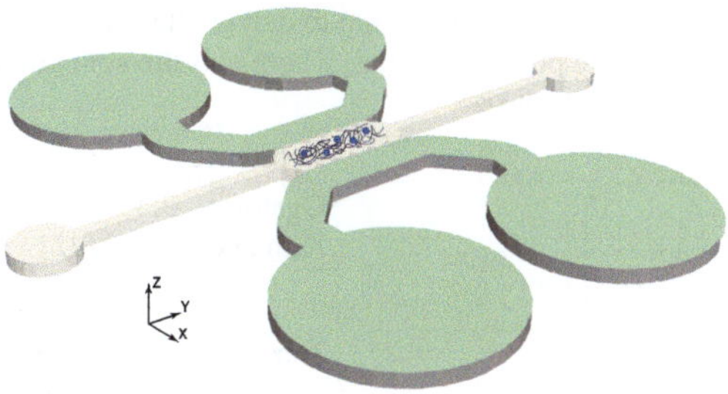

Fig. 10.9 Example of a type of hydrogel-based microfluidic chip detailing the different compartments: the channels (green volumes) and hydrogel-based chamber (grey volumes). The hydrogel is injected into the central cavity (grey channel), the main channels (green) are filled with culture media or PBS. When a biomolecule concentration is added to establish a chemical gradient, it is taken up by one of the channels (green) and diffuses through the hydrogel (grey) to the opposite channel (obtained from [11]). Open access article distributed under the terms and conditions of the Creative Commons Attribution (CC BY) license

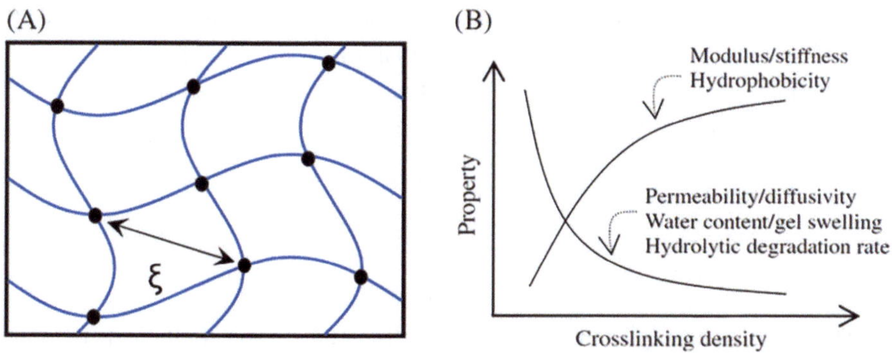

Fig. 10.10 (**a**) Simplified crosslinked hydrogel structure. Black dots represent crosslinking point; ξ represents mesh size of the gel. (**b**) Hydrogel property as a function of gel crosslinking density (obtained from [30]). Open access article distributed under the terms and conditions of the Creative Commons Attribution (CC BY) license

where k_B is the Boltzmann constant, T the absolute temperature, and η the viscosity of the fluid.

Certainly, this diffusion coefficient is determined in a free continuum medium, but when the molecules diffuse through one hydrogel or any kind of porous material, an effective diffusion coefficient has to be defined, which depends on the different geometrical characteristics of the porous material [24, 25]. Some authors [24, 26] uses as geometrical characteristics of the matrix the void ratio, but more recently other authors use the hydrogel mesh size [27, 28]. The *mesh size* of a hydrogel could be defined as the distance between two adjacent cross-links and has a critical role in determining the diffusion of agents within the hydrogel network [28, 29] (Fig. 10.10).

In Table 10.2 is presented a summary of these relationships according to some works in the literature.

Table 10.2 Relationships existent in the literature to define effective diffusivity in function of the geometrical characteristics of the porous material, where v_e is the void ratio, r_f is the fiber radius, r is the solute radius, and ξ is the mesh size

Relationship	References
$D_{eff} = D_\infty \exp\left(-\left[\sqrt{v_e}\left(1+\frac{r}{r_f}\right)\right]\right)$	Ogston et al. [31] Kim and Tarbell [32]
$D_{eff} = \frac{2D_\infty}{3-v_e}$	Tao and Nicholson [33]
$D_{eff} = D_\infty(1-(1-v_e)^{0.46})$	Mezedur et al. [34]
$D_{eff} = D_\infty \exp\left(-\pi\left[\frac{r+r_f}{\xi+2r_f}\right]\right)$	Amsden [27]

Table 10.3 Estimation of the specific diffusion coefficients particularized for each biochemical factor, together with the corresponding assumed radius, and for two different kind of hydrogels [24]

	Dextran	PDGF-BB	TGF-β_1
r (10^{-9}·m)	3.24	4.50	3.80
$D_\infty^1\left(10^{-11}\cdot\text{m}^2\cdot\text{s}^{-1}\right)$	1.75	1.26	1.49
$D_{\text{collagen}}^2\left(10^{-11}\cdot\text{m}^2\cdot\text{s}^{-1}\right)$	1.09	0.77	0.92
$D_{\text{fibrin}}^1\left(10^{-11}\cdot\text{m}^2\cdot\text{s}^{-1}\right)$	1.17	0.83	0.99

Therefore, as example, an estimation of different diffusion coefficients is shown in Table 10.3 for different growth factors and for different hydrogels. The effective diffusivity is altered when these molecules diffuse through the hydrogel instead of in a continuous fluid medium, using the relationship proposed by Ogston et al. [31].

Additionally, we have to keep in mind that while diffusion through the matrix pores in form of soluble molecules take place, the hydrogel (in the case of natural ones) may also serve as reservoir by offering binding sites to the GFs, but they can also interact with other molecules resulting in their degradation [24]. When this effect occurs, it is necessary to use ELISA assays in combination with 3D diffusion experiments in order to quantify the distribution of biomolecules across the hydrogel [24]. These equations have been solved numerically obtaining good agreement between the numerical results and the experimental data [24, 26]. Other authors have indicated that deviations from the model predictions can occur when the solute is deformable, and this effect is more pronounced for diffusion in highly concentrated hydrogels and in hydrogels possessing more flexible backbones [27].

10.6 The Role of Fluid Flow: Dynamic Cell Culture

Fluid flow plays an important role in many biological processes, such as, organ morphogenesis, homeostasis, and pathogenesis. Intuitively, it is clear that fluid flow can mechanically move cells, but it can also affect the transport of basic nutrients and signaling molecules, and it can even alter the ECM (see Fig. 10.11).

In the literature different strategies can be found to apply fluid flow in our 3D hydrogel-based microfluidic devices. These can be classified according to the fluid pressure origin in gravity-based (also known as static head) or pump-based systems. Their different characteristics determine, among other features, the preferential type of flow generated: luminal, interstitial, or transendothelial, or simple external agitation (see Fig. 10.12).

Depending on the system used to apply the fluid flow, different types of time-dependent fluid flow inputs are achieved, as shown in Table 10.4. This table also

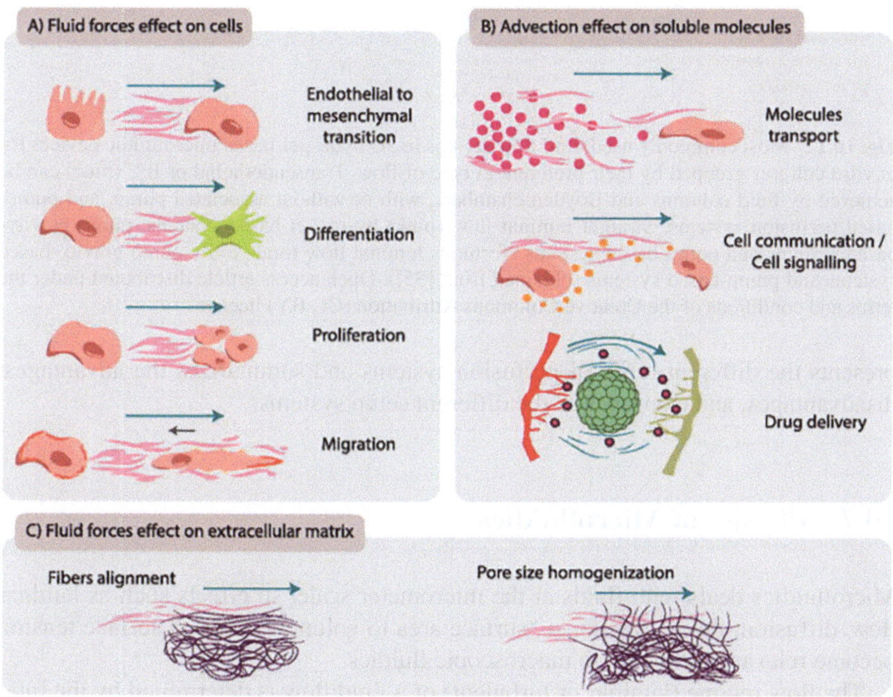

Fig. 10.11 Effect of fluid forces on cells, soluble molecules, and extracellular matrix (ECM). (**a**) Effect of fluid forces on cells, affecting endothelial-mesenchymal transition, differentiation, proliferation, and migration. (**b**) Advection effect on soluble molecules, regulation of nutrient transport, generation of chemical gradients, and control of cell communication/cell signaling and drug delivery (fluid flow is represented by blue lines). (**c**) Effect of fluid forces on the ECM, favoring alignment of ECM fibers and homogenization of pore size (obtained from [35]). Open access article distributed under the terms and conditions of the Creative Commons Attribution (CC BY) license

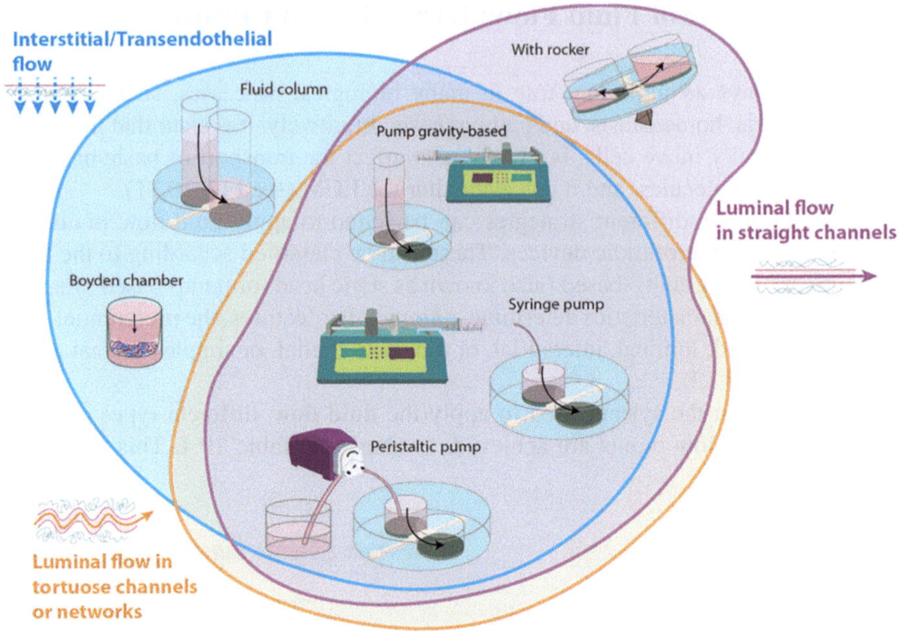

Fig. 10.12 Most commonly used fluid flow set-ups in 3D hydrogel-based microfluidic devices for in vitro cultures grouped by their preferential type of flow. Transendothelial or IFF (blue) can be achieved by fluid columns and Boyden chambers, with or without associated pump, and pump-based perfusion systems. Straight luminal flow (pink) by rocker-based systems, pump gravity-based systems and pump-based systems. Tortuous luminal flow (orange) by pump gravity-based systems and pump-based systems (obtained from [35]). Open access article distributed under the terms and conditions of the Creative Commons Attribution (CC BY) license

presents the different types of perfusion systems and summarizes the advantages, disadvantages, and examples of the different setup systems.

10.7 Physics of Microfluidics

Microfluidics deals with fluids at the micrometer scale, so effects such as laminar flow, diffusion, fluidic resistance, surface area to volume ratio, and surface tension become relevant compared to macroscopic fluidics.

The flow regime (laminar or turbulent) of a fluid flow is determined by the ratio of inertial forces to viscous forces in a fluidic system, which is described by the dimensionless Reynolds number (Re) which is given by [36]:

$$\mathrm{Re} = \frac{\rho v D_h}{\eta}$$

Table 10.4 Types of perfusion systems to induce flow in microfluidic-based cell cultures. This table groups the different types of perfusion platforms with their advantages, limitations and potential improvements as well as recommendations and examples of their use. * Depending on the aim of the experiment it may just be considered as an alternative. # Depending on the number or rollers and the presence of pulsation dampeners fluid flow could be either laminar or pulsatile-like (adapted from an open-access source: [35])

Perfusion System	Advantages	Disadvantages	Improvements	When to use *(preferentially)*	Quantitative data
Columns (or reservoirs) 	- High throughput - Simple handling - No need of bubble trap	- Limited in time and speed - Flow decreases over time - Low microscopy visualization in Boyden chambers	- Repetition over time *(although affecting the flow profile)* - Including external reservoirs - Including flow resistors *(affecting shear stress and advection)*	- Short experiments of interstitial or transendothelial fluid flow preferentially	- Initial pressure drop>1 to 25 mm H_2O - Manual pressure drop renewal for IFF>4 h to 24 h - Flow rate>2 to 15 µl/min - IFF speed>0.1 to 5 µm/s *Parameters highly dependent on gel permeability, gel dimensions and reservoir size*
Boyden chamber 					
Gravity-based with pump 	- Simple handling - Unlimited/long working time - No bubbles formation - Nutrients	- Low throughput - Limited in speed - Low sterility	- Including air filters to control sterility	- Very long experiments with renewal of nutrients	- Shear stress>0 to 2 $dyne/cm^2$ - Non-stop working time: weeks *Pressure drop and flow rates are similar to the rest of gravity-based systems.*
Rocker-based Bidirectional 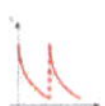	- High throughput - Simple handing - No need of bubble trap	- No renewal of nutrients - Low sterility	- Bypassing design to recirculate flow and perfuse unidirectionally *	- Shear stress stimulation of the cells without	- Angle>±7 to ±37º - Tilt interval>12 s to 8 min - Shear stress>0 to 6 $dyne/cm^2$
Pump-based *via* Peristaltic pump # 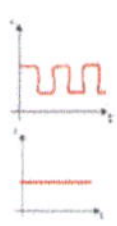	- Unlimited working time - Nutrients renewal - Allowed bidirectional flow *(although compromising medium renovation)*	- Compromised hydrogel stability when handling - Low throughput - Potential bubble formation	- Inclusion of bubble traps - Adapted system disposition - Pulsation dampeners	- Very long experiments with renewal of nutrients - Peristaltic stimulation	- Flow rate>0.5 to 4000 µl/min - Shear stress:~_0 to 20 $dyne/cm^2$ - Non-stop working time: weeks *Parameters highly dependent on the pump, tubing and microphysiological system*
Pump-based *via* Syringe pump 	- Long working time - Nutrients renewal - Allowed bidirectional flow *(although compromising medium renovation)*	- Compromised hydrogel stability when handling - Low throughput - Potential bubble formation	- Inclusion of bubble traps - Adapted system disposition	- Long experiments with renewal of nutrients	- Flow rate>1 to 50 µl/min - Shear stress:~_0 to 2 $dyne/cm^2$ - Non-stop working time: days *Parameters highly dependent on the pump, syringe and microphysiological system*

where ρ is the fluid density, v is the characteristic velocity of the fluid, η is the fluid viscosity, and D_h is the hydraulic diameter that depends on the channel's cross-sectional geometry.

Re < 2300 generally indicates a laminar flow. As Re approaches 2300, the fluid begins to show signs of turbulence, and as Re becomes greater than 2300 the flow is considered to be turbulent. Laminar flow is a condition in which the velocity of a particle in a fluid stream is not a random function of time. Because of the small size of microchannels, flow is almost always in the laminar flow regime [36].

In a Poiseuille flow the fluid is driven through a long, straight, and rigid channel by imposing a pressure difference between the two ends of the channel. The flow rate within a microchannel is given by

$$Q = \frac{dv}{dt} = \frac{\Delta P}{R_h}$$

where Q is the flow rate, ΔP is the pressure drop across the channel and R_h is the hydraulic resistance of the channel. For a circular section the hydraulic resistance is:

$$R_h = \frac{8\eta L}{AR^2}$$

The hydraulic resistance can be approximated for any section shape as

$$R_h \approx \frac{8\eta L}{Ar_h^2}$$

where r_h is the hydraulic radius of the channel

$$r_h = \frac{2A}{P}$$

being A the area and P the perimeter of the section.

In the case that in the middle of the tube there is a hydrogel or scaffold the hydraulic resistance is mainly due to the permeability of this hydrogel (see Fig. 10.13) through the equation:

$$R_h = \frac{\eta L}{AK}$$

where K is the intrinsic permeability of the matrix/hidrogel (m^2) and A is the cross section of the hydrogel.

This Hagen-Poiseuille law for fluid flow is analogous to Ohm's law for electrical circuits (see Fig. 10.14). Therefore, a circuit of microfluidic-based channels can be computed using the theory of electrical circuits [38].

The Hagen-Poiseuille law is insufficient if the pressure temporally is varying [39], which can normally occur in microfluidic devices when liquid containers are available in the design (see some examples in Figs. 10.9 and 10.12). In this case, it is necessary to define the hydraulic capacitance that describes the ability of a channel fluid system to store fluid (see Fig. 10.15).

Therefore, the volumetric rate of the fluid is defined through the hydraulic capacitance:

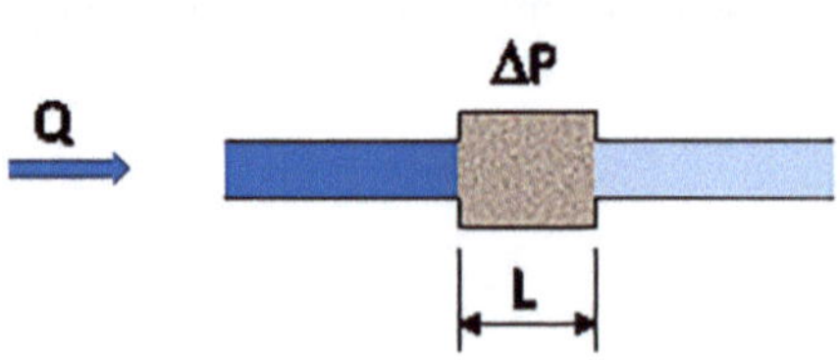

Fig. 10.13 Microfluidic-based system with a scaffold filling part of this channel

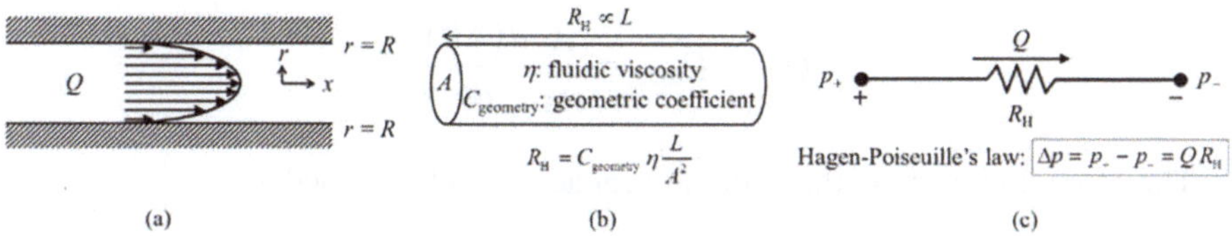

Fig. 10.14 The physical similarities between the flow of a fluid and the flow of electricity: (**a**) Hagen-Poiseuille flow in a circular channel, (**b**) the hydraulic resistance of the circular channel, (**c**) equivalent circuit symbol of a fluidic resistor for the hydraulic resistance and Hagen–Poiseuille's law analogous to a resistor for the electric resistance and Ohm's law (adapted from [37]). Open access article distributed under the terms and conditions of the Creative Commons Attribution (CC BY) license

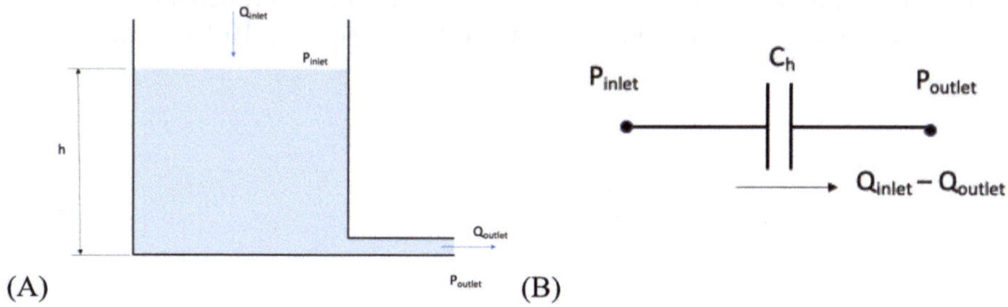

Fig. 10.15 (**a**) The hydraulic capacitance of one tank is the term used to describe energy storage with a liquid in the form of potential energy. (**b**) Symbol and pressure-flow-rate for a hydraulic capacitor (adapted from [39])

$$Q_{inlet} - Q_{outlet} = \frac{dV}{dt} = C_h \frac{d(\Delta P)}{dt}$$

where the hydraulic capacity can be defined as:

$$C_h = \frac{dV}{dP}$$

In the specific case shown in Fig. 10.15 of one container with a height of liquid h, the rate of change of the volume V is determined by the variation of this height:

$$Q_{inlet} - Q_{outlet} = \frac{dV}{dt} = \frac{d(Ah)}{dt} = A\frac{dh}{dt}$$

As the pressure difference between the input and output is *P*, and it is $\Delta P = h\rho g$ with ρ being the liquid density and *g* the acceleration due to gravity, this equation can be written as (if the liquid is assumed to be incompressible and the density does not change with pressure)

$$Q_{inlet} - Q_{outlet} = \frac{A}{\rho g} \frac{d(\Delta P)}{dt}$$

Therefore, in this case, the hydraulic capacity can be defined as

$$C_h = \frac{A}{\rho g}$$

Therefore, taking advantage of the analogy between electrical circuit methods and pressure-controlled microfluidic networks, with correlations between pressure and voltage, flow rate and current, and hydraulic and electrical resistance, any type of microfluidic network can be analyzed and the pressure-controlled laminar flow in the microchannels can be quickly estimated [38–40] (see Fig. 10.16 as example).

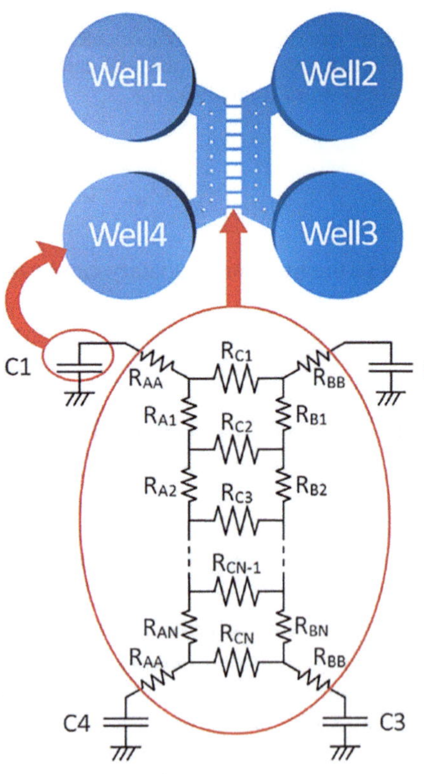

Fig. 10.16 Electrical circuit analogue of the device microfluidic network. The microfluidic device between the 4 open wells of the device was modelled as an electrical circuit comprising a resistive network connected to 4 capacitors. The height of the fluid in an open well determines the value of hydrostatic pressure (equivalent to an electrical potential) (adapted from an open-access source: [40]). Open access article distributed under the terms and conditions of the Creative Commons Attribution (CC BY) license

10.8 From 3D Cell Culture to Organoids

Current hydrogel-based microfluidic solutions enable the development of diverse and different 3D cell cultures to recreate the microenvironment at the tissue level in vivo with many applications in cancer [41, 42], tissue regeneration [43, 44] and drug screening [45]. Initially, most of these applications focus on understanding the processes associated with 3D migration [46–49]. But in addition, this technology provides new dynamic in vitro microenvironments that facilitate access to nutrients and oxygen, and remove waste products to maintain cellular physiological processes [50, 51]. All this makes this technology perfectly suitable for developing organoids in vitro, solving most of the limitations encountered when these organoids are cultured in 3D [52].

In any case, and regardless of the exact definition, organoid research requires the use of 3D culture technologies that provide a high level of control over the microenvironment with the possibility of applying interstitial fluid flow (Table 10.5). All of these needs can be met by organ-on-a-chip technology [56].

Therefore, it is very easy to find in the literature many examples of applications corresponding to organoids, especially in tumors [58], but also in tissue engineering applications, such as, cartilage [59], bone [60], kidney [61], liver [62], neural tissue [44].

Currently, the main challenge consists on the connection of multiples organs-on-chip to construct more complex multiorgan systems that allows to improve the understanding of how different organs interact between them [63]. The most optimistic aim will be to achieve a body-on-a-chip as many authors have proposed [35, 64] (see Fig. 10.17).

Table 10.5 Different definitions found in the literature about organoids

Definition of organoid	References
Simple tissue-engineered cell-based in vitro models that recapitulate many aspects of the complex structure and function of the corresponding in vivo tissue	Zhao et al. [53]
In vitro miniaturized and simplified model systems of organs that have gained enormous interest for modelling tissue development and disease, and for personalized medicine, drug screening and cell therapy	Hofer and Lutolf [54]
Multicellular structures that can be derived from adult organs or pluripotent stem cells	Takebe and Wells [55]
Array of stem cell–derived, self-organizing miniature organs that replicate the key structural and functional characteristics of their in vivo counterparts	Park et al. [56]
A 3D multicellular in vitro tissue construct that mimics its corresponding in vivo organ, such that it can be used to study aspects of that organ in the tissue culture dish	De Souza [57]

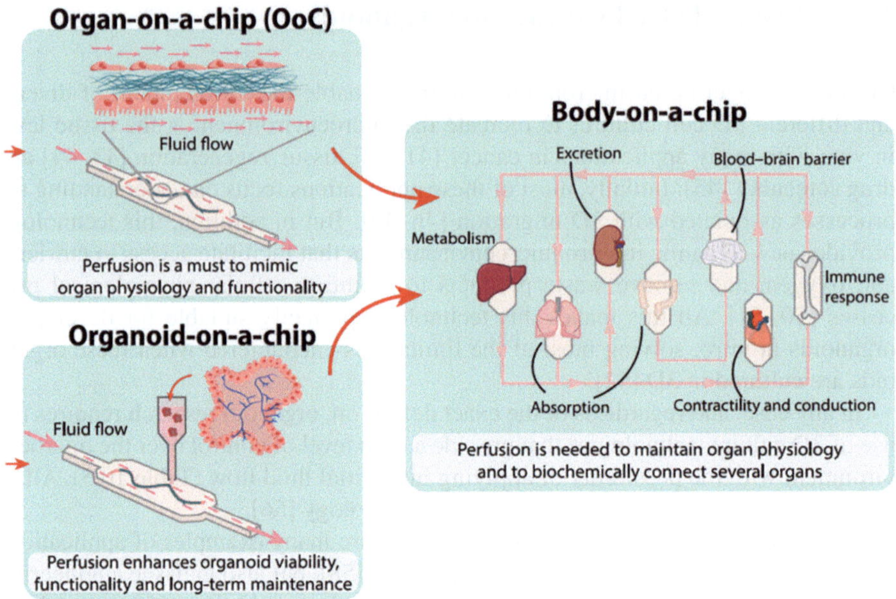

Fig. 10.17 Body-on-a-chip system combining different strategies: (1) Organ-on-a-chip (OoC) made of differentiated cell types grown under fluid flow are able to mimic some physiological functions of organs, and (2) Organoid-on-a-chip cultured under flow are able to model faithfully the intrinsic program of an organ (obtained from [35]). Open access article distributed under the terms and conditions of the Creative Commons Attribution (CC BY) license

10.9 Conclusions

At the end of this chapter, we can conclude that microfluidic technology offers many opportunities for a wide range of tissue engineering applications, from drug and biomaterial testing to the evaluation of various therapies. The development of more sophisticated in vitro models, such as those generated by microfluidics, offers many advantages. These models provide a more physiological and dynamic microenvironment, which can supply nutrients and oxygen and remove wastes in order to recreate cellular physiological conditions in healthy and/or pathological conditions.

In the case of human tissue regeneration, microfluidics-based technology allows working with cells and tissues from human patients, reducing and minimizing the number of animal experiments, as recommended by the 3Rs (replace, reduce, refine) ethical guidelines for animal experimentation. However, we have to keep in mind that although microfluidic 3D cell cultures offer much more possibilities in terms of quantitative analysis, they require the use of more powerful microscopes.

Questions

- Given the following microfluidic chip which is subjected to a pressure increase of 100 Pa. Calculate the average velocity in the channel of the chip, if it is a rectangular section of dimensions w = 20 μm and h = 10 mm, length L = 1 cm (the average V can be approximated as the flow rate divided by the channel area) and the channel is filled with water (viscosity = 1 mPa*s).

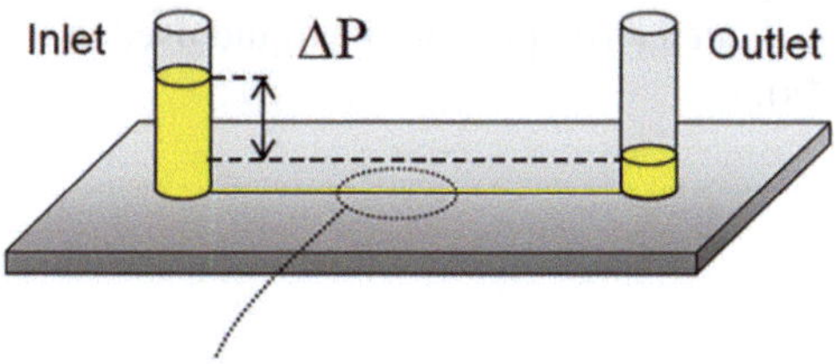

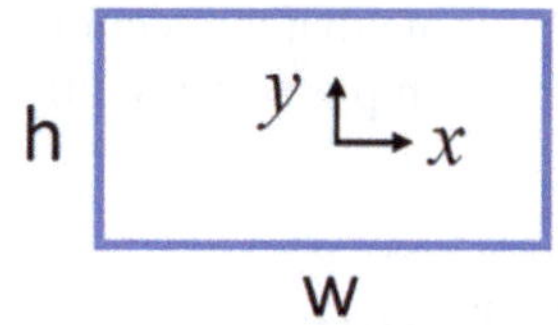

- A microfluidic chip consists of an inlet, an outlet, two 20 micron wide channel segments and a 500 micron wide reaction chamber, as shown in the image (top view). The channel depth throughout the chip is 80 μm and the lengths and widths of the three segments have been marked in the image. The liquid used in the chip is water at 20 °C (μ = 1 mPa*s).
- It is requested:

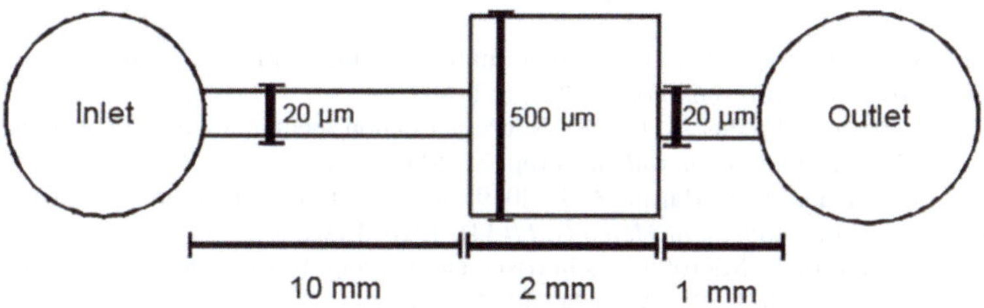

  - Calculate the microfluidic resistance of the chip using the hydraulic radius approach.
  - Calculate the volumetric flow rate (in μl/min) when a pressure difference of 1000 Pa is applied between the inlet and outlet.

- A microfluidic chip consists of a tank where there is fluid exerting a pressure at the channel inlet and an outlet, as shown in the figure [65]:

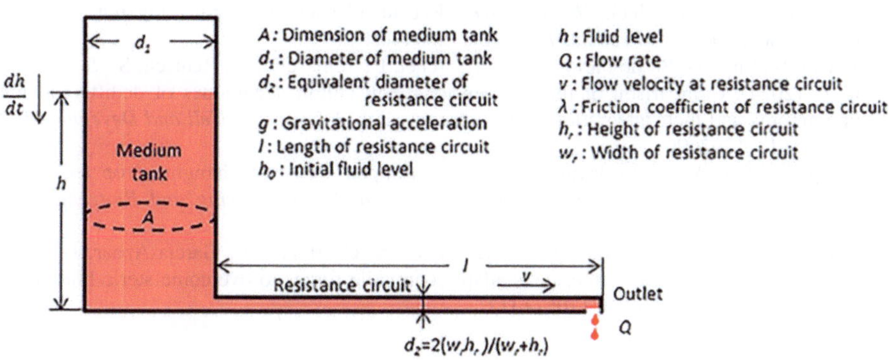

- Estimate whether the output flow rate Q increases or decreases as the length l of the resistive channel increases, justifying the answer.
- Calculate the microfluidic resistance of the chip as a function of the length l of the resistive channel, using the hydraulic radius approximation, considering that the section is circular with diameter 160 μm and the liquid used in the chip is water at 20 °C (μ = 1 mPa*s).

References

1. Yazhini, S., & Sreeja, B. (2024). Comprehensive overview of microfluidic flow sensors: Principles, materials, and fabrication. *Microsystem Technologies*. https://doi.org/10.1007/s00542-024-05787-9
2. Nguyen, N. T., Wereley, S. T., & Shaegh, S. A. M. (2019). *Fundamentals and applications of microfluidics*. Artech House.
3. Andersson, H., & Van Den Berg, A. (2004). Microfabrication and microfluidics for tissue engineering: State of the art and future opportunities. *Lab on a Chip, 4*(2), 98–103.
4. Zhao, Y., Landau, S., Okhovatian, S., Liu, C., Lu, R. X. Z., Lai, B. F. L., & Radisic, M. (2024). Integrating organoids and organ-on-a-chip devices. *Nature Reviews Bioengineering, 2*(7), 588–608.
5. Inamdar, N. K., & Borenstein, J. T. (2011). Microfluidic cell culture models for tissue engineering. *Current Opinion in Biotechnology, 22*(5), 681–689.
6. Silverio, V., & Cardoso de Freitas, S. (2018). Microfabrication techniques for microfluidic devices. In *Complex fluid-flows in microfluidics* (pp. 25–51). Springer.
7. Castiaux, A. D., Spence, D. M., & Martin, R. S. (2019). Review of 3D cell culture with analysis in microfluidic systems. *Analytical Methods, 11*(33), 4220–4232.
8. Sinha, S. K., & Bit, A. (2020). Microfluidics in tissue engineering. In *Biomaterials for organ and tissue regeneration* (pp. 567–598). Woodhead Publishing.
9. Niculescu, A. G., Chircov, C., Bîrcă, A. C., & Grumezescu, A. M. (2021). Fabrication and applications of microfluidic devices: A review. *International Journal of Molecular Sciences, 22*(4), 2011.
10. Gale, B. K., Jafek, A. R., Lambert, C. J., Goenner, B. L., Moghimifam, H., Nze, U. C., & Kamarapu, S. K. (2018). A review of current methods in microfluidic device fabrication and future commercialization prospects. *Inventions, 3*(3), 60.
11. Heckele, M., & Schomburg, W. K. (2003). Review on micro molding of thermoplastic polymers. *Journal of Micromechanics and Microengineering, 14*(3), R1.
12. Wu, J., & Gu, M. (2011). Microfluidic sensing: State of the art fabrication and detection techniques. *Journal of Biomedical Optics, 16*(8), 080901.
13. Bao, M., Xie, J., & Huck, W. T. (2018). Recent advances in engineering the stem cell microniche in 3D. *Advanced Science, 5*(8), 1800448.
14. Tahara, S., Sharma, S., de Faria, F. C. C., Sarchet, P., Tomasello, L., Rentsch, S., & Pollock, R. E. (2024). Comparison of three-dimensional cell culture techniques of dedifferentiated liposarcoma and their integration with future research. *Frontiers in Cell and Developmental Biology, 12*, 1362696.
15. Jeong, Y., Tin, A., & Irudayaraj, J. (2022). Flipped well-plate hanging-drop technique for growing three-dimensional tumors. *Frontiers in Bioengineering and Biotechnology, 10*, 898699.
16. Cóndor, M., Mark, C., Gerum, R. C., Grummel, N. C., Bauer, A., García-Aznar, J. M., & Fabry, B. (2019). Breast cancer cells adapt contractile forces to overcome steric hindrance. *Biophysical Journal, 116*(7), 1305–1312.

17. Sarama, R., Matharu, P. K., Abduldaiem, Y., Corrêa, M. P., Gil, C. D., & Greco, K. V. (2022). In vitro disease models for understanding psoriasis and atopic dermatitis. *Frontiers in Bioengineering and Biotechnology, 10*, 803218.
18. Lee, D. H., Bae, C. Y., Kwon, S., & Park, J. K. (2015). User-friendly 3D bioassays with cell-containing hydrogel modules: narrowing the gap between microfluidic bioassays and clinical end-users' needs. *Lab on a Chip, 15*(11), 2379–2387.
19. Shemesh, J., Jalilian, I., Shi, A., Yeoh, G. H., Tate, M. L. K., & Warkiani, M. E. (2015). Flow-induced stress on adherent cells in microfluidic devices. *Lab on a Chip, 15*(21), 4114–4127.
20. Halldorsson, S., Lucumi, E., Gómez-Sjöberg, R., & Fleming, R. M. (2015). Advantages and challenges of microfluidic cell culture in polydimethylsiloxane devices. *Biosensors and Bioelectronics, 63*, 218–231.
21. Choi, N. W., Cabodi, M., Held, B., Gleghorn, J. P., Bonassar, L. J., & Stroock, A. D. (2007). Microfluidic scaffolds for tissue engineering. *Nature Materials, 6*(11), 908–915.
22. Akther, F., Little, P., Li, Z., Nguyen, N. T., & Ta, H. T. (2020). Hydrogels as artificial matrices for cell seeding in microfluidic devices. *RSC Advances, 10*(71), 43682–43703.
23. Tang, Q., Yang, X., Xuan, C., Wu, K., Lai, C., & Shi, X. (2020). Generation of microfluidic gradients and their effects on cells behaviours. *Bio-Design and Manufacturing, 3*, 427–431.
24. Moreno-Arotzena, O., Mendoza, G., Cóndor, M., Rüberg, T., & García-Aznar, J. M. (2014). Inducing chemotactic and haptotactic cues in microfluidic devices for three-dimensional in vitro assays. *Biomicrofluidics, 8*(6), 064122.
25. Sainz-DeMena, D., Ye, W., Pérez, M. Á., & García-Aznar, J. M. (2022). A finite element based optimization algorithm to include diffusion into the analysis of DCE-MRI. *Engineering with Computers, 38*(5), 3849–3865.
26. Cóndor, M., Rüberg, T., Borau, C., Piles, J., & García-Aznar, J. M. (2018). A web-based application for automated quantification of chemical gradients induced in microfluidic devices. *Computers in Biology and Medicine, 95*, 118–128.
27. Amsden, B. G. (2022). Hydrogel mesh size and its impact on predictions of mathematical models of the solute diffusion coefficient. *Macromolecules, 55*(18), 8399–8408.
28. Campbell, K. T., Wysoczynski, K., Hadley, D. J., & Silva, E. A. (2019). Computational-based design of hydrogels with predictable mesh properties. *ACS Biomaterials Science & Engineering, 6*(1), 308–319.
29. Li, J., & Mooney, D. J. (2016). Designing hydrogels for controlled drug delivery. *Nature Reviews Materials, 1*(12), 1–17.
30. Lin, C. C., & Anseth, K. S. (2009). PEG hydrogels for the controlled release of biomolecules in regenerative medicine. *Pharmaceutical Research, 26*, 631–643.
31. Ogston, A. G., Preston, B. N., & Wells, J. D. (1973). On the transport of compact particles through solutions of chain-polymers. *Proceedings of the Royal Society of London A. Mathematical and Physical Sciences, 333*(1594), 297–316.
32. Kim, W. S., & Tarbell, J. M. (1994). Macromolecular transport through the deformable porous media of an artery wall. *Journal of Biomechanical Engineering, 116*, 156.
33. Tao, L., & Nicholson, C. (2004). Maximum geometrical hindrance to diffusion in brain extracellular space surrounding uniformly spaced convex cells. *Journal of Theoretical Biology, 229*(1), 59–68.
34. Mezedur, M. M., Kaviany, M., & Moore, W. (2002). Effect of pore structure, randomness and size on effective mass diffusivity. *AICHE Journal, 48*(1), 15–24.
35. Juste-Lanas, Y., Hervas-Raluy, S., García-Aznar, J. M., & González-Loyola, A. (2023). Fluid flow to mimic organ function in 3D in vitro models. *APL Bioengineering, 7*(3), 031501.
36. Beebe, D. J., Mensing, G. A., & Walker, G. M. (2002). Physics and applications of microfluidics in biology. *Annual Review of Biomedical Engineering, 4*(1), 261–286.
37. Ereku, L. T., Mackay, R. E., Ajayi, K., & Balachandran, W. (2014). Continuous flow pressure driven microfluidic techniques for point of care testing. *Sensors & Transducers, 170*(5), 1.
38. Oh, K. W., Lee, K., Ahn, B., & Furlani, E. P. (2012). Design of pressure-driven microfluidic networks using electric circuit analogy. *Lab on a Chip, 12*(3), 515–545.

39. Kirby, B. J. (2010). *Micro-and nanoscale fluid mechanics: Transport in microfluidic devices*. Cambridge University Press.
40. Robertson, G., Bushell, T. J., & Zagnoni, M. (2014). Chemically induced synaptic activity between mixed primary hippocampal co-cultures in a microfluidic system. *Integrative Biology, 6*(6), 636–644.
41. Zervantonakis, I. K., Hughes-Alford, S. K., Charest, J. L., Condeelis, J. S., Gertler, F. B., & Kamm, R. D. (2012). Three-dimensional microfluidic model for tumor cell intravasation and endothelial barrier function. *Proceedings of the National Academy of Sciences, 109*(34), 13515–13520.
42. Bersini, S., Jeon, J. S., Dubini, G., Arrigoni, C., Chung, S., Charest, J. L., & Kamm, R. D. (2014). A microfluidic 3D in vitro model for specificity of breast cancer metastasis to bone. *Biomaterials, 35*(8), 2454–2461.
43. Movilla, N., Gonçalves, I. G., Borau, C., & García-Aznar, J. M. (2022). A novel integrated experimental and computational approach to unravel fibroblast motility in response to chemical gradients in 3D collagen matrices. *Integrative Biology, 14*(8–12), 212–227.
44. Karimi, M., Bahrami, S., Mirshekari, H., Basri, S. M. M., Nik, A. B., Aref, A. R., & Hamblin, M. R. (2016). Microfluidic systems for stem cell-based neural tissue engineering. *Lab on a Chip, 16*(14), 2551–2571.
45. Zhai, J., Liu, Y., Ji, W., Huang, X., Wang, P., Li, Y., & Jia, Y. (2024). Drug screening on digital microfluidics for cancer precision medicine. *Nature Communications, 15*(1), 4363.
46. Shin, Y., Han, S., Jeon, J. S., Yamamoto, K., Zervantonakis, I. K., Sudo, R., & Chung, S. (2012). Microfluidic assay for simultaneous culture of multiple cell types on surfaces or within hydrogels. *Nature Protocols, 7*(7), 1247–1259.
47. Del Amo, C., Borau, C., Movilla, N., Asín, J., & García-Aznar, J. M. (2017). Quantifying 3D chemotaxis in microfluidic-based chips with step gradients of collagen hydrogel concentrations. *Integrative Biology, 9*(4), 339–349.
48. Chung, S., Sudo, R., Mack, P. J., Wan, C. R., Vickerman, V., & Kamm, R. D. (2009). Cell migration into scaffolds under co-culture conditions in a microfluidic platform. *Lab on a Chip, 9*(2), 269–275.
49. Anguiano, M., Castilla, C., Maška, M., Ederra, C., Peláez, R., Morales, X., & Ortiz-de-Solorzano, C. (2017). Characterization of three-dimensional cancer cell migration in mixed collagen-matrigel scaffolds using microfluidics and image analysis. *PLoS One, 12*(2), e0171417.
50. Limongi, T., Guzzi, F., Parrotta, E., Candeloro, P., Scalise, S., Lucchino, V., & Perozziello, G. (2022). Microfluidics for 3D cell and tissue cultures: Microfabricative and ethical aspects updates. *Cells, 11*(10), 1699.
51. Vickerman, V., Blundo, J., Chung, S., & Kamm, R. (2008). Design, fabrication and implementation of a novel multi-parameter control microfluidic platform for three-dimensional cell culture and real-time imaging. *Lab on a Chip, 8*(9), 1468–1477.
52. Garreta, E., Kamm, R. D., de Sousa, C., Lopes, S. M., Lancaster, M. A., Weiss, R., Trepat, X., & Montserrat, N. (2021). Rethinking organoid technology through bioengineering. *Nature Materials, 20*(2), 145–155.
53. Zhao, Z., Chen, X., Dowbaj, A. M., Sljukic, A., Bratlie, K., Lin, L., & Yu, H. (2022). Organoids. *Nature Reviews Methods Primers, 2*(1), 94.
54. Hofer, M., & Lutolf, M. P. (2021). Engineering organoids. *Nature Reviews Materials, 6*(5), 402–420.
55. Takebe, T., & Wells, J. M. (2019). Organoids by design. *Science, 364*(6444), 956–959.
56. Park, S. E., Georgescu, A., & Huh, D. (2019). Organoids-on-a-chip. *Science, 364*(6444), 960–965.
57. de Souza, N. (2018). Organoids. *Nature Methods, 15*(1), 23–23.
58. Leung, C. M., De Haan, P., Ronaldson-Bouchard, K., Kim, G. A., Ko, J., Rho, H. S., & Toh, Y. C. (2022). A guide to the organ-on-a-chip. *Nature Reviews Methods Primers, 2*(1), 33.

59. Upadhyay, U., Maredupaka, S., Kancherla, R., Srinivasulu, K., & Chelluri, L. K. (2025). Bioengineered articular cartilage biomimetic organ-on-a-chip using microfluidics. *npj Biomedical Innovations, 2*(1), 6.
60. Nasello, G., Alamán-Díez, P., Schiavi, J., Pérez, M. Á., McNamara, L., & García-Aznar, J. M. (2020). Primary human osteoblasts cultured in a 3D microenvironment create a unique representative model of their differentiation into osteocytes. *Frontiers in Bioengineering and Biotechnology, 8*, 336.
61. Musah, S., Bhattacharya, R., & Himmelfarb, J. (2024). Kidney disease modeling with organoids and organs-on-chips. *Annual Review of Biomedical Engineering, 26*, 383–414.
62. Wang, J., Huang, D., Yu, H., Cheng, Y., Ren, H., & Zhao, Y. (2022). Developing tissue engineering strategies for liver regeneration. *Engineered Regeneration, 3*(1), 80–91.
63. Nguyen, V. V., Ye, S., Gkouzioti, V., van Wolferen, M. E., Yengej, F. Y., Melkert, D., & van Balkom, B. W. (2022). A human kidney and liver organoid-based multi-organ-on-a-chip model to study the therapeutic effects and biodistribution of mesenchymal stromal cell-derived extracellular vesicles. *Journal of Extracellular Vesicles, 11*(11), 12280.
64. Huh D., Matthews B. D., Mammoto A., Montoya-Zavala M., Hsin H. Y., & Ingber D. E. (2010). Reconstituting organ-level lung functions on a chip. *Science, 328*(5986), 1662–1668. https://doi.org/10.1126/science.1188302
65. Komeya, M., Hayashi, K., Nakamura, H., Yamanaka, H., Sanjo, H., Kojima, K., & Ogawa, T. (2017). Pumpless microfluidic system driven by hydrostatic pressure induces and maintains mouse spermatogenesis in vitro. *Scientific Reports, 7*(1), 15459.

Chapter 11
Computational Modelling of the Mechanobiology of Scaffolds and Their Interaction with Cells and Tissues

Abstract In this chapter, we present how computational modelling can be used as an additional tool for working in tissue engineering and regenerative medicine. The ability to use computational models in combination with other technologies, such as in vitro and in vivo experiments, is very useful because it provides the opportunity to study many different conditions from different perspectives, obtaining different data that can be compared and integrated to achieve a deep understanding of the biological mechanisms involved. Moreover, computer simulations allow to reduce the number of in vivo experiments, providing us with an alternative methodology. As it will be shown here, computational predictions can assist in final design decisions in bioreactors, microfluidic systems, or organ-on-chip devices, in optimizing culture conditions, or in estimating the long-term behavior of engineered tissues in vivo. Here, several modelling strategies are shown to present its potential applicability in tissue engineering and regenerative medicine.

11.1 Introduction

The aim of this chapter is to present the computational methodology as an additional method to work in tissue engineering [1], showing different examples of application that can inspire and illustrate the future work of students and researchers that are not familiarized with this technology.

The first idea that we have to keep in mind is that computational tools cannot be used alone, but they have to be used in combination with other methodologies, such as in-vitro and in-vivo experiments (Fig. 11.1). Although from computational simulations we can obtain a lot of information, this model is very far from reality and validation is always required, which implies developing a quantitative comparison between the in-vitro or in-vivo results with the computational simulations in order to calibrate our model. Special emphasis on this front has been developed in the last section of this chapter.

© The Author(s), under exclusive license to Springer Nature Switzerland AG 2026

E. García-Gareta, J. M. García-Aznar, *Contemporary Tissue Engineering and Regenerative Medicine*, Series in BioEngineering,
https://doi.org/10.1007/978-3-032-03441-0_11

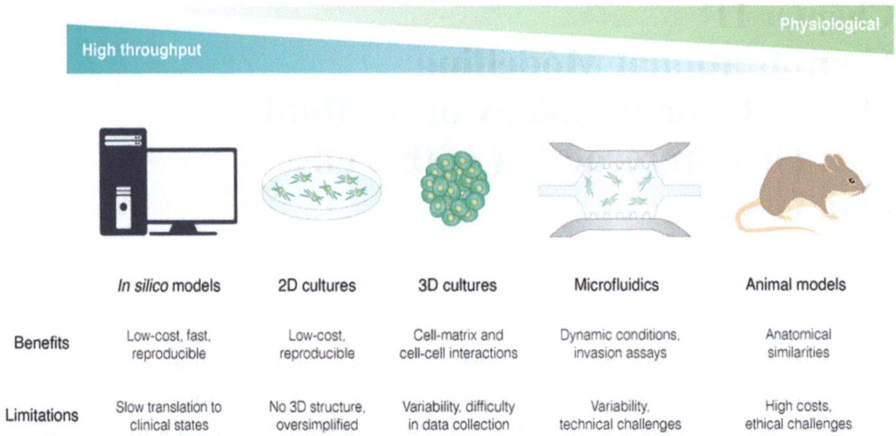

Fig. 11.1 Different kinds of models in mechanobiology research. Comparative analysis of benefits and limitations for different in silico, in vitro and in vivo models. In vitro models are well-established as they are less expensive and faster than in vivo models, which are also known to raise ethical concerns. Despite being less biologically accurate, in silico models have been used consistently in combination with experimental assays to understand biological data and test hypotheses about its causality (Obtained from [2])

Therefore, the development of these computational models must be designed to facilitate their validation with experimental data and to test hypotheses in a quantitative, predictive, and reproducible manner [3]. As pointed out by Markowetz [4] in the literature, there are many predictive models, but few of them have been validated. Hence validation is a crucial aspect to be considered when a computational model is built.

Thus, to develop a computational model that allows us to simulate a specific problem in the context of Tissue Engineering and Regenerative Medicine (TERM) is required to define some specific aspects or steps [5] that have to be clearly defined and specified:

1. Modelling approach. Normally, the most common approach to simulate tissue dynamics is to consider the tissue as a continuous material and apply classical techniques from continuum mechanics and thermodynamics, where the cell populations and biochemical factors are described by a macroscopic continuum variable. However, it is also usual to simulate tissue dynamics by means of agent-based approaches, where normally one agent simulates one cell. The first approach allows to simulate large multicellular tissues, whereas the second one has to focus on smaller multicellular tissues, due to the computational cost. Certainly, it is also possible to find different hybrid approaches combining both strategies. In Table 11.1 (based on the work of [6]) is described the most common modelling approaches used in biological models, clearly showing the main advantages and disadvantages for each model approach.

Table 11.1 Different modelling approaches used in biological models, based on Butner et al. [6]

Model type	Common uses	Advantages	Disadvantages
Continuum PDE	Modelling spatiotemporal behaviors and feedback between populations	• Computationally low cost • Relatively easier to develop • Model large-scale processes that are impractical for discrete representations • Allows data-driven empirical description or mechanistic formalism of the system	• Difficult to characterize heterogeneity • Deterministic solutions • Phenomenological model parameters (if used) may not have direct biological translation • Parameters may be difficult to measure experimentally
Continuum ODE, compartmental modelling	Modelling time-dependent population shifts and feedback between populations	• Computationally low cost • Relatively easier to develop • Model large-scale processes that are impractical for discrete representations • Allows data-driven empirical description or mechanistic formalism of the system • High potential for prospective clinical use or supporting regulatory approval applications	• Does not describe spatial effects • Difficult to characterize heterogeneity • Deterministic solutions • Phenomenological model parameters (if used) may not have direct biological translation • Parameters may be difficult to measure experimentally
Agent-based modelling	Describe spatiotemporal dynamics of heterogeneous entities, such as molecules, cells, tissue and organs	• Explicit description of heterogeneous properties and relevant underlying mechanisms (biological and physical) • Allows stochasticity and are non-deterministic	• Difficult to develop and implement • Computationally intensive • May be impractical for describing large-scale systems • Multiscale modelling may be difficult
Hybrid modelling	Describe spatiotemporal dynamics of both continuous and heterogeneous entities and feedback between populations and across scales	• Explicit description of heterogeneous properties and relevant underlying mechanisms (biological and physical) • Allows stochasticity • Explores complexity and diversity of cancer • Multiscale in nature	• Difficult to design • Computationally intensive • Extensive numerical techniques requires • Model parameterization may be difficult

2. Model construction. The model construction requires to define the initial conditions and geometry of the problem of study, being therefore known data. Normally the geometry can be reconstructed from patient image data in an automatic manner. A strong effort has been addressed in the literature to construct models from these image data, it is explained in detail in the Sect. 11.2 known as CATE (Computer Aided Tissue Engineering). Moreover, in computational biology, the modelling assumptions must be carefully tailored to the biological system of study and the available data [5], facing many different challenges and difficulties from a mathematical modelling perspective that is discussed in Sect. 11.3. Computational modelling in tissue engineering and regenerative medicine requires a multiscale (Sect. 11.4) and multiphysics analysis (Sect. 11.5) per se. At the end, this decision depends on the level of description that the model is able to provide and the possibility to compare and validate the computational results with the available data.
3. Numerical solver. The numerical model that is constructed must be solved numerically. Different numerical techniques are commonly used to numerically solve this kind of models, such as finite element, finite volume, finite difference and lattice Boltzmann methods. In any case, this numerical method should be chosen to ensure convergence and accuracy.
4. Verification. Verification is the process of confirming that the developed model is valid to simulate the problem of study. For this aim, it is necessary to define benchmark problems with a known solution to evaluate if the model is able to predict this solution.
5. Validation. Validation is the process of assessing the degree to which the computational model is an appropriate representation of the problem of study [7]. For this evaluation it is required to compare the computational model results with those obtained from the in-vitro/in-vivo/clinical data. In V&V 40 (the American Society of Mechanical Engineers Verification and Validation 40 standard and FDA proposed AI/ML model life cycle management framework for Software as a Medical Device in biopharmaceutical manufacturing use cases) proposed different levels of comparison according to the rigorous of this comparison. From a level 1 very qualitative, such as a visual comparison, to level 5 (see Table 11.2) where the model not only has to predict the average data, but also the uncertainty of the data.
6. Assessment of predictive capacity. After validation, it is important to evaluate the predictive capacity of the model using additional data from in-vitro/in-vivo/clinical to confirm the model potential.

Overall, this chapter provides some examples of how different strategies in computer modelling can be used to analyze and understand complex problems in tissue engineering. We explicitly show how modelling may represent tissue matrix, scaffolds, cells and their interaction. In particular, we present in different sections the

Table 11.2 Different levels of comparison between computational results and experimental data according to V&V 40 criteria [8]

Level 1	Visual comparison concludes good agreement
Level 2	Comparison by measuring the difference between computational results and experimental data. Differences are less than 20%
Level 3	Comparison by measuring the difference between computational results and experimental data. Differences are less than 10%
Level 4	Comparison with uncertainty estimated and incorporated from the comparator or computational model. Differences between computational results and experimental data are less than 5%. Includes consideration of uncertainty, but statistical distributions for uncertainty quantification are unknown
Level 5	Comparison with uncertainties estimated and incorporated from both the comparator and the computational model, including comparison error. Differences between computational results and experimental data are less than 5%. Statistical distributions for uncertainty quantification are known

application of computational models to simulate in-vivo experiments with animals (Sect. 11.6), in-vitro experiments in bioreactors (Sect. 11.7) and in-vitro experiments in organs-on-chips (Sect. 11.8). Obviously, the computational modelling of all these experiments require validation and, therefore, Sect. 11.9 describes different strategies that can be used for validating these models.

11.2 Computer-Aided Tissue Engineering (CATE)

The concept of CATE was originally proposed by Sun et al. [9] as a consequence of the combination of computer-aided design (CAD), image processing, manufacturing and solid free-form fabrication (SFF) for modelling, designing, simulation and manufacturing of biological tissue and organ substitutes. This work describes some salient technologies for the development of high-resolution non-invasive imaging and image-based three-dimensional reconstructions, and various reconstructive techniques for CAD-based tissue modelling generation.

More recently, the book of Rainer and Moroni [10] shows different examples of computer aided design and manufacturing of 3D scaffolds and biofabricated constructs for tissue engineering applications. Most of these works focus on the computational design of scaffolds [11–13] and their 3D printing [14, 15].

It is important to remark that most of these models focus on the short-term tissue response after scaffold implantation, although now computational models are focusing on the long-term response. Therefore, next section of this chapter mainly presents computational models to predict the long-term response of the tissues and the biomaterial.

11.3 Challenges and Difficulties in Mechanobiological Modelling in Tissue Engineering

Mechanobiological model requires to evaluate not only the mechanical alteration (stresses and strains) that tissues suffer after scaffold implantation (short-term response), but also how cells associated to these tissues respond accordingly to these alterations, producing new tissue or removing tissue. Therefore, the application of these models to simulate different biological processes involved in tissue engineering and regenerative medicine, require to face some relevant challenges that we have to take into account:

- Multiphysics modelling. Multiple fields of physics are coupled in all the biological processes involved in regeneration, such as, fluid flow, mass transfer, biochemistry, solid mechanics among others (see, for instance, Fig. 2.11 to show multiphysics in bone mechanobiology). But additionally, biological tissues are multiphysics materials by nature. For example, collagen is a piezoelectric material, where mechanical stimuli are coupled with the electrical stimuli [16, 17]. This fact has motivated that many works are using piezoelectric materials to design and fabricate scaffolds [18, 19]. Therefore, multiphysics modelling is compulsory when a model is constructed in TERM, so coupling between biomechanics and biological factors is the minimum approach to be considered.
- Multiscale modelling. Tissues are multiscale in nature, where their morphogenesis, development and regeneration are governed by the cell response according to the local stimuli that they are sensing, which can be biochemical or mechanical signals. How these signals are transmitted through the tissues from the organ scale to the cell level require a multiscale analysis (see Fig. 2.11 as an example in bone multiscale). If a multiscale computational approach is implemented, the main disadvantage is its high computational cost that requires some simplifications, and thus different strategies can be found in the literature (see Sect. 11.5).
- Modelling cell response. Cells are one of the main responsible of tissue adaptation. Therefore, modelling how cells perceive the different regulatory stimuli from their extracellular microenvironment and how they respond is one of the most relevant assumptions that have to be defined. But we also have to keep in mind that cell response is also strongly influenced by the interactions with the ECM and with their surrounding cells. All these aspects should be considered when a model is defined for TERM.
- Very complex and difficult validation that requires image quantification. Validation is not an easy task, because it requires a lot of quantitative data from experiments. One possible strategy to quantify experiments is the use of image quantification.
- Patient-specific simulations. If simulation of a clinical case is developed, computational models are personalized using data from a specific patient, for example, through image-based biomarkers [20].

11.4 Multiphysics Modelling in Tissue Engineering

In biological processes, mechanical forces and biochemical reactions are strongly coupled at subcellular level, as shown in Fig. 11.2. So, when a cell protein is deformed by a mechanical force, signaling domains become available to capture different signaling molecules that can activate new biochemical reactions. Therefore, multiphysics analysis is always involved in computational modelling in TERM, which is usually known as computational mechanobiology or mechano-chemo-biology [21]. Therefore, many different multiphysics works can be found in the literature [21], which are normally divided into three main categories (see Fig. 11.3) according to the biophysical stimuli that regulate the different biological events: solid mechanics, fluid mechanics, and mass transport. Of course, these stimuli are not used in isolation, but as a combination that establish regulatory theories in function of them [21–24].

But multiphysics is not only occurring at subcellular or cellular level, but also at tissue level. Piezoelectricity has been normally simulated computationally to simulate bone regeneration [17, 25] or scaffold behaviour [26]. In addition, more recently, Rennekamp et al. [27] showed that collagen is more than a passive material because when it is supporting mechanical stresses, it behaves as a huge mechano-"enzyme", producing mechano-radicals and subsequently reactive oxygen species (ROS), which are essential biological signaling molecules [28]. Therefore, multiphysics analysis is also required at tissue level.

11.5 Multiscale Modelling in Tissue Engineering

According to the detail level at spatio-temporal description of the model we can work in one scale or in different linked scales, what is normally known as a multi-scale approach. This modelling approach is becoming increasingly popular in

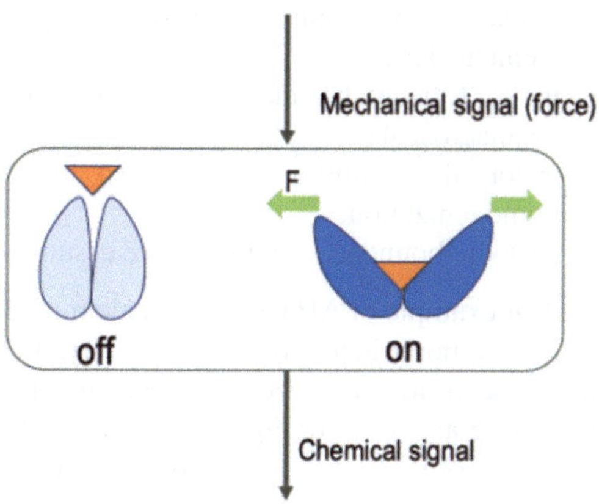

Fig. 11.2 Mechano-chemo-biology coupling clearly defines the multiphysics analysis required for modelling in tissue engineering and regenerative medicine. When a protein is deformed by a mechanical force F, signaling domains are available for capturing different signaling molecules that can active new biochemical reactions

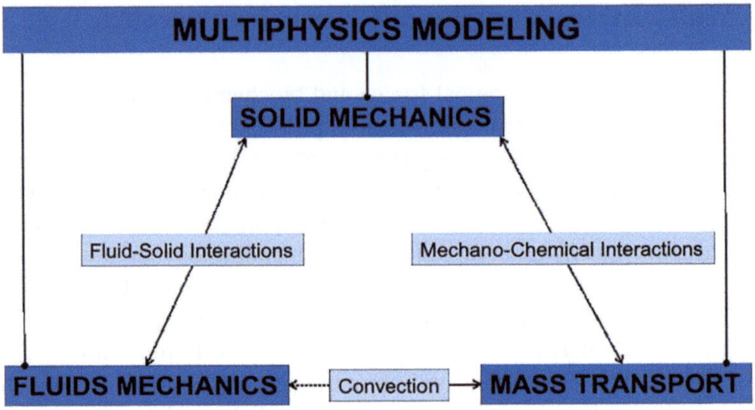

Fig. 11.3 Multiphysics models are classified according to different fields of physics (based on [21]): computational mechano-chemo-biology

mechanobiology and TERM [29, 30]. A multiscale approach can include different mathematical modelling approaches. For example, continuum models treat cells as concentrations of cell types, whereas discrete models (such as agent-based or particle models) consider discrete cells; the cell behaviors, including interactions between cells, can be described as deterministic or stochastic [31]. Normally, in a multiscale approach, a continuum description is used at tissue scale or macroscale, while a discrete approach is used at lower scales for the cells, but keeping a continuum approach for molecular species, such as glucose, oxygen, growth factors, chemokines and drugs. In Fig. 11.4, it is shown a possible scale organization in multiscale modelling in TERM, specially focused on the validation techniques and the modelling approaches normally used in each scale.

A critical aspect in the construction of a multiscale model is the definition of the scale, or how the scale separation will be defined [30, 32]. One common theoretical strategy to define a multiscale approach is the use of asymptotic homogenization theory (AHT) [33–35], which is based on the following hypotheses:

- a macroscopic point is microscopically represented by a representative volume element (RVE);
- the RVE has to be selected according to the length scale separation (microscale << macroscale);
- averaged quantities are defined in the RVE in a process known as homogenization;
- periodic boundary conditions are assumed during localization process;

A clear example of AHT in multiscale modelling for bone tissue engineering applications is shown in the work of Sanz-Herrera et al. [33, 36, 37], where bone ingrowth in one scaffold is analyzed macroscopically and microscopically as is indicated in Figs. 11.5 and 11.6. In Fig. 11.5 is shown the multiscale approach, where the macroscopic (organ) level is solved by means of a continuum approach using PDEs

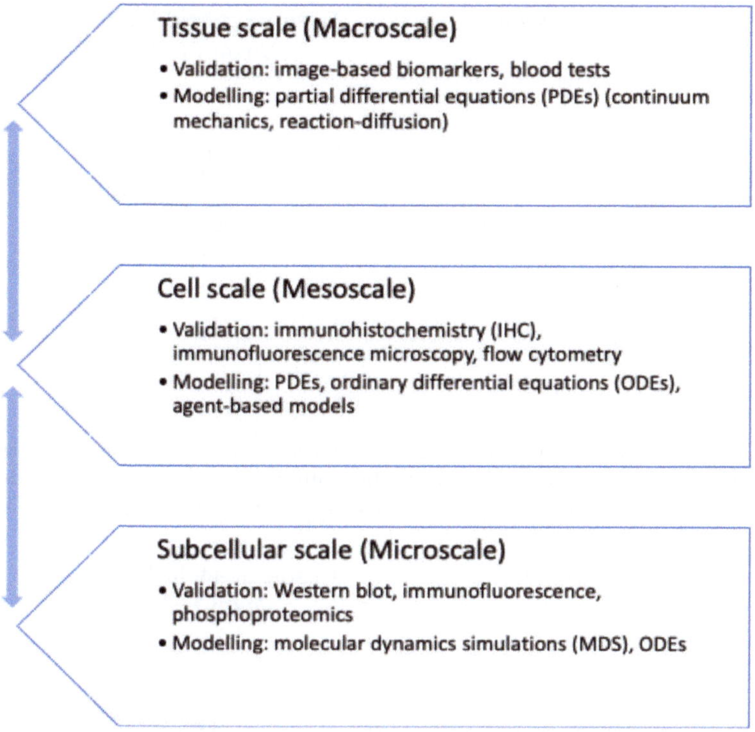

Fig. 11.4 Hierarchy of spatial scales in multiscale analysis with associated examples of validation technologies and modelling approaches for each scale (Adapted from [5, 31])

solved by finite element method (FEM), with a different approach in the bone tissue and in the bone scaffold replacement. In the bone tissue a macroscopic bone remodelling model is used to simulate bone density evolution, while in the bone scaffold a multiscale approach is solved. In the macroscopic scale of the scaffold, a mechanical problem is numerically solved to estimate energy density distribution in combination with reaction-diffusion equations to evaluate the invasion of cells inside the scaffold [33, 36, 37]. Each point of analysis in the macroscopic scaffold corresponds to a computational model at the tissue level, where homogenization techniques are used to compute the stiffness and diffusion tensor (see Fig. 11.5).

After the numerical solution of the macroscopic problem (mechanical and diffusion analysis), the localization process is calculated by estimating the mechanical and diffusion variables locally in the scaffold surface (Fig. 11.6). From these local variables, a phenomenological law is proposed to evaluate bone formation in the scaffold surface [33, 36, 37] and scaffold degradation. Numerical results of bone formation in scaffolds with different initial microarchitectures are shown as an example in Fig. 11.7, where clearly this initial microarchitecture of the scaffold is guiding the growth of the bone.

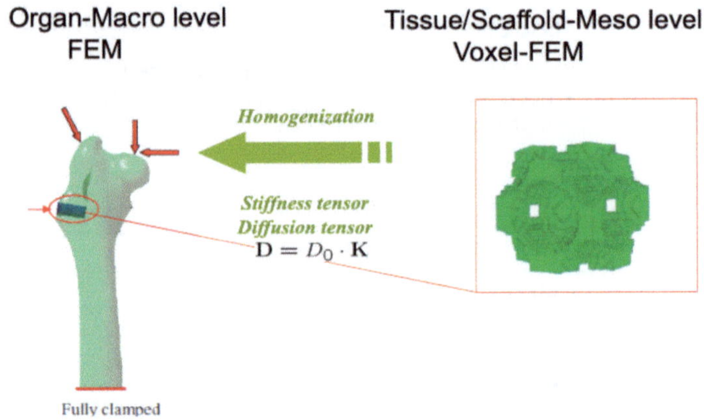

Fig. 11.5 Computational homogenization analysis of one scaffold where multiphysics and multiscale approach is used. A mechanical and reaction-diffusion macroscopy analysis is developed, therefore, stiffness and diffusion tensor are averaged at the tissue scale (Adapted from [33, 36, 37])

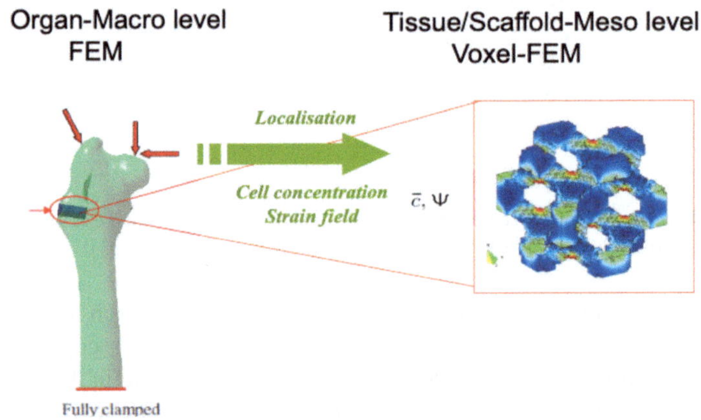

Fig. 11.6 Computational localization analysis of one scaffold where multiphysics and multiscale approach is used. The mechanical strain stimulus and the cell surface distribution is computed at scaffold surface level, where a phenomenological law is evaluated to estimate if scaffold degradation occurs and it is replaced by bone tissue (Adapted from [33, 36, 37])

11.6 Modelling Macroscopic In-Vivo Conditions

In this section, several examples of macroscopic modelling are shown, which have been compared with experimental results. Certainly, it is possible to find many other different examples in the literature, but we have focused on two of them with which we are more familiarized. On the one hand, a macroscopic model to simulate would healing closure is presented, where a quantitative comparison with experiments in animals is shown based on a previous computational work [38]. On the other hand,

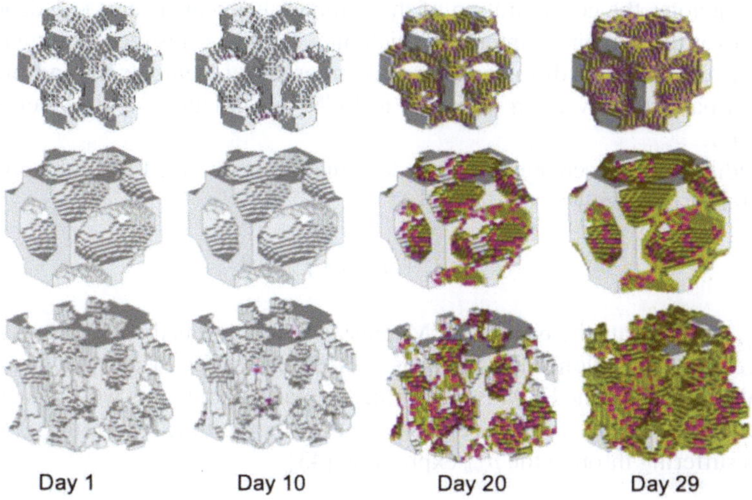

Fig. 11.7 Numerical results obtained from the computational model with scaffolds of different initial architecture (in the top isotropic distribution, in the middle orthotropic distribution and in the bottom anisotropic distribution) and how bone is formed after 29 days in the simulation (obtained from the work of [37])

we show the simulation of one scaffold implantation and how bone ingrowth can occur inside this scaffold [39].

11.6.1 Macroscopic Modelling of Wound Healing Closure

In this first example a macroscopic model is presented for the simulation of the wound closure event. Wound healing is a natural process in which the damaged skin layers regenerate autonomously through several complex mechano-biochemical processes that ends with the wound closure. The different processes taking place during wound healing overlap in time and can be divided into different stages: inflammatory, proliferative and remodelling [40]. Just after wound formation, initially the wound area slightly increases for later start to contract until the wound is closed [41, 42]. To simulate this mechanobiological process, Valero et al. [38] proposed a computational-based model considering the temporal evolution of different cellular species (fibroblasts and myofibroblasts) and the growth factors (a generic growth factor with the combined effects of PDGF and TGF-b on (myo)fibroblasts and the collagen density), which are described mathematically through the conservation law:

$$\frac{\partial \rho_i}{\partial t} + \nabla \cdot \boldsymbol{J}_i = G_i$$

where r_i denotes the concentration of the cellular/chemical species i, $\boldsymbol{J}_i$ denotes its net flux over the domain of interest (which may include terms representing random dispersal -migration or diffusion-, directed migration (chemotaxis), and may also include a passive convection term due to ECM deformation), and G_i denotes net production.

In addition, the mechanical matrix deformation is obtained from the conservation of linear momentum

$$\nabla \cdot \left(\sigma_{ecm} + \sigma_{cell} \right) + \boldsymbol{f}_{ext} = \mathbf{0}$$

where s_{ecm} represents the passive ECM stress, s_{cell} represents the ECM stress due to the fibroblast and myofibroblast forces exerted on the ECM and $\boldsymbol{f}_{\text{ext}}$ denotes the external stresses that the ECM is supporting. In particular, in this work, Valero et al. [38] assumed that cell forces are dependent on the volumetric strain (q) that the ECM is suffering through the p_{cell} expression [43]:

$$\sigma_{cell} = p_{cell}\left(\theta\right)\left(n + \xi m\right)\mathbf{1}$$

where n is the cellular density of fibroblasts and m is the cellular density of myofibroblasts, being x the ratio that quantifies the myofibroblast-enhanced traction force.

Using this model, Valero et al. simulated the wound geometries provided by Roy et al. [41] and McGrath and Simon [42] in animal models. Roy et al. [41] considered a circular wound area of 0.5 cm^2 in pigs, however, McGrath and Simon [42] considered square wounds with areas of 6.25cm^2 and 12.54cm^2 in rats.

The numerical predictions about the time evolution of the wound area (normalized with respect to its initial size) is presented in Fig. 11.8, where clearly the model is able to qualitatively predict the tendency of the wound closure.

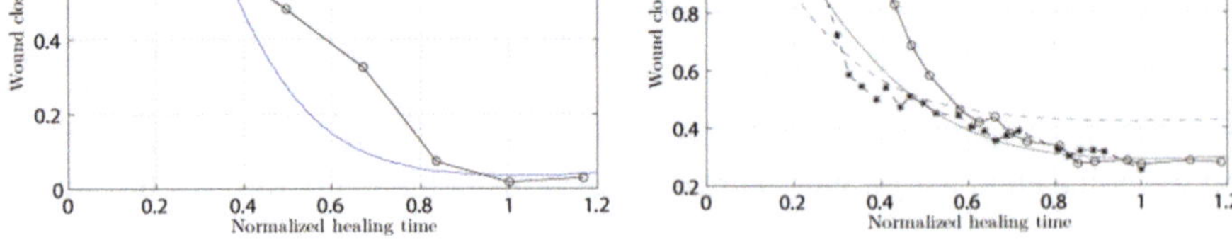

Fig. 11.8 Normalized wound area as a function of the normalized healing time for different conditions: (left) a circular wound with a radius of 0.4 cm and (right) two square wounds with areas of 6.25 cm^2 and 12.54 cm^2 (obtained from [38]). Open access article distributed under the terms and conditions of the Creative Commons Attribution (CC BY) license

11.6.2 Macroscopic Modelling of Bone Ingrowth Inside One Scaffold

In this case, it is shown an application example of a computational model to simulate bone ingrowth in scaffolds using animal experiments for validation [39]. In particular, a computational macroscopic approach is used to simulate bone formation in porous titanium scaffolds simulated in the distal femur and the proximal tibia of three goats and compared to experimental results (Fig. 11.9).

In this example, a macroscopic continuum approach based on partial differential equations solved by the Finite Element Method (FEM) is used. It is assumed that cells can invade the scaffold through the granulation domain, differentiating to bone cells and forming bone tissue. This cell differentiation and bone formation phase is assumed to be regulated by the mechanical stimulus through a phenomenological regulatory law [39]. Therefore, a couple of mechanical and diffusion analysis are implemented in the model.

Despite the difficulties in quantifying and measuring the experimental data, a qualitative validation was developed as one of the main results of this work. On the one hand, Fig. 11.10 shows the distribution of bone ingrowth at 12 weeks after surgery within the scaffold placed in the goat femoral epiphysis, where bone formation occurred mostly in the outer pores of the scaffold for all samples, with limited growth in the core and different distribution depending on the anatomical location. On the other hand, Fig. 11.11 shows a quantitative comparison between experimental and numerical results focusing on the temporal evolution of bone formation.

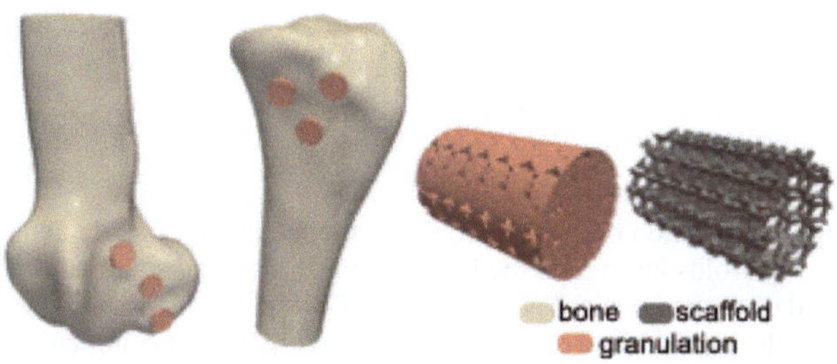

Fig. 11.9 Finite element model of the femur and tibia of goats and the scaffold implantation in different locations, distinguishing between different materials: bone, scaffold and granulation tissue (obtained from [39]). Open access article distributed under the terms and conditions of the Creative Commons Attribution (CC BY) license

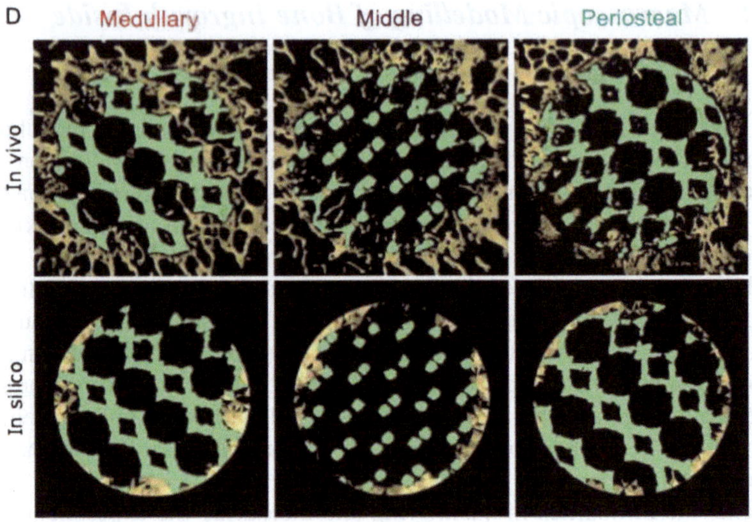

Fig. 11.10 Representative *in vivo* and *in silico* microCT slices in the scaffold located in the femoral epiphysis of the medullary, middle and periosteal subregions where bone is labeled in yellow and metal is labeled in green. Obtained from Nasello et al. [39]. Open access article distributed under the terms and conditions of the Creative Commons Attribution (CC BY) license

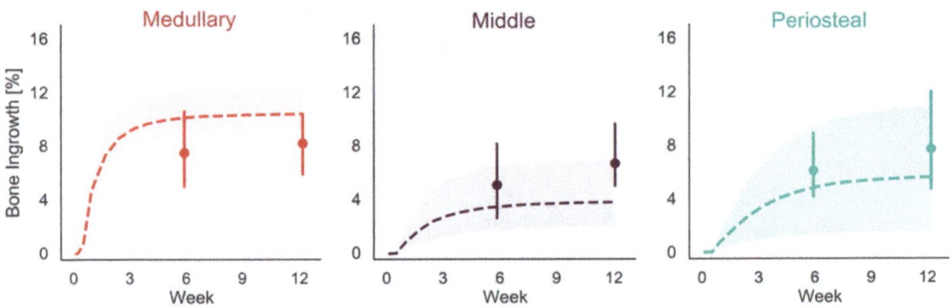

Fig. 11.11 Temporal evolution of bone ingrowth in the medullary, middle and periosteal regions of titanium porous scaffolds inserted in the femoral epiphysis of goat animal models. Comparison between in silico (dashes) and in vivo (points) results. Point plots and dashed lines denote the mean of n = 3 samples, with error bars and error bands representing 95% of confidence interval. Obtained from Nasello et al. [39]. Open access article distributed under the terms and conditions of the Creative Commons Attribution (CC BY) license

11.7 Modelling Bioreactor Conditions

Numerical simulation of bioreactors is one of the most relevant topics in computational tissue engineering existing multiple and different approaches that go from computational fluid dynamics (CFD) to cell-based simulations. The most common approaches used to simulate the functional behavior of bioreactors are CFD simulations that focus on biochemical processes, mainly in mechanical mixing with diffusion and bioreaction kinetics [44–47]. As conclusion, these works show that computer simulations are supportive tools for testing and are useful for the design, optimization and control of large-scale bioreactors.

However, in this section, our focus is based on those models that simulate cell populations and their interaction with the biomechanical and biochemical microenvironment defined in the bioreactor. From a macroscopic perspective, computational fluid dynamics (CFD) approach continuous being the most common modelling framework used to estimate flow-induced shear stresses throughout tissue-engineered constructs. This approach has been used for investigating the influence of different conditions: to optimize the location of the regular pore scaffolds within the perfusion bioreactor chamber [48]; to analyze the influence of cylindrical bioreactor dimensions (length and diameter) on the fluid flow and scaffold shear stresses [49]; to study the bioreactor performance (e.g., predicting shear stress, turbulence, and oxygen transfer rate) for better bioreactors design so that cell growth can be optimized [50]; to analyze the process of cell seeding and adhesion inside a scaffold [51, 52]; among others.

Therefore, this kind of cell analysis in a bioreactor, requires the combination and integration of CFD analyses with dynamic cell reaction kinetic (CRK) models to simulate perfusion bioreactor performance [40]. In this section, it is reviewed the different strategies found in the literature to couple CFD and CRK methods, distinguishing between eulerian and lagrangian approaches. The eulerian approach is also known as the Euler–Euler (EE) approach, as the fluid and the cells are not individually tracked. In this approach, all the phases are considered as a continuum medium described in terms of their volume fractions [53–55]. The assumption of a biological entity as a continuum is an oversimplification and does not consider the individual nature of each cell. Thus, this methodology is unable to account for the individual response of the cells, but their global behavior. A clear example of this approach is presented by Osborne et al. [54], who present a continuum macroscopic formulation based on a three-phase model for the growth of a tissue construct in a perfusion bioreactor, where the cell population (and associated extracellular matrix), culture medium, and porous scaffold are treated as distinct phases. The bioreactor system is represented by a two-dimensional channel containing a cell-seeded rigid porous scaffold (tissue construct), which is perfused with a culture medium. Through the definition of different functions for cell proliferation and extracellular matrix deposition rates, the model is used to compare the influence of cell density-, pressure-, and culture medium shear stress-regulated growth on the composition of the engineered tissue. Nevertheless, more examples can be found in the literature, the work of Singh et al. [40] presents a review with many of these examples.

The Lagrangian approach emphasizes tracking each particle individually for all phases. Consequently, this approach results in a high computational cost. Therefore, a more practical modelling approach is called Euler-Lagrange simulations, where the Lagrangian approach is normally applied to the cells and the Eulerian approach is used for the rest of components (nutrients, fluids, among others). Singh et al. [40] provides a review showing different examples. As example, I would like to mention

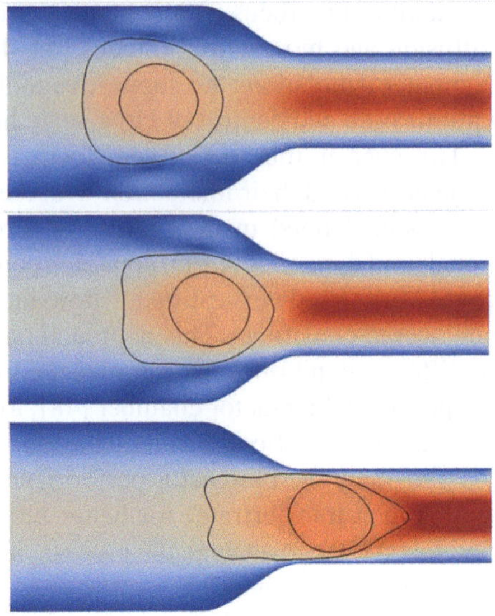

Fig. 11.12 Deformed shape of a cell and its nucleus moving through one channel, showing the fluid velocity profile (obtained using the model proposed by Rüberg and Aznar [64])

the work of Olivares and Lacroix [51] where they used dynamic cell particles to simulate the cell-seeding process within a scaffold under perfusion conditions. In this work, cells are treated as spherical particles dragged by the fluid media, where the physical parameters are computed through a Lagrangian formulation.

In most of these models, cells are simulated by rigid particles, however the mechanical deformation of cells by fluid flow is one critical aspect in all perfusion bioreactors during many processes such as, cell expansion [56], seeding [52], adhesion [57], differentiation [58], matrix formation [59, 60]. Therefore, it is necessary to simulate cell deformation during fluid flow due to perfusion in the bioreactor, which requires a fully coupled fluid-solid interaction. The most common technique for this fully-coupled analysis of fluid–solid interaction is the Arbitrary Lagrangian-Eulerian (ALE) technique [61, 62]. For example, Serrano-Alcalde et al. [63] used this technique to simulate how one single cell is deformed by a creeping flow inside a channel with a variable cross section and estimate the strains and stresses that the cell nucleus is supporting.

Another interesting approach to solve the solid–fluid interaction has been proposed by Rüberg and García-Aznar [21], where an immersed finite element method is presented. Cell deformation is shown in Fig. 11.12 together with the fluid pressure when the cell is moving through a channel with a variable cross section.

11.8 Modelling Organoids for TERM Applications

Different numerical approaches have been used to simulate in vitro organoids, mainly ranging from a continuum to a discrete (or an agent-based model, ABM) approach. In a continuum approach, normally ordinary differential equations (ODEs) or partial differential equations (PDEs) are used to simulate organoids at the cell population level [65, 66]. However, discrete models describe cells at an individual level and are better suited to reproduce the heterogeneous behavior of organoids. Discrete models can be defined as on-lattice or off-lattice. On-lattice models are computationally efficient, while off-lattice models provide more realistic representations of cells. In the work of Van Liedekerke et al. [67], there is an extensive review of the different types of agent-based models, analyzing and comparing the main advantages and disadvantages between them.

In both model types, the diffusion of nutrients and oxygen are normally considered through continuum reaction diffusion equations, even modelling glucose metabolism [68]. Therefore, continuum models normally follow an eulerian approach because both nutrients, oxygen and cells are modelled following a continuum approach [65, 66]. However, in the case of agent-based models normally follow a Euler-Lagrange approach, where the lagrangian approach is normally applied to the cells and the Eulerian approach is used for the nutrients and oxygen [69–71] (Fig. 11.13).

Most of these models have focused on the simulation of tumor organoids, because their growth is one of the aspects that have been more studied [65, 66, 69, 72]. To simulate the growth of the organoid is crucial to take into account the regulatory

Fig. 11.13 Classification of discrete, or agent-based, models, according to the space location of the cells: on-lattice and off-lattice models, based on whether the cells are bound to occupy fixed positions defined by a lattice or if they are able to freely move through the domain. In the CP model, different shades of the same color represent different cells since a single cell may occupy more than one voxel. From Gonçalves and García-Aznar [68], open access article distributed under the terms and conditions of the Creative Commons Attribution (CC BY) license

role of the biophysical properties of the surrounding matrix like the stiffness. Thus, it is known that tumour spheroid growth is inhibited when it is embedded in agarose matrices due to compressive stresses [73–75]. However, and although the stiffness of the surrounding matrix is very relevant, it is also crucial to know the porosity and permeability of this matrix, because some cells can migrate and escape from the organoid clearly reducing its size ("Go or Grow Hypothesis" is dependent on the surrounding ECM) [76, 77]. Therefore, clearly, the extracellular matrix (ECM) properties drive the organoid morphogenesis, finding strong differences if the ECM is Matrigel or collagen type I. For example, in Matrigel, normal or transformed epithelial cell types, including pancreatic cancer epithelial cells, self-organize into spherical structures with hollow lumen and polarized epithelial lining [78]. However, the same cells seeded in matrices of collagen type I induce branched tubulogenesis, and the formation of both lumen and terminal buds for organoids growing [78].

Although most of existent numerical models simulate solid organoids or spheroids, the most frequent morphological feature of organoids is the presence of one or multiple lumens [79, 80]. Two main computational works have focused on this aspect. On the one hand, Camacho-Gómez et al. proposed a 3D centre-based model 3D for lumen morphogenesis of pancreatic organoids that includes cells' fluid secretion and the density of the ECM, where the lumen hydrostatic pressure and the density of the ECM are considered the main regulators of the lumen formation [81]. On the other hand, van Liedekerke et al. [82] proposed a computational model of biliary lumen formation using a deformable cell model (DCM) that considers the deformation of the cell surface.

Nevertheless, there still many challenges to face in the numerical modelling of organoids that is interesting to comment:

- The morphogenesis of organoids requires the deformation of cells, therefore, although some models simplify this effect it is very important to take into account its deformability as it is proposed by van Liedekerke et al. [82].
- Organoids not always present a monolayer architecture, but sometimes it is necessary to simulate the formation of a cell multilayer [79].
- It is important to consider complex cell-cell interactions, such as the osmoregulation as it was simulated by McEvoy et al. [70].
- Organoids can also present an anisotropic shape [83], so it is important to simulate all kind of shapes like complex branched phenotypes [78]. Most of models simulate spherical or mostly isotropic structures.
- Even in the same conditions of the in-vitro experiments, organoids present a high variability, because different cell types are present and the fate of the organoid can vary from one to another. It is really interesting to see the work of Camacho-Gomez et al. [71] where they achieved to explain why different organoids present a different size according to the times where cells exert a different functionality. The use of AI algorithms opens new possibilities in this direction.

11.9 Integration of Experimental and Numerical Results

To achieve models that could be useful [4] it is necessary to validate their numerical predictions with experimental data from in-vitro, ex-vivo or in-vivo. To facilitate this task is necessary to combine and integrate the experimental data in our models. Computational models simulate experiments using some input parameters, however, normally these parameters cannot be measured and have to be calibrated. For this fitting process, the most common approach consists on solving a least-square minimization problem with a regularization term [84]. However, this approach may lead to ill-posed problems. Therefore, an alternative is to solve the calibration problem as a Bayesian inverse problem, where the uncertainty of the experimental data can be taken into account. Rather than obtaining a specific value of the parameters, the Bayesian solution provides a probability distribution for the uncertain parameters given the variability characteristics of the measurements [85]. For practically relevant problems, the inference task cannot be solved analytically. Nevertheless, the distribution can be approximated using sampling, particle, or variational methods [84]. Although these strategies provide good results [85, 86] with some particular computational models with a low computational cost, they can be prohibitive for those expensive computer models (in the case of sampling and particle methods, which require a large number of simulations) or they can require to compute the gradient of the model outputs with respect to the uncertain parameters, which is quite complex (in the case of variational methods). Therefore, in these cases, a different strategy is currently being used, which is based on the application of different surrogate modelling approaches [66, 87], which ranging from polynomial chaos expansion, adaptive sparse grid collocation methods, Bayesian neural networks, Gaussian process regression, to physics-constrained surrogate models [84].

11.10 Conclusions

When a new computer model is designed for simulating a biological application, it is especially important to take into account that the model has to be focused on facilitating validation, since without validation its impact and utility is very low [4]. Therefore, all the steps required to develop a computational model, which have been briefly described in Sect. 11.1, have to be defined according to facilitate the comparison with the experimental data and consequently its validation (Fig. 11.14). Thus, the computational model can be adjusted to the experimental case is investigated and quantified.

In this chapter, different numerical approaches are presented to study and investigate different mechanisms that regulate tissue regeneration for both in-vivo and in-vitro experiments. Nevertheless, despite the potential of the current numerical simulations as tool for investigation in TERM, they require a very high computational cost, above all, when multiphysics and multiscale simulations are involved,

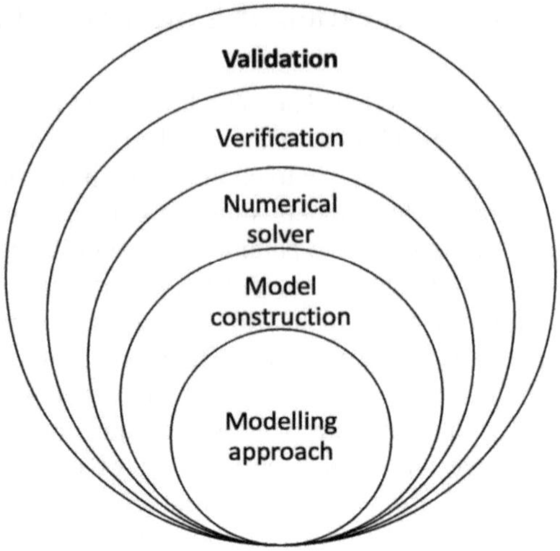

Fig. 11.14 Validation drives the modelling strategy followed during the development of computational models in TERM

which is crucial in TERM. Hence, novel numerical strategies are necessary that allow to link scales and reduce this computational cost.

Questions

- Which is the main difference between a center-based model and a deformable cell model in a discrete approach?
- Please, cite the most relevant steps that have to be followed to construct a computational model in TERM.
- Which is the main difference between an eulerian and a lagrangian approach?
- If you would like to simulate a cluster of cells that is moving collectively, where you are interested on following each cell individually to understand the role of each cell, which kind of numerical approach you would follow? Please, justify.
- Which is the main difference between verification and validation of a computational model?

References

1. Geris, L., Lambrechts, T., Carlier, A., & Papantoniou, I. (2018). The future is digital: In silico tissue engineering. *Current Opinion in Biomedical Engineering, 6*, 92–98.
2. Gonçalves, M. I. G. (2023). *Computational modelling of cancer systems: From individual to collective cell behaviour*. PhD Thesis. https://zaguan.unizar.es/record/127606/files/TESIS-2023-152.pdf
3. Sharpe, J. (2017). Computer modeling in developmental biology: Growing today, essential tomorrow. *Development, 144*(23), 4214–4225.
4. Markowetz, F. (2024). All models are wrong and yours are useless: making clinical prediction models impactful for patients. *npj Precision Oncology, 8*(1), 54.
5. Fletcher, A. G., & Osborne, J. M. (2022). Seven challenges in the multiscale modeling of multicellular tissues. *WIREs Mechanisms of Disease, 14*(1), e1527.

6. Butner, J. D., Dogra, P., Chung, C., Pasqualini, R., Arap, W., Lowengrub, J., & Wang, Z. (2022). Mathematical modeling of cancer immunotherapy for personalized clinical translation. *Nature Computational Science, 2*(12), 785–796.
7. Morrison, T. M., Pathmanathan, P., Adwan, M., & Margerrison, E. (2018). Advancing regulatory science with computational modeling for medical devices at the FDA'S office of science and engineering laboratories. *Frontiers in Medicine, 5*, 241.
8. ASME. (2018). *V&V40-assessing credibility of computational modeling through verification and validation: Application to medical devices*. ASME.
9. Sun, W., Darling, A., Starly, B., & Nam, J. (2004). Computer-aided tissue engineering: Overview, scope and challenges. *Biotechnology and Applied Biochemistry, 39*(1), 29–47.
10. Rainer, A., & Moroni, L. (2021). *Computer-aided tissue engineering* (Vol. 2147). Springer.
11. Ramaraju, H., Sferra, S. R., Kunisaki, S. M., & Hollister, S. J. (2022). Finite element analysis of esophageal atresia repair with biodegradable polymer sleeves. *Journal of the Mechanical Behavior of Biomedical Materials, 133*, 105349.
12. Nyirjesy, S. C., Heller, M., von Windheim, N., Gingras, A., Kang, S. Y., Ozer, E., & VanKoevering, K. K. (2022). The role of computer aided design/computer assisted manufacturing (CAD/CAM) and 3-dimensional printing in head and neck oncologic surgery: A review and future directions. *Oral Oncology, 132*, 105976.
13. Geris, L. (2013). *Computational modeling in tissue engineering*. Springer.
14. Rahatuzzaman, M., Mahmud, M., Rahman, S., & Hoque, M. E. (2024). Design, fabrication, and characterization of 3D-printed ABS and PLA scaffolds potentially for tissue engineering. *Results in Engineering, 21*, 101685.
15. Lieberthal, T. J., Sahakyants, T., Szabo-Wexler, N. R., Hancock, M. J., Spann, A. P., Oliver, M. S., & Vacanti, J. P. (2024). Implantable 3D printed hydrogels with intrinsic channels for liver tissue engineering. *Proceedings of the National Academy of Sciences, 121*(47), e2403322121.
16. Fukada, E., & Yasuda, I. (1964). Piezoelectric effects in collagen. *Japanese Journal of Applied Physics, 3*(2), 117.
17. Fernández, J. R., García-Aznar, J. M., & Martínez, R. (2012). Piezoelectricity could predict sites of formation/resorption in bone remodelling and modelling. *Journal of Theoretical Biology, 292*, 86–92.
18. Zaszczynska, A., Sajkiewicz, P., & Gradys, A. (2020). Piezoelectric scaffolds as smart materials for neural tissue engineering. *Polymers, 12*(1), 161.
19. Rajabi, A. H., Jaffe, M., & Arinzeh, T. L. (2015). Piezoelectric materials for tissue regeneration: A review. *Acta Biomaterialia, 24*, 12–23.
20. Hervas-Raluy, S., Sainz-DeMena, D., Gomez-Benito, M. J., & García-Aznar, J. M. (2024). Image-based biomarkers for engineering neuroblastoma patient-specific computational models. *Engineering with Computers, 40*(5), 3215–3231.
21. García-Aznar, J. M., Valero, C., Borau, C., & Garijo, N. (2016). Computational mechano-chemo-biology: A tool for the design of tissue scaffolds. *Biomanufacturing Reviews, 1*, 1–10.
22. Perier-Metz, C., Duda, G. N., & Checa, S. (2022). A mechanobiological computer optimization framework to design scaffolds to enhance bone regeneration. *Frontiers in Bioengineering and Biotechnology, 10*, 980727.
23. Loerakker, S., & Humphrey, J. D. (2023). Computer model-driven design in cardiovascular regenerative medicine. *Annals of Biomedical Engineering, 51*(1), 45–57.
24. Valero, C., Javierre, E., García-Aznar, J. M., Gómez-Benito, M. J., & Menzel, A. (2015). Modeling of anisotropic wound healing. *Journal of the Mechanics and Physics of Solids, 79*, 80–91.
25. Mohammadkhah, M., Marinkovic, D., Zehn, M., & Checa, S. (2019). A review on computer modeling of bone piezoelectricity and its application to bone adaptation and regeneration. *Bone, 127*, 544–555.
26. Badali, V., Checa, S., Zehn, M. M., Marinkovic, D., & Mohammadkhah, M. (2024). Computational design and evaluation of the mechanical and electrical behavior of a piezoelectric scaffold: A preclinical study. *Frontiers in Bioengineering and Biotechnology, 11*, 1261108.

27. Rennekamp, B., Karfusehr, C., Kurth, M., Ünal, A., Monego, D., Riedmiller, K., & Gräter, F. (2023). Collagen breaks at weak sacrificial bonds taming its mechanoradicals. *Nature Communications, 14*(1), 2075.
28. Zapp, C., Obarska-Kosinska, A., Rennekamp, B., Kurth, M., Hudson, D. M., Mercadante, D., & Gräter, F. (2020). Mechanoradicals in tensed tendon collagen as a source of oxidative stress. *Nature Communications, 11*(1), 2315.
29. Post, J. N., Loerakker, S., Merks, R. M., & Carlier, A. (2022). Implementing computational modeling in tissue engineering: where disciplines meet. *Tissue Engineering Part A, 28*(11–12), 542–554.
30. Bhattacharya, P., Li, Q., Lacroix, D., Kadirkamanathan, V., & Viceconti, M. (2021). A systematic approach to the scale separation problem in the development of multiscale models. *PLoS One, 16*(5), e0251297.
31. Norton, K. A., Gong, C., Jamalian, S., & Popel, A. S. (2019). Multiscale agent-based and hybrid modeling of the tumor immune microenvironment. *PRO, 7*(1), 37.
32. de Melo Quintela, B., Hervas-Raluy, S., Garcia-Aznar, J. M., Walker, D., Wertheim, K. Y., & Viceconti, M. (2022). A theoretical analysis of the scale separation in a model to predict solid tumour growth. *Journal of Theoretical Biology, 547*, 111173.
33. Sanz-Herrera, J. A., García-Aznar, J. M., & Doblaré, M. (2009). A mathematical approach to bone tissue engineering. *Philosophical Transactions of the Royal Society A: Mathematical, Physical and Engineering Sciences, 367*(1895), 2055–2078.
34. Geers, M. G., Kouznetsova, V. G., Matouš, K., & Yvonnet, J. (2017). Homogenization methods and multiscale modeling: Nonlinear problems. In *Encyclopedia of computational mechanics* (2nd ed., pp. 1–34). Wiley.
35. Zohdi, T. I. (2004). Homogenization methods and multiscale modeling. Encyclopedia of computational mechanics. *Solids and Structures, 2*, 357–383.
36. Sanz-Herrera, J. A., García-Aznar, J. M., & Doblaré, M. (2008). Micro–macro numerical modelling of bone regeneration in tissue engineering. *Computer Methods in Applied Mechanics and Engineering, 197*(33-40), 3092–3107.
37. Sanz-Herrera, J. A., Doblaré, M., & García-Aznar, J. M. (2010). Scaffold microarchitecture determines internal bone directional growth structure: A numerical study. *Journal of Biomechanics, 43*(13), 2480–2486.
38. Valero, C., Javierre, E., García-Aznar, J. M., & Gómez-Benito, M. J. (2014). A cell-regulatory mechanism involving feedback between contraction and tissue formation guides wound healing progression. *PLoS One, 9*(3), e92774.
39. Nasello, G., Vautrin, A., Pitocchi, J., Wesseling, M., Kuiper, J. H., Pérez, M. Á., & García-Aznar, J. M. (2021). Mechano-driven regeneration predicts response variations in large animal model based on scaffold implantation site and individual mechano-sensitivity. *Bone, 144*, 115769.
40. Singh, V. K., Jiménez del Val, I., Glassey, J., & Kavousi, F. (2024). Integration approaches to model bioreactor hydrodynamics and cellular kinetics for advancing bioprocess optimisation. *Bioengineering, 11*(6), 546.
41. Roy, S., Biswas, S., Khanna, S., Gordillo, G., Bergdall, V., et al. (2009). Characterization of a preclinical model of chronic ischemic wound. *Physiological Genomics, 37*, 211–224.
42. McGrath, M., & Simon, R. (1983). Wound geometry and the kinetics of wound contraction. *Plastic and Reconstructive Surgery, 72*, 66–72.
43. Moreo, P., Garcia-Aznar, J. M., & Doblare, M. (2008). Modeling mechanosensing and its effect on the migration and proliferation of adherent cells rid f-8256-2010. *Acta Biomaterialia, 4*, 613–621.
44. Theodoropoulos, C., & Sun, C. (2019). Bioreactor models and modeling approaches. In M. Butler (Ed.), *Comprehensive biotechnology* (3rd ed., pp. 663–680). Elsevier BV.
45. Losoi, P., Konttinen, J., & Santala, V. (2024). Modeling large-scale bioreactors with diffusion equations. Part I: Predicting axial dispersion coefficient and mixing times. *Biotechnology and Bioengineering, 121*(3), 1060–1075.

46. Cho, B. A., Teke, G. M., Gakingo, G. K., Pott, R. W. M., & Zhang, D. (2025). CFD predictive simulations of miniature bioreactor mixing dynamics coupled with photo-bioreaction kinetics in transitional flow regime. *Biochemical Engineering Journal, 214*, 109585.
47. Kleinstreuer, C., & Poweigha, T. (2005). Modeling and simulation of bioreactor process dynamics. In *Bioprocess parameter control* (pp. 91–146). Springer.
48. Papantoniou, I., Guyot, Y., Sonnaert, M., Kerckhofs, G., Luyten, F. P., Geris, L., & Schrooten, J. (2014). Spatial optimization in perfusion bioreactors improves bone tissue-engineered construct quality attributes. *Biotechnology and Bioengineering, 111*(12), 2560–2570.
49. Pereira, R. F., Freitas, D., Tojeira, A., Almeida, H. A., Alves, N., & Bártolo, P. J. (2014). Computer modelling and simulation of a bioreactor for tissue engineering. *International Journal of Computer Integrated Manufacturing, 27*(10), 946–959.
50. Rivera, F. J. C., & Chen, J. (2022). Computational fluid dynamics modeling of cell cultures in bioreactors and its potential for cultivated meat production—A mini-review. *Future Foods, 6*, 100195.
51. Olivares, A. L., & Lacroix, D. (2012). Simulation of cell seeding within a three-dimensional porous scaffold: A fluid-particle analysis. *Tissue Engineering Part C: Methods, 18*(8), 624–631.
52. Zhang, Z., Zhu, J., Liu, Y., Shao, J., & Xie, S. (2023). Effects of cell deformability and adhesion strength on dynamic cell seeding: Cell-scale investigation via mesoscopic modeling. *Journal of Biomechanics, 153*, 111589.
53. Schmalzriedt, S., Jenne, M., Mauch, K., & Reuss, M. (2003). Integration of physiology and fluid dynamics. *Advances in Biochemical Engineering/Biotechnology, 80*, 19–68.
54. Osborne, J. M., O'Dea, R. D., Whiteley, J. P., Byrne, H. M., & Waters, S. L. (2010). The influence of bioreactor geometry and the mechanical environment on engineered tissues. *Journal of Biomechanical Engineering, 132*(5), 051006.
55. Prokharau, P. A., Vermolen, F. J., & García-Aznar, J. M. (2014). A mathematical model for cell differentiation, as an evolutionary and regulated process. *Computer Methods in Biomechanics and Biomedical Engineering, 17*(10), 1051–1070.
56. King, J. A., & Miller, W. M. (2007). Bioreactor development for stem cell expansion and controlled differentiation. *Current Opinion in Chemical Biology, 11*(4), 394–398.
57. Escribano, J., Sánchez, M. T., & García-Aznar, J. M. (2015). Modeling the formation of cell-matrix adhesions on a single 3D matrix fiber. *Journal of Theoretical Biology, 384*, 84–94.
58. Verhoeff, K., Cuesta-Gomez, N., Maghera, J., Dadheech, N., Pawlick, R., Smith, N., & Shapiro, A. J. (2025). Scalable bioreactor-based suspension approach to generate stem cell–derived islets from healthy donor-derived iPSCs. *Transplantation, 109*(1), e22–e35.
59. Jungreuthmayer, C., Jaasma, M. J., Al-Munajjed, A. A., Zanghellini, J., Kelly, D. J., & O'Brien, F. J. (2009). Deformation simulation of cells seeded on a collagen-GAG scaffold in a flow perfusion bioreactor using a sequential 3D CFD-elastostatics model. *Medical Engineering & Physics, 31*(4), 420–427.
60. Guyot, Y., Luyten, F. P., Schrooten, J., Papantoniou, I., & Geris, L. (2015). A three-dimensional computational fluid dynamics model of shear stress distribution during neotissue growth in a perfusion bioreactor. *Biotechnology and Bioengineering, 112*(12), 2591–2600.
61. Kalliontzis, D. (2022). Fluid–structure interaction with ALE formulation and skeleton-based structural models. *Journal of Fluids and Structures, 110*, 103513.
62. Arena, G., Groh, R., Theunissen, R., Weaver, P., & Pirrera, A. (2016). Adaptive nonlinear structures for flow regulation: Modelling fluid-structure interactions with coupled Eulerian-Lagrangian meshes. In *2016 SIMULIA regional user meeting*. Simuleon.
63. Serrano-Alcalde, F., García-Aznar, J. M., & Gómez-Benito, M. J. (2017). The role of nuclear mechanics in cell deformation under creeping flows. *Journal of Theoretical Biology, 432*, 25–32.
64. Rüberg, T., & Aznar, J. M. G. (2016). Numerical simulation of solid deformation driven by creeping flow using an immersed finite element method. *Advanced Modeling and Simulation in Engineering Sciences, 3*(1), 9.

65. Walker, B. J., Celora, G. L., Goriely, A., Moulton, D. E., & Byrne, H. M. (2023). Minimal morphoelastic models of solid tumour spheroids: A tutorial. *Bulletin of Mathematical Biology, 85*(5), 38.
66. Hervas-Raluy, S., Wirthl, B., Guerrero, P. E., Rei, G. R., Nitzler, J., Coronado, E., & Wall, W. A. (2023). Tumour growth: An approach to calibrate parameters of a multiphase porous media model based on in vitro observations of neuroblastoma spheroid growth in a hydrogel microenvironment. *Computers in Biology and Medicine, 159*, 106895.
67. Van Liedekerke, P., Palm, M. M., Jagiella, N., & Drasdo, D. (2015). Simulating tissue mechanics with agent-based models: Concepts, perspectives and some novel results. *Computational Particle Mechanics, 2*, 401–444.
68. Gonçalves, I. G., & García-Aznar, J. M. (2023). Hybrid computational models of multicellular tumour growth considering glucose metabolism. *Computational and Structural Biotechnology Journal, 21*, 1262–1271.
69. González-Valverde, I., Semino, C., & García-Aznar, J. M. (2016). Phenomenological modelling and simulation of cell clusters in 3D cultures. *Computers in Biology and Medicine, 77*, 249–260.
70. McEvoy, E., Han, Y. L., Guo, M., & Shenoy, V. B. (2020). Gap junctions amplify spatial variations in cell volume in proliferating tumor spheroids. *Nature Communications, 11*(1), 6148.
71. Camacho-Gomez, D., Sorzabal-Bellido, I., Ortiz-de-Solorzano, C., Garcia-Aznar, J. M., & Gomez-Benito, M. J. (2023). A hybrid physics-based and data-driven framework for cellular biological systems: Application to the morphogenesis of organoids. *Iscience, 26*(7), 107164.
72. Gerlee, P. (2013). The model muddle: In search of tumor growth laws. *Cancer Research, 73*, 2407–2411.
73. Helmlinger, G., Netti, P. A., Lichtenbeld, H. C., Melder, R. J., & Jain, R. K. (1997). Solid stress inhibits the growth of multicellular tumor spheroids. *Nature Biotechnology, 15*(8), 778–783.
74. Cheng, G., Tse, J., Jain, R. K., & Munn, L. L. (2009). Micro-environmental mechanical stress controls tumor spheroid size and morphology by suppressing proliferation and inducing apoptosis in cancer cells. *PLoS One, 4*(2), e4632.
75. Montel, F., Delarue, M., Elgeti, J., Vignjevic, D., Cappello, G., & Prost, J. (2012). Isotropic stress reduces cell proliferation in tumor spheroids. *New Journal of Physics, 14*(5), 055008.
76. Gonçalves, I. G., & Garcia-Aznar, J. M. (2021). Extracellular matrix density regulates the formation of tumour spheroids through cell migration. *PLoS Computational Biology, 17*(2), e1008764.
77. Plou, J., Juste-Lanas, Y., Olivares, V., Del Amo, C., Borau, C., & García-Aznar, J. M. (2018). From individual to collective 3D cancer dissemination: Roles of collagen concentration and TGF-β. *Scientific Reports, 8*(1), 12723.
78. Randriamanantsoa, S., Papargyriou, A., Maurer, H. C., Peschke, K., Schuster, M., Zecchin, G., & Bausch, A. R. (2022). Spatiotemporal dynamics of self-organized branching in pancreas-derived organoids. *Nature Communications, 13*(1), 5219.
79. Lee, B. H., Seijo-Barandiaran, I., & Grapin-Botton, A. (2022). Epithelial morphogenesis in organoids. *Current Opinion in Genetics & Development, 72*, 30–37.
80. Lee, B. H., Fuji, K., Petzold, H., Seymour, P., Yennek, S., Schewin, C., & Grapin-Botton, A. (2024). Control of lumen geometry and topology by cell proliferation rate and pressure.
81. Camacho-Gómez, D., García-Aznar, J. M., & Gómez-Benito, M. J. (2022). A 3D multi-agent-based model for lumen morphogenesis: The role of the biophysical properties of the extracellular matrix. *Engineering with Computers, 38*(5), 4135–4149.
82. Van Liedekerke, P., Gannoun, L., Loriot, A., Johann, T., Lemaigre, F. P., & Drasdo, D. (2022). Quantitative modeling identifies critical cell mechanics driving bile duct lumen formation. *PLoS Computational Biology, 18*(2), e1009653.
83. Koike, H., Iwasawa, K., Ouchi, R., Maezawa, M., Giesbrecht, K., Saiki, N., & Takebe, T. (2019). Modelling human hepato-biliary-pancreatic organogenesis from the foregut–midgut boundary. *Nature, 574*(7776), 112–116.

84. Dinkel, M., Geitner, C. M., Rei, G. R., Nitzler, J., & Wall, W. A. (2024). Solving Bayesian inverse problems with expensive likelihoods using constrained Gaussian processes and active learning. *Inverse Problems, 40*(9), 095008.
85. Merino-Casallo, F., Gomez-Benito, M. J., Juste-Lanas, Y., Martinez-Cantin, R., & Garcia-Aznar, J. M. (2018). Integration of in vitro and in silico models using Bayesian optimization with an application to stochastic modeling of mesenchymal 3D cell migration. *Frontiers in Physiology, 9*, 1246.
86. Sainz-DeMena, D., Ye, W., Pérez, M. Á., & García-Aznar, J. M. (2022). A finite element based optimization algorithm to include diffusion into the analysis of DCE-MRI. *Engineering with Computers, 38*(5), 3849–3865.
87. Wirthl, B., Brandstaeter, S., Nitzler, J., Schrefler, B. A., & Wall, W. A. (2023). Global sensitivity analysis based on Gaussian-process metamodelling for complex biomechanical problems. *International Journal for Numerical Methods in Biomedical Engineering, 39*(3), e3675.

Chapter 12
The Future of Tissue Engineering and Regenerative Medicine

Abstract Tissue engineering and regenerative medicine (TERM) offer immense potential for treating organ failure, traumatic injury, and multiple diseases like cancer. Significant breakthroughs are upcoming in key areas such as bioprinting, biomaterial innovation, stem cell therapies, immunoengineering, organ-on-a-chip models, artificial intelligence, or gene editing. Nevertheless, important challenges must be addressed to realize the full potential of TERM, which include vascularization and innervation of large constructs, immune rejection, ethics, interdisciplinary collaboration, and standardization and regulation. This last chapter discusses the areas where pivotal breakthroughs for TERM are envisaged and ends with a brief reminder and discussion of the challenges still to overcome for advancing the field towards clinical translation.

12.1 The Future of Tissue Engineering and Regenerative Medicine: A Transformative Horizon

Tissue engineering and regenerative medicine (TERM) stands as a beacon of hope in modern medicine, promising revolutionary solutions for organ failure, tissue damage, traumatic injury, and a myriad of debilitating diseases [1, 2]. By combining the three elements of the TERM paradigm, i.e. cells, biomaterial scaffolds, and suitable biochemical and biophysical factors, this interdisciplinary field of research aims to restore, maintain, improve, or replace biological tissues [3]. The coming years are poised for significant breakthroughs, driven by advancements in several key areas and contingent on overcoming persistent challenges. These areas include bioprinting, biomaterial innovation, stem cell therapies, immunoengineering, organ-on-a-chip models, artificial intelligence and machine learning, and gene editing (Fig. 12.1).

© The Author(s), under exclusive license to Springer Nature Switzerland AG 2026

E. García-Gareta, J. M. García-Aznar, *Contemporary Tissue Engineering and Regenerative Medicine*, Series in BioEngineering,
https://doi.org/10.1007/978-3-032-03441-0_12

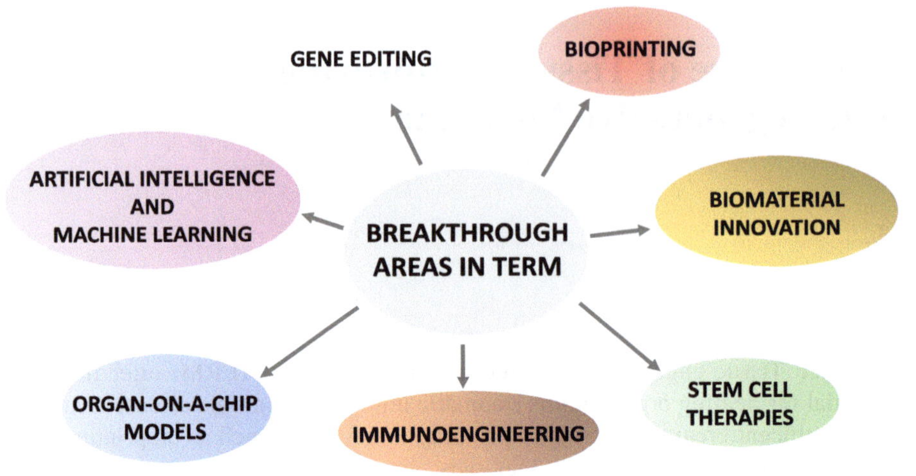

Fig. 12.1 Visual representation of the areas poised for significant breakthrough in the TERM field

12.1.1 Bioprinting

Bioprinting, an additive manufacturing process, is at the forefront of TERM, enabling the precise layer-by-layer construction of three-dimensional (3D) tissue constructs [4–6]. The future of bioprinting lies in enhancing resolution, speed, and the ability to print complex, multi-cellular structures with integrated vascular networks. Researchers are exploring 4D bioprinting, where the printed structures can change shape or function over time in response to specific stimuli, mimicking natural tissue development [7, 8]. The ultimate goal is the on-demand printing of fully functional organs, which could alleviate organ transplant shortages and eliminate the need for lifelong immunosuppression by using a patient's own cells. Continued innovation in bio-inks, which are the printable materials containing living cells and biocompatible scaffolds, is crucial for achieving these ambitious targets. The convergence of bioprinting and nanotechnology, by incorporating nanomaterials in bio-inks, holds particular promise as seen in Chap. 6 [5]. Furthermore, integrating bioprinting with advanced imaging and computational modelling will allow for patient-specific designs and optimized tissue maturation in bioreactors that simulate physiological conditions [9].

12.1.2 Biomaterial Innovation

The development of novel biomaterials is a cornerstone of TERM's future progress. Next-generation biomaterials are being designed to be not just biocompatible and biodegradable, but also able to instruct cellular behaviour and guide tissue formation [10]. Innovations include smart biomaterials that can respond to physiological cues, release therapeutic agents in a controlled manner, or possess self-healing or shape-memory capabilities [11]. Researchers are increasingly focusing on materials that mimic the native extracellular matrix (ECM) more closely, incorporating specific biological signals and biomechanical properties, or imitating its architecture [12–16]. Decellularized ECM from natural tissues is also gaining traction as a scaffold material for clinical applications, offering inherent bioactivity, as well as native architectural and biochemical cues [17]. Hydrogels that incorporate decellularized ECM material are particularly promising for in vitro models, since they add biomimicry in terms of native composition. Decellularized ECM from diseased tissues is also of particular interest since the ECM of pathological tissues differs in terms of composition and architectural organization from that of healthy ones [18]. This is especially relevant in tumour models used in cancer research, as decellularized tumoural ECM best recreates the ECM component of the tumour microenvironment [18]. Research into smart hydrogels, which have abilities like sensing, stimuli responsiveness, shape memory, self-assembly, or electroactivity will allow the clinical application of this particular class of scaffolds [19]. Nanomaterials are also at the forefront of biomaterial research for TERM applications, since cells in our body interact with nano-cues [20].

Looking into the future, the synergy between synthetic polymer chemistry, nanotechnology, and an understanding of cell-biomaterial interactions will drive the creation of sophisticated scaffolds that promote enhanced tissue integration and functionality.

12.1.3 Stem Cell Therapies

Stem cells, with their remarkable ability to differentiate into various cell types and self-renew, remain central to the future of TERM. In Chap. 4 of this book, we saw the different types of stem cells in terms of potency and sources and the associated ethical considerations of each type. The field is moving beyond the use of adult stem cells towards the broader application of induced pluripotent stem cells (iPSCs), which can be generated directly from a patient's adult cells, thereby circumventing ethical concerns associated with embryonic stem cells and reducing the risk of immune rejection [21]. Future advancements will focus on refining differentiation protocols to generate pure and mature cell populations efficiently and safely [21]. Furthermore, combining stem cell technology with gene editing offers the potential to correct genetic defects before tissue construction [21, 22]. Moreover, understanding and manipulating the stem cell niche—the microenvironment that governs stem cell behaviour—will be critical for controlling cell fate and enhancing the therapeutic efficacy of stem cell-based TERM products, as well as increasing the complexity of tissue models to study tissue development and disease [23–25].

12.1.4 Immunoengineering

Overcoming the host immune response to engineered tissues represents a critical and multifaceted hurdle in the advancement of TERM. The moment an engineered construct is introduced into the body, it is recognized as foreign, triggering a complex cascade of immune reactions that can range from acute inflammation to chronic rejection, ultimately leading to graft failure. This inherent biological defence mechanism poses a significant barrier to the successful integration and long-term functionality of engineered tissues and organs.

Immunoengineering has emerged as a pivotal and rapidly evolving field, specifically focused on developing sophisticated strategies to intelligently modulate the immune system [26]. The goal is not simply to suppress the immune response, which can leave the patient vulnerable to infections and other complications, but rather to guide it towards a state of tolerance and active support for tissue regeneration [26]. This sophisticated balancing act involves designing biomaterials that are inherently less immunogenic, perhaps by mimicking natural host tissues or by presenting surfaces that discourage immune cell adhesion and activation [27]. Beyond passive non-immunogenicity, cutting-edge approaches aim to create biomaterials that actively engage with the immune system, for instance, by releasing signals that calm inflammatory responses or that recruit and activate pro-regenerative immune cell populations, such as M2 macrophages or regulatory T cells, which are known to contribute to tissue repair and remodelling [27]. One promising avenue involves the incorporation of immunomodulatory drugs or biological agents directly into the tissue scaffolds themselves [27]. These could be released in a controlled and localized manner, minimizing systemic side effects while maximizing therapeutic impact at the site of implantation. Furthermore, the development of "off-the-shelf" universal donor cells, potentially derived from iPSCs that have been genetically edited for immune compatibility, could revolutionize the field by providing readily available, rejection-resistant cellular building blocks for a wide range of patients [28].

Looking ahead, a deeper understanding of the complex interplay between implanted constructs and the host immune system at the cellular and molecular level is needed. This involves deciphering the precise signalling pathways, cell types, and temporal dynamics that govern both destructive and constructive immune responses to engineered tissues.

12.1.5 Organ-on-a-Chip Models

Organ-on-a-chip (OOC) models represent a significant advancement in TERM, integrating microfluidic technologies with cell biology and biomaterial science to create miniaturized systems that recapitulate the dynamic microenvironments and functional units of human organs [29–33]. As previously seen in Chap. 10, these platforms typically consist of micro-fabricated channels and chambers where cells, either patient-derived or from established lines, are cultured under precisely controlled conditions, including fluid flow, mechanical stimuli, and biochemical gradients [29, 34]. By engineering tissue-tissue interfaces and multicellular architectures, OOCs aim to mimic organ-level physiology and pathophysiology with greater fidelity than conventional 2D or static 3D culture systems. This capability is pivotal in TERM for investigating cell-cell and cell-matrix interactions, understanding tissue development and regeneration mechanisms, and evaluating the efficacy and safety of novel therapeutic strategies and engineered constructs in a more physiologically relevant context prior to in vivo studies. Thus, OOC models are revolutionizing drug discovery, disease modelling, and personalized medicine.

The future will see the development of more complex, multi-organ systems on a chip, allowing the study of systemic effects and interactions between different tissue types. Integrating sensors for real-time monitoring of cellular behaviour and tissue maturation within these chips will provide invaluable data. As these models become more sophisticated and validated, they will reduce the reliance on animal testing and accelerate the development of new TERM therapies [29].

12.1.6 Artificial Intelligence and Machine Learning

Artificial intelligence (AI) and machine learning (ML) are increasingly being applied to TERM research to analyze complex biological data, simulate regenerative processes, optimize scaffold designs, predict material properties, and streamline manufacturing processes [35–37]. AI algorithms can identify patterns and correlations that may not be apparent to human researchers, thereby accelerating the discovery of new biomaterials and optimal culture conditions. In the future, AI could play a crucial role in personalized tissue engineering by designing patient-specific constructs based on individual medical imaging and biological data. ML models can also be used to predict the in vivo performance of engineered tissues, reducing the need for extensive preclinical testing [36]. The integration of AI into bioreactors could enable real-time monitoring and adaptive control of tissue development, leading to more consistent and high-quality products [35].

12.1.7 Gene Editing

Gene editing technologies, particularly the CRISPR-Cas9 system, represent a significant advancement in the field of TERM [38, 39]. These tools enable precise and efficient genomic modifications in cells used for tissue constructs, thereby facilitating the enhancement of cellular functionality, the directed differentiation of stem cells into specific lineages, and the correction of deleterious mutations underlying genetic disorders. Emerging applications include the engineering of cells to express therapeutic molecules such as growth factors or immunomodulatory proteins that support tissue regeneration and improve graft integration. Additionally, gene editing can be employed to reduce the immunogenicity of allogeneic or xenogeneic cells, thereby decreasing the likelihood of host immune rejection and expanding the applicability of off-the-shelf tissue-engineered products. Beyond therapeutic constructs, gene editing offers powerful capabilities for disease modelling [40]. The introduction of pathogenic mutations or the modification of regulatory sequences can produce physiologically relevant in vitro models that better mimic the molecular and cellular pathology of human disease [40]. These models are instrumental for high-throughput drug screening, functional genomics studies, and the investigation of fundamental processes underlying tissue development and repair.

As gene editing platforms continue to advance—with improvements in targeting accuracy, reduced off-target activity, and enhanced genomic stability—their integration into TERM workflows is anticipated to expand therapeutic possibilities. This integration will likely accelerate clinical translation by enabling the development of more sophisticated, functional, and personalized tissue-engineered products [38, 39].

12.2 Challenges on the Path to Clinical Translation

As seen throughout this book, the TERM field has achieved remarkable progress in the few decades of its existence. Nevertheless, several significant challenges must be addressed to realize the full potential of TERM, especially for clinical translation. These challenges include vascularization and innervation of large constructs, immune rejection, ethical considerations, interdisciplinary collaboration, and standardization and regulation (Fig. 12.2).

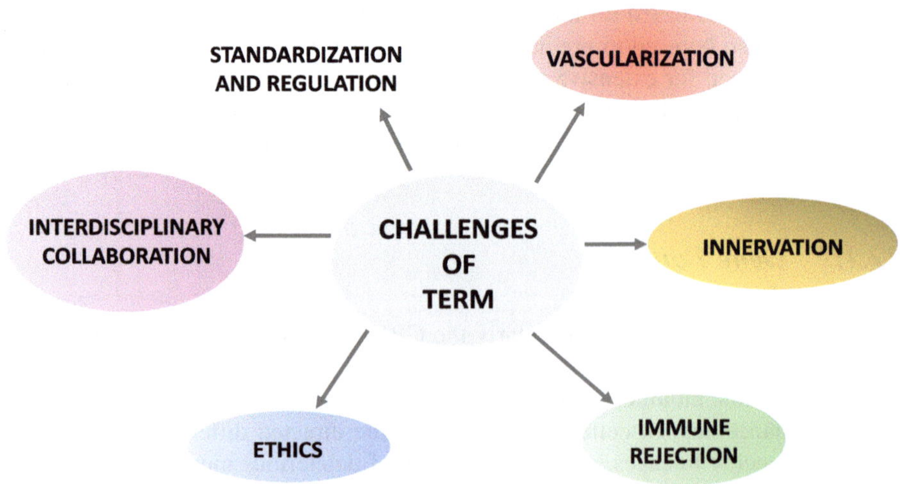

Fig. 12.2 Visual representation of the challenges of TERM

Supplying engineered tissues, especially larger and more complex constructs, with an adequate blood supply remains a major hurdle, as seen in Chap. 8 of this book. Without rapid and efficient vascularization, cells in at the core of the construct will die due to lack of oxygen and nutrients. Developing strategies to promote the formation of stable and functional vascular networks within engineered tissues is paramount for their survival and integration with host tissues. The scenario is similar with nerves, since they play an important role in tissue formation and regeneration, as seen in Chap. 8. Therefore, the TERM field needs to develop strategies that promote innervation of large implantable constructs.

Even with the use of a patient's own cells, components of biomaterials or complex tissue structures can trigger an adverse immune response. For allogeneic (donor-derived) or xenogeneic (animal-derived) tissues, immune rejection is an even greater concern. Finding ways to induce immune tolerance or create universally compatible tissues is crucial for the widespread clinical application of TERM products. In this regard, immunoengineering becomes a powerful ally in TERM.

Importantly, as TERM technologies advance, they raise complex ethical questions. These include concerns about the source of cells (particularly embryonic stem cells), the potential for enhancement versus therapy, equitable access to expensive new treatments, and the moral implications of creating highly sophisticated human-like tissues or even partial organs. Open public discourse and robust ethical frameworks are necessary to guide responsible innovation.

The TERM field is, since its inception, inherently multidisciplinary, requiring expertise from cell biology, materials science, engineering, chemistry, immunology, surgery, and regulatory affairs, among others. Fostering effective collaboration and communication between these diverse fields is essential for translating research discoveries into clinical realities. Breaking down silos and creating integrated research and development environments will accelerate progress.

Finally, the path from laboratory research to clinical application for TERM products is complex and requires clear regulatory pathways. Establishing standardized protocols for manufacturing, quality control, and preclinical testing is crucial for ensuring the safety and efficacy of these novel therapies. International cooperation will be important for harmonizing regulatory approaches and facilitating global access to these transformative technologies.

12.3 Conclusions

We can conclude that the future of TERM is exceptionally bright, with ongoing innovations poised to revolutionize healthcare. Addressing the existing challenges through continued research, interdisciplinary collaboration, and careful ethical consideration will be key to translating these promising technologies into life-changing therapies for patients worldwide.

References

1. Sahakyants, T., & Vacanti, J. P. (2020). Tissue engineering: From the bedside to the bench and back to the bedside. *Pediatric Surgery International, 36*(10), 1123–1133.
2. Ashammakhi, N., GhavamiNejad, A., Tutar, R., Fricker, A., Roy, I., Chatzistavrou, X., et al. (2022). Highlights on advancing frontiers in tissue engineering. *Tissue Engineering. Part B, Reviews, 28*(3), 633–664.
3. Langer, R., & Vacanti, J. P. (1993). Tissue engineering. *Science, 260*(5110), 920. http://science.sciencemag.org/content/260/5110/920.abstract
4. Liu, S., Cheng, L., Liu, Y., Zhang, H., Song, Y., Park, J.-H., et al. (2023). 3D bioprinting tissue analogs: Current development and translational implications. *Journal of Tissue Engineering, 1*(14), 20417314231187110. https://doi.org/10.1177/20417314231187113
5. Zhang, S., Chen, X., Shan, M., Hao, Z., Zhang, X., Meng, L., et al. (2023). Convergence of 3D bioprinting and nanotechnology in tissue engineering scaffolds. *Biomimetics, 8*(1), 94. https://www.mdpi.com/2313-7673/8/1/94
6. Dell, A. C., Wagner, G., Own, J., & Geibel, J. P. (2022). 3D Bioprinting using hydrogels: Cell inks and tissue engineering applications. *Pharmaceutics, 14*, 2596.
7. Arif, Z. U., Khalid, M. Y., Ahmed, W., & Arshad, H. (2022). A review on four-dimensional (4D) bioprinting in pursuit of advanced tissue engineering applications. *Bioprinting, 27*, e00203.
8. Abolhassani, S., Fattahi, R., Safshekan, F., Saremi, J., & Hasanzadeh, E. (2025). Advances in 4D bioprinting: The next Frontier in regenerative medicine and tissue engineering applications. *Advanced Healthcare Materials, 14*(4), e2403065.
9. Moheb Afzali, A., Kheradmand, M. A., & Naghib, S. M. (2024). Bioreactor design-assisted bioprinting of stimuli-responsive materials for tissue engineering and drug delivery applications. *Bioprinting, 37*, e00325.
10. Montoya, C., Du, Y., Gianforcaro, A. L., Orrego, S., Yang, M., & Lelkes, P. I. (2021). On the road to smart biomaterials for bone research: Definitions, concepts, advances, and outlook. *Bone Research, 9*(1), 12.
11. Cecen, B., Hassan, S., Li, X., & Zhang, Y. S. (2024). Smart biomaterials in biomedical applications: Current advances and possible future directions. *Macromolecular Bioscience, 24*(3), e2200550.
12. Owida, H. A., Al-Nabulsi, J. I., Alnaimat, F., Al-Ayyad, M., Turab, N. M., Al Sharah, A., et al. (2022). Recent applications of electrospun nanofibrous scaffold in tissue engineering. *Applied Bionics and Biomechanics, 2022*, 1953861.
13. Lecina-Tejero, Ó., Pérez, M. Á., García-Gareta, E., & Borau, C. (2023). The rise of mechanical metamaterials: Auxetic constructs for skin wound healing. *Journal of Tissue Engineering, 14*, 20417314231177838.
14. Carter, A., Popowski, K., Cheng, K., Greenbaum, A., Ligler, F. S., & Moatti, A. (2021). Enhancement of bone regeneration through the converse piezoelectric effect, A novel approach for applying mechanical stimulation. *Bioelectricity, 3*(4), 255–271. https://doi.org/10.1089/bioe.2021.0019
15. Jin, M., Shi, J., Zhu, W., Yao, H., & Wang, D.-A. (2020). Polysaccharide-based biomaterials in tissue engineering: A review. *Tissue Engineering Reviews (Part B), 27*(6), 604–626. https://doi.org/10.1089/ten.teb.2020.0208
16. García-Gareta, E. (2023). Protein-based materials for wound healing. In F. R. A. Maia, J. M. Oliveira, & R. L. Reis (Eds.), *Handbook of the extracellular matrix: Biologically-derived materials* (pp. 1–28). Springer International Publishing. https://doi.org/10.1007/978-3-030-92090-6_20-1
17. García-Gareta, E., Abduldaiem, Y., Sawadkar, P., Kyriakidis, C., Lali, F., & Greco, K. V. (2020). Decellularised scaffolds: Just a framework? Current knowledge and future directions. *Journal of Tissue Engineering, 11*, 2041731420942903.
18. García-Gareta, E., Pérez, M. Á., & García-Aznar, J. M. (2022). Decellularization of tumours: A new frontier in tissue engineering. *Journal of Tissue Engineering, 13*, 20417314221091682.

19. Khan, M. U. A., Stojanović, G. M., Bin, A. M. F., Dolatshahi-Pirouz, A., Marei, H. E., Ashammakhi, N., et al. (2024). Fundamental properties of smart hydrogels for tissue engineering applications: A review. *International Journal of Biological Macromolecules, 254*, 127882.
20. Sharma, R., Kumar, S., Bhawna, G. A., Dheer, N., Jain, P., et al. (2022). An insight of nanomaterials in tissue engineering from fabrication to applications. *Tissue Engineering and Regenerative Medicine, 19*(5), 927–960. https://doi.org/10.1007/s13770-022-00459-z
21. Cerneckis, J., Cai, H., & Shi, Y. (2024). Induced pluripotent stem cells (iPSCs): Molecular mechanisms of induction and applications. *Signal Transduction and Targeted Therapy, 9*(1), 112.
22. Greşiţă, A., Hermann, D. M., Boboc, I. K. S., Doeppner, T. R., Petcu, E., Semida, G. F., et al. (2025). Glial cell reprogramming in ischemic stroke: A review of recent advancements and translational challenges. *Translational Stroke Research, 16*, 1811–1835.
23. Mao, H., & Ito, Y. (2017). Engineering niches for embryonic and induced pluripotent stem cells. In *Biology and engineering of stem cell niches* (pp. 445–457). Elsevier.
24. Ding, Z., Pan, X., Wang, X., Xie, H., & Ye, Q. (2021). Interactions between induced pluripotent stem cells and stem cell niche augment osteogenesis and bone regeneration. *Smart Materials in Medicine, 2*, 196–208. https://www.sciencedirect.com/science/article/pii/S2590183421000144
25. Frenz-Wiessner, S., Fairley, S. D., Buser, M., Goek, I., Salewskij, K., Jonsson, G., et al. (2024). Generation of complex bone marrow organoids from human induced pluripotent stem cells. *Nature Methods, 21*(5), 868–881.
26. McBride, D. A., Jones, R. M., Bottini, N., & Shah, N. J. (2024). The therapeutic potential of immunoengineering for systemic autoimmunity. *Nature Reviews Rheumatology, 20*(4), 203–215.
27. Yousefpour, P., Ni, K., & Irvine, D. J. (2023). Targeted modulation of immune cells and tissues using engineered biomaterials. *Nature Reviews Bioengineering, 1*(2), 107–124. https://www.nature.com/articles/s44222-022-00016-2
28. Simpson, A., Hewitt, A. W., & Fairfax, K. A. (2023). Universal cell donor lines: A review of the current research. *Stem Cell Reports, 18*(11), 2038–2046.
29. Ingber, D. E. (2022). Human organs-on-chips for disease modelling, drug development and personalized medicine. *Nature Reviews. Genetics, 23*(8), 467–491.
30. Sontheimer-Phelps, A., Hassell, B. A., & Ingber, D. E. (2019). Modelling cancer in microfluidic human organs-on-chips. *Nature Reviews Cancer, 19*(2), 65–81. https://www.nature.com/articles/s41568-018-0104-6
31. Zhang, S., Wan, Z., & Kamm, R. D. (2021). Vascularized organoids on a chip: Strategies for engineering organoids with functional vasculature. *Lab on Chip, 21*(3), 473–488. https://doi.org/10.1039/D0LC01186J
32. Risue, I., Valencia, L., Jorcano, J. L., & Velasco, D. (2021). Skin-on-a-chip models: General overview and future perspectives. *APL Bioengineering, 5*, 30901. https://doi.org/10.1063/5.0046376
33. Fan, Y., Nguyen, D. T., Akay, Y., Xu, F., & Akay, M. (2016). Engineering a brain cancer chip for high-throughput drug screening. *Scientific Reports, 6*(1), 25062.
34. Juste-Lanas, Y., Hervas-Raluy, S., García-Aznar, J. M., & González-Loyola, A. (2023). Fluid flow to mimic organ function in 3D in vitro models. *APL Bioengineering, 7*(3), 031501.
35. Gharibshahian, M., Torkashvand, M., Bavisi, M., Aldaghi, N., & Alizadeh, A. (2024). Recent advances in artificial intelligent strategies for tissue engineering and regenerative medicine. *Skin Research and Technology, 30*(9), e70016.
36. Bagherpour, R., Bagherpour, G., & Mohammadi, P. (2025). Application of artificial intelligence in tissue engineering. *Tissue Engineering. Part B, Reviews, 31*(1), 31–43.
37. Asadi Sarabi, P., Shabanpouremam, M., Eghtedari, A. R., Barat, M., Moshiri, B., Zarrabi, A., et al. (2024). AI-Based solutions for current challenges in regenerative medicine. *European Journal of Pharmacology, 984*, 177067. https://www.sciencedirect.com/science/article/pii/S001429992400757X

38. Dubey, A. K., & Mostafavi, E. (2023). Biomaterials-mediated CRISPR/Cas9 delivery: Recent challenges and opportunities in gene therapy. *Frontiers in Chemistry, 28*, 11.
39. Farag, V. E., Devey, E. A., & Leong, K. W. (2025). The interface of gene editing with regenerative medicine. *Engineering, 46*, 73–100. https://www.sciencedirect.com/science/article/pii/S2095809924006726
40. Geurts, M. H., & Clevers, H. (2023). CRISPR engineering in organoids for gene repair and disease modelling. *Nature Reviews Bioengineering, 1*(1), 32–45.

MIX
Papier aus verantwortungsvollen Quellen
Paper from responsible sources
FSC® C105338

If you have any concerns about our products,
you can contact us on
ProductSafety@springernature.com

In case Publisher is established outside the EU,
the EU authorized representative is:
Springer Nature Customer Service Center GmbH
Europaplatz 3, 69115 Heidelberg, Germany

Printed by Libri Plureos GmbH
in Hamburg, Germany